## Teacher Edition

# Eureka Math Grade 5 Module 2

Special thanks go to the Gordon A. Cain Center and to the Department of Mathematics at Louisiana State University for their support in the development of *Eureka Math*.

For a free *Eureka Math* Teacher Resource Pack, Parent Tip Sheets, and more please visit www.Eureka.tools

**Published by the non-profit Great Minds**



Printed in the U.S.A.
This book may be purchased from the publisher at eureka-math.org
10 9 8

ISBN 978-1-63255-379-9

***Eureka Math: A Story of Units* Contributors**

Katrina Abdussalaam, Curriculum Writer
Tiah Alphonso, Program Manager—Curriculum Production
Kelly Alsup, Lead Writer / Editor, Grade 4
Catriona Anderson, Program Manager—Implementation Support
Debbie Andorka-Aceves, Curriculum Writer
Eric Angel, Curriculum Writer
Leslie Arceneaux, Lead Writer / Editor, Grade 5
Kate McGill Austin, Lead Writer / Editor, Grades PreK–K
Adam Baker, Lead Writer / Editor, Grade 5
Scott Baldridge, Lead Mathematician and Lead Curriculum Writer
Beth Barnes, Curriculum Writer
Bonnie Bergstresser, Math Auditor
Bill Davidson, Fluency Specialist
Jill Diniz, Program Director
Nancy Diorio, Curriculum Writer
Nancy Doorey, Assessment Advisor
Lacy Endo-Peery, Lead Writer / Editor, Grades PreK–K
Ana Estela, Curriculum Writer
Lessa Faltermann, Math Auditor
Janice Fan, Curriculum Writer
Ellen Fort, Math Auditor
Peggy Golden, Curriculum Writer
Maria Gomes, Pre-Kindergarten Practitioner
Pam Goodner, Curriculum Writer
Greg Gorman, Curriculum Writer
Melanie Gutierrez, Curriculum Writer
Bob Hollister, Math Auditor
Kelley Isinger, Curriculum Writer
Nuhad Jamal, Curriculum Writer
Mary Jones, Lead Writer / Editor, Grade 4
Halle Kananak, Curriculum Writer
Susan Lee, Lead Writer / Editor, Grade 3
Jennifer Loftin, Program Manager—Professional Development
Soo Jin Lu, Curriculum Writer
Nell McAnelly, Project Director

Ben McCarty, Lead Mathematician / Editor, PreK–5
Stacie McClintock, Document Production Manager
Cristina Metcalf, Lead Writer / Editor, Grade 3
Susan Midlarsky, Curriculum Writer
Pat Mohr, Curriculum Writer
Sarah Oyler, Document Coordinator
Victoria Peacock, Curriculum Writer
Jenny Petrosino, Curriculum Writer
Terrie Poehl, Math Auditor
Robin Ramos, Lead Curriculum Writer / Editor, PreK–5
Kristen Riedel, Math Audit Team Lead
Cecilia Rudzitis, Curriculum Writer
Tricia Salerno, Curriculum Writer
Chris Sarlo, Curriculum Writer
Ann Rose Sentoro, Curriculum Writer
Colleen Sheeron, Lead Writer / Editor, Grade 2
Gail Smith, Curriculum Writer
Shelley Snow, Curriculum Writer
Robyn Sorenson, Math Auditor
Kelly Spinks, Curriculum Writer
Marianne Strayton, Lead Writer / Editor, Grade 1
Theresa Streeter, Math Auditor
Lily Talcott, Curriculum Writer
Kevin Tougher, Curriculum Writer
Saffron VanGalder, Lead Writer / Editor, Grade 3
Lisa Watts-Lawton, Lead Writer / Editor, Grade 2
Erin Wheeler, Curriculum Writer
MaryJo Wieland, Curriculum Writer
Allison Witcraft, Math Auditor
Jessa Woods, Curriculum Writer
Hae Jung Yang, Lead Writer / Editor, Grade 1

This page intentionally left blank

Table of Contents

# GRADE 5 • MODULE 2

## Multi-Digit Whole Number and Decimal Fraction Operations

# Grade 5 • Module 2

# Multi-Digit Whole Number and Decimal Fraction Operations

## OVERVIEW

In Module 1, students explored the relationships of adjacent units on the place value chart to generalize whole number algorithms to decimal fraction operations. In Module 2, students apply the patterns of the base ten system to mental strategies and the multiplication and division algorithms.

Topics A through D provide a sequential study of multiplication. To link to prior learning and set the foundation for understanding the standard multiplication algorithm, students begin at the concrete–pictorial level in Topic A. They use place value disks to model multi-digit multiplication of place value units, for example, 42 × 10, 42 × 100, 42 × 1,000, leading to problems such as 42 × 30, 42 × 300, and 42 × 3,000 (**5.NBT.1**, **5.NBT.2**). They then round factors in Lesson 2 and discuss the reasonableness of their products. Throughout Topic A, students evaluate and write simple expressions to record their calculations using the associative property and parentheses to record the relevant order of calculations (**5.OA.1**).

In Topic B, place value understanding moves toward understanding the distributive property via area models, which are used to generate and record the partial products (**5.OA.1**, **5.OA.2**) of the standard algorithm (**5.NBT.5**). Topic C moves students from whole numbers to multiplication with decimals, again using place value as a guide to reason and make estimations about products (**5.NBT.7**). In Topic D, students explore multiplication as a method for expressing equivalent measures. For example, they multiply to convert between meters and centimeters or ounces and cups with measurements in both whole number and decimal form (**5.MD.1**).

Topics E through H provide a similar sequence for division. Topic E begins concretely with place value disks as an introduction to division with multi-digit whole numbers (**5.NBT.6**).

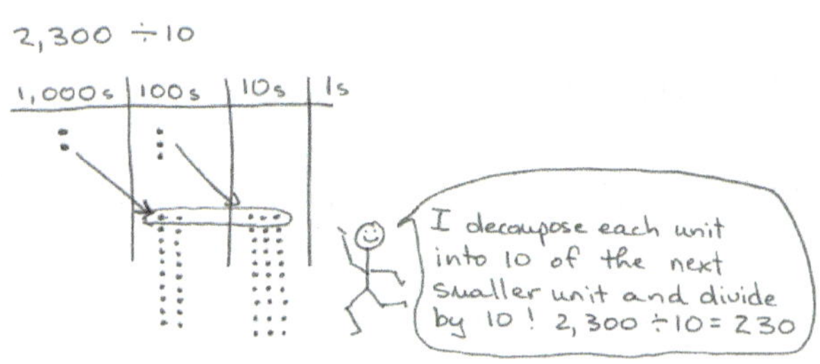

In the same lesson, 420 ÷ 60 is interpreted as 420 ÷ 10 ÷ 6. Next, students round dividends and two-digit divisors to nearby multiples of 10 in order to estimate single-digit quotients (e.g., 431 ÷ 58 ≈ 420 ÷ 60 = 7) and then multi-digit quotients. This work is done horizontally, outside the context of the written vertical method. The series of lessons in Topic F lead students to divide multi-digit dividends by two-digit divisors using the written vertical method. Each lesson moves to a new level of difficulty with a sequence beginning with divisors that are multiples of 10 to non-multiples of 10. Two instructional days are devoted to single-digit quotients with and without remainders before progressing to two- and three-digit quotients (**5.NBT.6**).

In Topic G, students use their understanding to divide decimals by two-digit divisors in a sequence similar to that of Topic F with whole numbers (**5.NBT.7**). In Topic H, students apply the work of the module to solve multi-step word problems using multi-digit division with unknowns representing either the group size or number of groups. In this topic, an emphasis on checking the reasonableness of their answers draws on skills learned throughout the module, including refining their knowledge of place value, rounding, and estimation.

## Notes on Pacing for Differentiation

If pacing is a challenge, consider the following modifications and omissions. Depending on students' strengths, consider consolidating Lessons 5 and 6. In Lesson 5, omit Problem 1 of the Concept Development, and move directly into renaming with the algorithm after Problem 2. Use the Problem Set from Lesson 6 for independent student practice. Consider consolidating Lessons 7 and 8 as well. Ask students to estimate the product beginning with the Concept Development of Lesson 7, and then use the Problem Set from Lesson 8 for student practice. Similarly, Lessons 11 and 12 can also be consolidated. Use estimation from the outset, and have students practice with the Problem Set from Lesson 12.

It is not recommended to omit any lessons from Topic D as it is a foundation for work later in the year. Students convert measurement units from small to large and from large to small using multiplication. This significantly expedites their understanding of and fluency with conversion and fraction multiplication as the year continues. In Lesson 14, students multiply whole numbers by unit fractions, which they learned to do in Grade 4 Module 5. If necessary, consider moving the fluency activity, "Multiply Unit Fractions," from Lesson 14 to Topic C to provide a few extra days of practice prior to beginning Lesson 14.

## Distribution of Instructional Minutes

This diagram represents a suggested distribution of instructional minutes based on the emphasis of particular lesson components in different lessons throughout the module.

- Fluency Practice
- Concept Development
- Application Problems
- Student Debrief

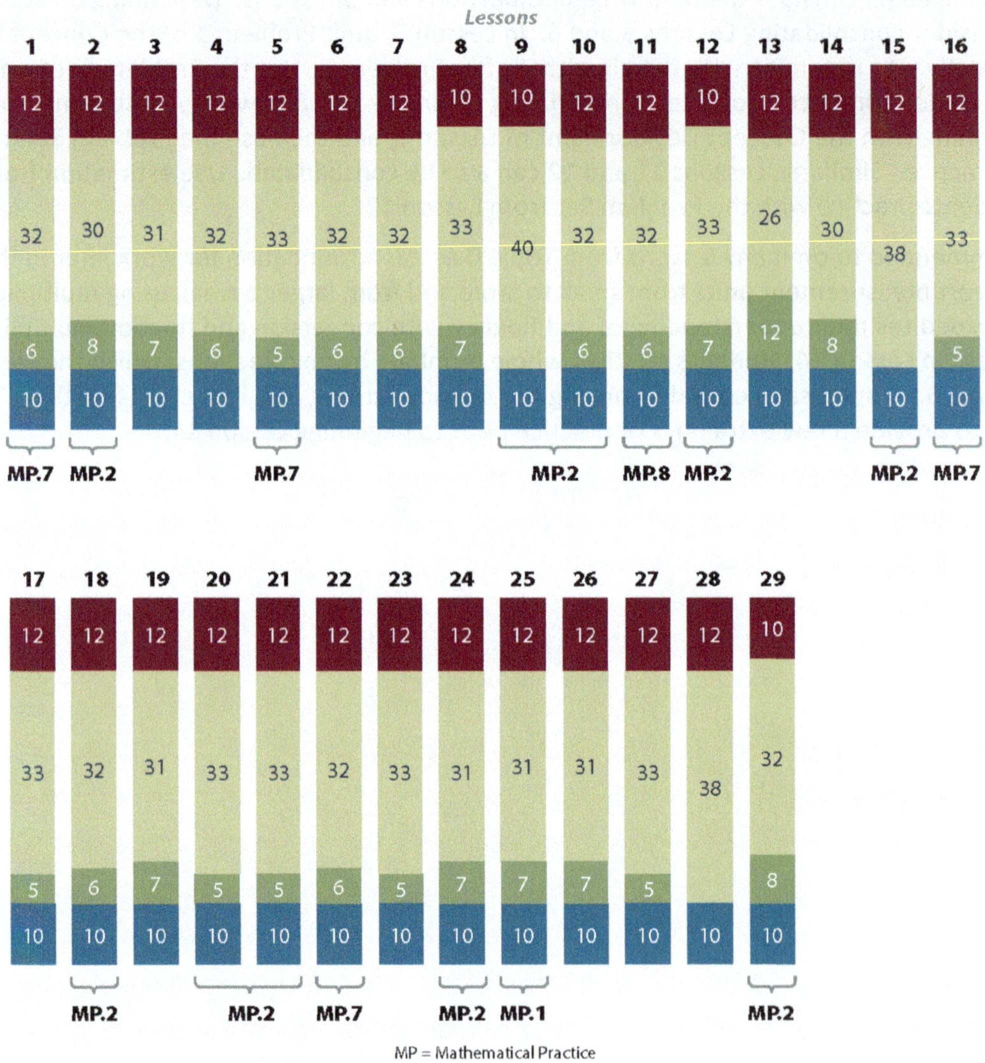


MP = Mathematical Practice

# Focus Grade Level Standards

## Write and interpret numerical expressions.

**5.OA.1** Use parentheses, brackets, or braces in numerical expressions, and evaluate expressions with these symbols.

**5.OA.2** Write simple expressions that record calculations with numbers, and interpret numerical expressions without evaluating them. *For example, express the calculation "add 8 and 7, then multiply by 2" as 2 × (8 + 7). Recognize that 3 × (18932 + 921) is three times as large as 18932 + 921, without having to calculate the indicated sum or product.*

## Understand the place value system.[1]

**5.NBT.1** Recognize that in a multi-digit number, a digit in one place represents 10 times as much as it represents in the place to its right and 1/10 of what it represents in the place to its left.

**5.NBT.2** Explain patterns in the number of zeros of the product when multiplying a number by powers of 10, and explain patterns in the placement of the decimal point when a decimal is multiplied or divided by a power of 10. Use whole-number exponents to denote power of 10.

## Perform operations with multi-digit whole numbers and with decimals to hundredths.

**5.NBT.5** Fluently multiply multi-digit whole numbers using the standard algorithm.

**5.NBT.6** Find whole-number quotients of whole numbers with up to four-digit dividends and two-digit divisors, using strategies based on place value, the properties of operations, and/or the relationship between multiplication and division. Illustrate and explain the calculation by using equations, rectangular arrays, and/or area models.

**5.NBT.7** Add, subtract, multiply, and divide decimals to hundredths, using concrete models or drawings and strategies based on place value, properties of operations, and/or the relationship between addition and subtraction; relate the strategy to a written method and explain the reasoning used.[2]

## Convert like measurement units within a given measurement system.

**5.MD.1** Convert among different-sized standard measurement units within a given measurement system (e.g., convert 5 cm to 0.05 m), and use these conversions in solving multi-step, real world problems.

---

[1]The balance of this cluster is addressed in Module 1.

[2]Focus on decimal multiplication of a single-digit whole number factor times a multi-digit number with up to two decimal places (e.g., 3 × 64.98). Restrict decimal division to a single digit whole number divisor with a multi-digit dividend with up to two decimal places (e.g., 64.98 ÷ 3). The balance of the standard is taught in Module 4.

## Foundational Standards

**4.OA.1** Interpret a multiplication equation as a comparison, e.g., interpret 35 = 5 × 7 as a statement that 35 is 5 times as many as 7 and 7 times as many as 5. Represent verbal statements of multiplicative comparisons as multiplication equations.

**4.OA.3** Solve multistep word problems posed with whole numbers and having whole-number answers using the four operations, including problems in which remainders must be interpreted. Represent these problems using equations with a letter standing for the unknown quantity. Assess the reasonableness of answers using mental computation and estimation strategies including rounding.

**4.NBT.4** Fluently add and subtract multi-digit whole numbers using the standard algorithm.

**4.NBT.5** Multiply a whole number of up to four digits by a one-digit whole number, and multiply two two-digit numbers, using strategies based on place value and the properties of operations. Illustrate and explain the calculation by using equations, rectangular arrays, and/or area models.

**4.NBT.6** Find whole-number quotients and remainders with up to four-digit dividends and one-digit divisors, using strategies based on place value, the properties of operations, and/or the relationship between multiplication and division. Illustrate and explain the calculation by using equations, rectangular arrays, and/or area models.

## Focus Standards for Mathematical Practice

**MP.1** **Make sense of problems and persevere in solving them.** Students make sense of problems when they use place value disks and area models to conceptualize and solve multiplication and division problems.

**MP.2** **Reason abstractly and quantitatively.** Students make sense of quantities and their relationships when they use both mental strategies and the standard algorithms to multiply and divide multi-digit whole numbers. Students also *decontextualize* when they represent problems symbolically and *contextualize* when they consider the value of the units used and understand the meaning of the quantities as they compute.

**MP.7** **Look for and make use of structure.** Students apply the *times 10, 100, 1,000* and the *divide by 10* patterns of the base ten system to mental strategies and the multiplication and division algorithms as they multiply and divide whole numbers and decimals.

**MP.8** **Look for and express regularity in repeated reasoning.** Students express the regularity they notice in repeated reasoning when they apply the partial quotients algorithm to divide two-, three-, and four-digit dividends by two-digit divisors. Students also check the reasonableness of the intermediate results of their division algorithms as they solve multi-digit division word problems.

## Overview of Module Topics and Lesson Objectives

| Standards | | Topics and Objectives | | Days |
|---|---|---|---|---|
| **5.NBT.1**<br>**5.NBT.2**<br>5.OA.1 | A | **Mental Strategies for Multi-Digit Whole Number Multiplication** | | 2 |
| | | Lesson 1: | Multiply multi-digit whole numbers and multiples of 10 using place value patterns and the distributive and associative properties. | |
| | | Lesson 2: | Estimate multi-digit products by rounding factors to a basic fact and using place value patterns. | |
| **5.OA.1**<br>**5.OA.2**<br>**5.NBT.5** | B | **The Standard Algorithm for Multi-Digit Whole Number Multiplication** | | 7 |
| | | Lesson 3: | Write and interpret numerical expressions, and compare expressions using a visual model. | |
| | | Lesson 4: | Convert numerical expressions into unit form as a mental strategy for multi-digit multiplication. | |
| | | Lesson 5: | Connect visual models and the distributive property to partial products of the standard algorithm without renaming. | |
| | | Lessons 6–7: | Connect area models and the distributive property to partial products of the standard algorithm with renaming. | |
| | | Lesson 8: | Fluently multiply multi-digit whole numbers using the standard algorithm and using estimation to check for reasonableness of the product. | |
| | | Lesson 9: | Fluently multiply multi-digit whole numbers using the standard algorithm to solve multi-step word problems. | |
| **5.NBT.7**<br>5.OA.1<br>5.OA.2<br>5.NBT.1 | C | **Decimal Multi-Digit Multiplication** | | 3 |
| | | Lesson 10: | Multiply decimal fractions with tenths by multi-digit whole numbers using place value understanding to record partial products. | |
| | | Lesson 11: | Multiply decimal fractions by multi-digit whole numbers through conversion to a whole number problem and reasoning about the placement of the decimal. | |
| | | Lesson 12: | Reason about the product of a whole number and a decimal with hundredths using place value understanding and estimation. | |
| **5.NBT.5**<br>**5.NBT.7**<br>**5.MD.1**<br>5.NBT.1<br>5.NBT.2 | D | **Measurement Word Problems with Whole Number and Decimal Multiplication** | | 3 |
| | | Lesson 13: | Use whole number multiplication to express equivalent measurements. | |
| | | Lesson 14: | Use fraction and decimal multiplication to express equivalent measurements. | |
| | | Lesson 15: | Solve two-step word problems involving measurement conversions. | |

| Standards | Topics and Objectives | | Days |
|---|---|---|---|
| | | Mid-Module Assessment: Topics A–D (assessment ½ day, return ½ day, remediation or further applications 2 days) | 3 |
| 5.NBT.1<br>5.NBT.2<br>5.NBT.6 | E | **Mental Strategies for Multi-Digit Whole Number Division**<br>Lesson 16: Use *divide by 10* patterns for multi-digit whole number division.<br>Lessons 17–18: Use basic facts to approximate quotients with two-digit divisors. | 3 |
| 5.NBT.6 | F | **Partial Quotients and Multi-Digit Whole Number Division**<br>Lesson 19: Divide two- and three-digit dividends by multiples of 10 with single-digit quotients, and make connections to a written method.<br>Lessons 20–21: Divide two- and three-digit dividends by two-digit divisors with single-digit quotients, and make connections to a written method.<br>Lessons 22–23: Divide three- and four-digit dividends by two-digit divisors resulting in two- and three-digit quotients, reasoning about the decomposition of successive remainders in each place value. | 5 |
| 5.NBT.2<br>5.NBT.7 | G | **Partial Quotients and Multi-Digit Decimal Division**<br>Lesson 24: Divide decimal dividends by multiples of 10, reasoning about the placement of the decimal point and making connections to a written method.<br>Lesson 25: Use basic facts to approximate decimal quotients with two-digit divisors, reasoning about the placement of the decimal point.<br>Lessons 26–27: Divide decimal dividends by two-digit divisors, estimating quotients, reasoning about the placement of the decimal point, and making connections to a written method. | 4 |
| 5.NBT.6<br>5.NBT.7 | H | **Measurement Word Problems with Multi-Digit Division**<br>Lessons 28–29: Solve division word problems involving multi-digit division with group size unknown and the number of groups unknown. | 2 |
| | | End-of-Module Assessment: Topics A–H (assessment ½ day, return ½ day, remediation or further application 2 days) | 3 |
| **Total Number of Instructional Days** | | | **35** |

# Terminology

## New or Recently Introduced Terms

- Conversion factor (the factor in a multiplication sentence that renames one measurement unit as another equivalent unit, e.g., 14 x (1 in) = 14 x ($\frac{1}{12}$ ft); 1 in and $\frac{1}{12}$ ft are the conversion factors)
- Decimal fraction (a proper fraction whose denominator is a power of 10)
- Multiplier (a quantity by which a given number—a multiplicand—is to be multiplied)
- Parentheses (the symbols used to relate order of operations)

## Familiar Terms and Symbols[3]

- Decimal (a fraction whose denominator is a power of ten and whose numerator is expressed by figures placed to the right of a decimal point)
- Digit (a symbol used to make numbers: 0, 1, 2, 3, 4, 5, 6, 7, 8, 9)
- Divisor (the number by which another number is divided)
- Equation (a statement that the values of two mathematical expressions are equal)
- Equivalence (a state of being equal or equivalent)
- Equivalent measures (e.g., 12 inches = 1 foot; 16 ounces = 1 pound)
- Estimate (approximation of the value of a quantity or number)
- Exponent (the number of times a number is to be used as a factor in a multiplication expression)
- Multiple (a number that can be divided by another number without a remainder like 15, 20, or any multiple of 5)
- Pattern (a systematically consistent and recurring trait within a sequence)
- Product (the result of multiplying numbers together)
- Quotient (the answer of dividing one quantity by another)
- Remainder (the number left over when one integer is divided by another)
- Renaming (decomposing or composing a number or units within a number)
- Rounding (approximating the value of a given number)
- Unit form (place value counting, e.g., 34 stated as 3 tens 4 ones)

[3]These are terms and symbols students have used or seen previously.

## Suggested Tools and Representations

- Area models (e.g., an array)
- Number bond
- Place value disks

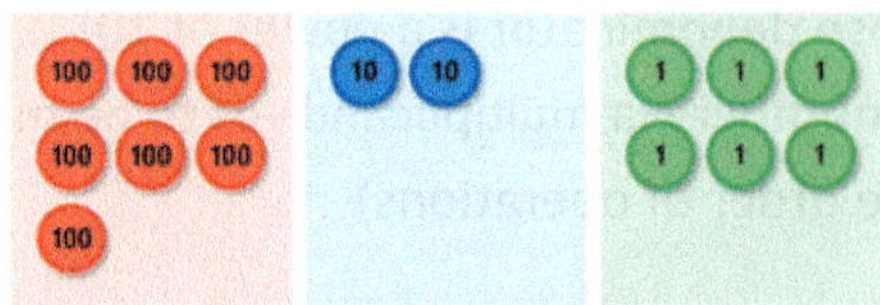


Unit form modeled with place value disks:
7 hundreds 2 tens 6 ones = 72 tens 6 ones = 726 ones

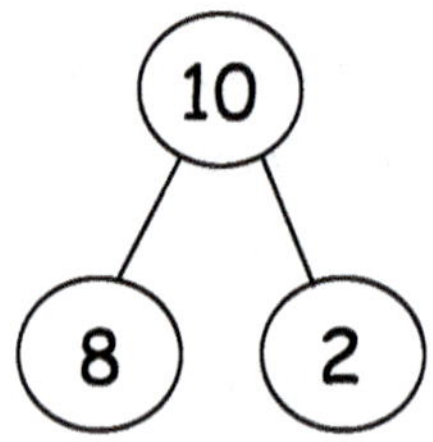


Number Bond

- Partial product (an algorithmic method that takes base ten decompositions of factors, makes products of all pairs, and adds all products together)
- Partial quotient (an algorithmic method using successive approximation)

## Scaffolds[4]

The scaffolds integrated into *A Story of Units* give alternatives for how students access information as well as express and demonstrate their learning. Strategically placed margin notes are provided within each lesson, elaborating on the use of specific scaffolds at applicable times. They address many needs presented by English language learners, students with disabilities, students performing above grade level, and students performing below grade level. Many of the suggestions are organized by Universal Design for Learning (UDL) principles and are applicable to more than one population. To read more about the approach to differentiated instruction in *A Story of Units,* please refer to "How to Implement *A Story of Units*."

[4]Students with disabilities may require Braille, large print, audio, or special digital files. Please visit the website, www.p12.nysed.gov/specialed/aim, for specific information on how to obtain student materials that satisfy the National Instructional Materials Accessibility Standard (NIMAS) format.

## Assessment Summary

| Type | Administered | Format | Standards Addressed |
|---|---|---|---|
| Mid-Module Assessment Task | After Topic D | Constructed response with rubric | 5.OA.1<br>5.OA.2<br>5.NBT.1<br>5.NBT.2<br>5.NBT.5<br>5.NBT.7<br>5.MD.1 |
| End-of-Module Assessment Task | After Topic H | Constructed response with rubric | 5.OA.1<br>5.OA.2<br>5.NBT.1<br>5.NBT.2<br>5.NBT.5<br>5.NBT.6<br>5.NBT.7<br>5.MD.1 |

# Topic A

# Mental Strategies for Multi-Digit Whole Number Multiplication

**5.NBT.1, 5.NBT.2,** 5.OA.1

| | | |
|---|---|---|
| **Focus Standards:** | 5.NBT.1 | Recognize that in a multi-digit number, a digit in one place represents 10 times as much as it represents in the place to its right and 1/10 of what it represents in the place to its left. |
| | 5.NBT.2 | Explain patterns in the number of zeros of the product when multiplying a number by powers of 10, and explain patterns in the placement of the decimal point when a decimal is multiplied or divided by a power of 10. Use whole-number exponents to denote power of 10. |
| **Instructional Days:** | 2 | |
| **Coherence -Links from:** | G4–M3 | Multi-Digit Multiplication and Division |
| **-Links to:** | G5–M5 | Addition and Multiplication with Volume and Area |
| | G6–M5 | Area, Surface Area, and Volume Problems |

Topic A begins a sequential study of multiplication that culminates in Topic D. In order to link prior learning from Grade 4 Module 1 and Grade 5 Module 1 and to set the stage for solidifying the standard multiplication algorithm, students begin at the concrete–pictorial level. They use place value disks to model multi-digit multiplication of place value units, for example, 42 × 10, 42 × 100, 42 × 1,000, leading quickly to problems such as 42 × 30, 42 × 300, and 42 × 3,000 (**5.NBT.1**, **5.NBT.2**). Students then round factors in Lesson 2 and discuss the reasonableness of their products. Throughout Topic A, students evaluate and write simple expressions to record their calculations using the associative property and parentheses to record the relevant order of calculations (**5.OA.1**).

## A Teaching Sequence Toward Mastery of Mental Strategies for Multi-Digit Whole Number Multiplication

**Objective 1: Multiply multi-digit whole numbers and multiples of 10 using place value patterns and the distributive and associative properties.
(Lesson 1)**

**Objective 2: Estimate multi-digit products by rounding factors to a basic fact and using place value patterns.
(Lesson 2)**

# Lesson 1

## Objective: Multiply multi-digit whole numbers and multiples of 10 using place value patterns and the distributive and associative properties.

### Suggested Lesson Structure

| | |
|---|---|
| Fluency Practice | (12 minutes) |
| Application Problem | (6 minutes) |
| Concept Development | (32 minutes) |
| Student Debrief | (10 minutes) |
| **Total Time** | **(60 minutes)** |

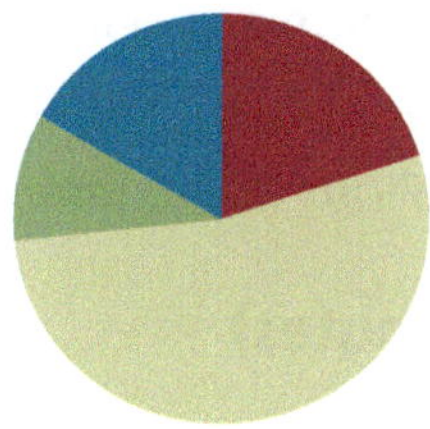

## Fluency Practice (12 minutes)

- Multiply by 10, 100, and 1,000 **5.NBT.2** (3 minutes)
- Place Value **5.NBT.3** (4 minutes)
- Round to Different Place Values **5.NBT.4** (5 minutes)

### Multiply by 10, 100, and 1,000 (3 minutes)

Note: This fluency activity reviews Module 1 skills and lays the groundwork for today's lesson in which both factors are multiples of 10.

T: (Write 3 × 10.) Say the product.

S: 30.

Repeat the process using the following possible sequence: 3 × 100, 3 × 1,000, 5 × 1,000, 0.005 × 1,000, 50 × 100, 0.05 × 100, 30 × 100, 30 × 1,000, 32 × 1,000, 0.32 × 1,000, 52 × 100, 5.2 × 100, 4 × 10, 0.4 × 10, 0.45 × 1,000, 30.45 × 1,000, 7 × 100, 72 × 100, and 7.002 × 100.

**NOTES ON MULTIPLE MEANS FOR ACTION AND EXPRESSION:**

Scaffold the Multiply by 10, 100, and 1,000 fluency activity for students working below grade level and others. Students may benefit from the aid of a place value chart or concrete place value disks, for example. Gradually decrease these scaffolds and encourage independence and strategies through pattern analysis, for example.

### Place Value (4 minutes)

Note: This fluency activity reviews composing and decomposing units, crucial to multiplying multiples of 10 in Lesson 2.

Materials: (S) Personal white board, millions to thousandths place value chart (Template)

T: (Project place value chart. Draw 4 tens disks in the tens column.) How many tens do you see?

S: 4 tens.

T: (Write 4 underneath the disks.) There are 4 tens and how many ones?

S: Zero ones.

T: (Write 0 in the ones column. Below it, write 4 tens = ____.) Fill in the blank.

S: 4 tens = 40.

Repeat the process for 4 ten thousands, 4 hundred thousands, 7 millions, and 2 thousands.

T: (Write 5 hundreds = ____.) Show the answer in your place value chart.

S: (Students write 5 in the hundreds column and 0 in the tens and ones columns.)

Repeat the process for 3 tens, 53 tens, 6 ten thousands, 36 ten thousands, 8 hundred thousands 36 ten thousands, 8 millions 24 ten thousands, 8 millions 17 hundred thousands, and 1,034 hundred thousands.

## Round to Different Place Values (5 minutes)

Note: Practicing rounding to different place values in isolation helps students when they estimate to find products in Lesson 2.

Materials: (S) Personal white board

T: (Project 8,735.) Say the number.

S: 8,735.

T: Let's round to the thousands, hundreds, and tens places.

T: Draw a vertical number line on your personal white board with two points and a midpoint between them.

T: Between which two thousands is 8,735?

S: 8 thousand and 9 thousand.

T: Label the two outside points with these values.

S: (Label.)

T: What's the midpoint for 8,000 and 9,000?

S: 8,500.

T: Label your number line. 8,500 is the same as how many hundreds?

S: 85 hundreds.

T: How many hundreds are in 8,735?

S: 87 hundreds.

T: (Write 8,735 ≈ ____.) Show 8,735 on your number line, and write the number sentence.

S: (Label 8,735 between 8,500 and 9,000 on the number line, and write 8,735 ≈ 9,000.)

Students round to the hundreds and tens. Follow the same process and procedure for 7,458.

## Application Problem (6 minutes)

The top surface of a desk has a length of 5.6 feet. The length is 4 times its width. What is the width of the desk?

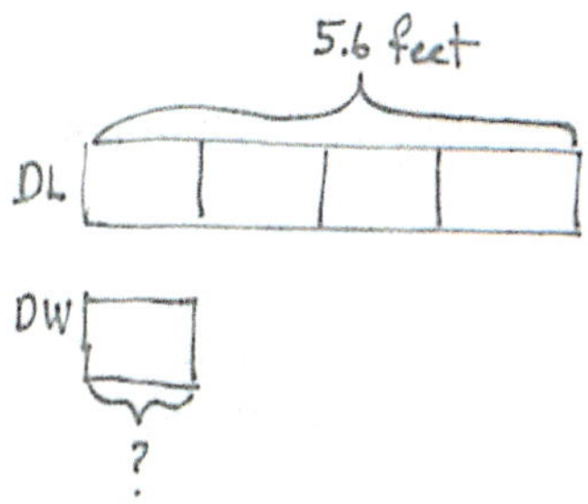


56 tenths ÷ 4 = 14 tenths
14 tenths = 1.4
The width of the desk is 1.4 ft.

Note: This is a review of Grade 5 Module 1 Topic F, dividing decimals by single-digit whole numbers. Allow students to share their approaches with the class. Accept any valid approach.

**NOTES ON MULTIPLE MEANS OF ENGAGEMENT:**

As the Application Problem is time-based rather than task-based, increase student engagement and decrease possible frustration by making specific goals for students who may need longer than 6 minutes to find their solutions. Monitor student use of Read, Draw, and Write to solve. If students, for example, find drawing accurate and relevant representations challenging, make that a goal. Celebrate every step toward success.

## Concept Development (32 minutes)

Materials: (S) Personal white board, millions to thousandths place value chart (Template)

**Problems 1–4**

4 × 30

40 × 30

40 × 300

4,000 × 30

T: (Write 4 × 30. Below it, write 4 × 3 tens = ____.) To find the product, start by multiplying the whole numbers, remembering to state the unit in your product.

S: 12 tens.

T: Show 12 tens on your place value chart. What is 12 tens in standard form?

S: 120.

T: (Write 4 tens × 3 tens = ____.) Solve with a partner.

S: (Solve.)

T: How did you use the previous problem to help you solve 4 tens × 3 tens?

S: The only difference was the place value unit of the first factor, so it was 12 hundreds. → It's the same as 4 threes times 10 times 10, which is 12 hundreds. → I multiplied 4 × 3, which is 12. I then multiplied tens by tens, so my new units are hundreds. Now, I have 12 hundreds, or 1,200.

T: Let me record what I hear you saying. (Write (4 × 3) × 100.)

T: (Write 4 tens × 3 hundreds = ___ on the board.) How is this problem different than the last problem?

S: We are multiplying tens and hundreds, not ones and hundreds or tens and tens.

T: 4 tens is the same as 4 times 10. (Write 4 × 10 on the board). 3 hundreds is the same as 3 times what?

S: 100.

T: (Write 3 × 100 next to 4 × 10 on the board.) So, another way to write our problem would be (4 × 10) × (3 × 100). (Now, write (4 × 3) × (10 × 100) on the board.) Are these expressions equal? Why or why not? Turn and talk.

S: Yes, they are the same. → We can multiply in any order, so they are the same.

T: What is 4 × 3?

S: 12.

T: (Record 12 under 4 × 3.) What is 10 × 100?

S: 1,000.

T: (Record 1,000 under 10 × 100.)

T: What is the product of 12 and 1,000?

S: 12,000.

Repeat the sequence with 4,000 × 30.

**Problems 5–8**

60 × 5

60 × 50

60 × 500

60 × 5,000

T: (Write 60 × 5.)

T: (Underneath the expression above, write (6 × 10) × 5 and (6 × 5) × 10.) Are both of these equivalent to 60 × 5? Why or why not? Turn and talk.

T: When we change the order of the factors, we are using the commutative (any-order) property. When we group the factors differently (point to the board), we are using the associative property of multiplication.

MP.7

T: Let's solve (6 × 5) × 10.

S: (Solve 30 × 10 = 300.)

T: For the next problem, use the properties and what you know about multiplying multiples of 10 to help you solve.

T: (Write 60 × 50 = ___.) Work with a partner to solve, and then explain.

MP.7

S: I thought of 60 as 6 × 10 and 50 as 5 × 10. I rearranged the factors to see (6 × 5) × (10 × 10). I got 30 × 100 = 3,000. → I first multiplied 6 times 5 and got 30. Then, I multiplied by 10 to get 300 and then multiplied by 10 to get 3,000.

T: I notice that in Problems 5–8 the number of zeros in the product was exactly the same as the number of zeros in our factors. That doesn't seem to be the case here. Why is that?

S: Because 6 × 5 is 30, then we have to multiply by 100. So, 30 ones × 100 is 30 hundreds, or 3,000.

T: Think about that as you solve 60 × 500 and 60 × 5,000 independently.

**Problems 9–12**

451 × 8

451 × 80

4,510 × 80

4,510 × 800

T: Find the product, 451 × 8, using any method.

S: (Solve to find 3,608.)

T: How did you solve?

S: I used the vertical algorithm. → I used the distributive property. I multiplied 400 × 8, then 50 × 8, and then 1 × 8. I added those products together.

T: What makes the distributive property useful here? Why does it help here when we didn't really use it in our other problems? Turn and talk.

S: There are different digits in three place values instead of all zeros. If I break the number apart by unit, then I can use basic facts to get the products.

T: Turn and talk to your partner about how you can use 451 × 8 to help you solve the 451 × 80, 4,510 × 80, and 4,510 × 800. Then, evaluate these expressions.

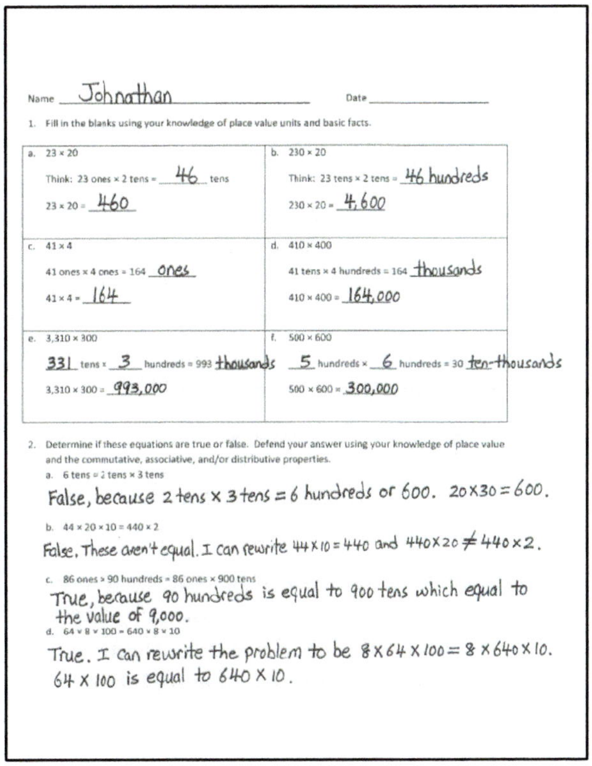

Name Johnathan Date

1. Fill in the blanks using your knowledge of place value units and basic facts.

| a. 23 × 20<br>Think: 23 ones × 2 tens = 46 tens<br>23 × 20 = 460 | b. 230 × 20<br>Think: 23 tens × 2 tens = 46 hundreds<br>230 × 20 = 4,600 |
|---|---|
| c. 41 × 4<br>41 ones × 4 ones = 164 ones<br>41 × 4 = 164 | d. 410 × 400<br>41 tens × 4 hundreds = 164 thousands<br>410 × 400 = 164,000 |
| e. 3,310 × 300<br>331 tens × 3 hundreds = 993 thousands<br>3,310 × 300 = 993,000 | f. 500 × 600<br>5 hundreds × 6 hundreds = 30 ten-thousands<br>500 × 600 = 300,000 |

2. Determine if these equations are true or false. Defend your answer using your knowledge of place value and the commutative, associative, and/or distributive properties.

a. 6 tens = 2 tens × 3 tens

False, because 2 tens × 3 tens = 6 hundreds or 600. 20×30 = 600.

b. 44 × 20 × 10 = 440 × 2

False. These aren't equal. I can rewrite 44×10 = 440 and 440×20 ≠ 440×2.

c. 86 ones × 90 hundreds = 86 ones × 900 tens

True, because 90 hundreds is equal to 900 tens which equal to the value of 9,000.

d. 64 × 8 × 100 = 640 × 8 × 10

True. I can rewrite the problem to be 8×64×100 = 8×640×10. 64×100 is equal to 640×10.

## Problem Set (10 minutes)

Students should do their personal best to complete the Problem Set within the allotted 10 minutes. Some problems do not specify a method for solving. This is an intentional reduction of scaffolding that invokes MP.5, Use Appropriate Tools Strategically. Students should solve these problems using the RDW approach used for Application Problems.

For some classes, it may be appropriate to modify the assignment by specifying which problems students should work on first. With this option, let the purposeful sequencing of the Problem Set guide the selections so that problems continue to be scaffolded. Balance word problems with other problem types to ensure a range of practice. Consider assigning incomplete problems for homework or at another time during the day.

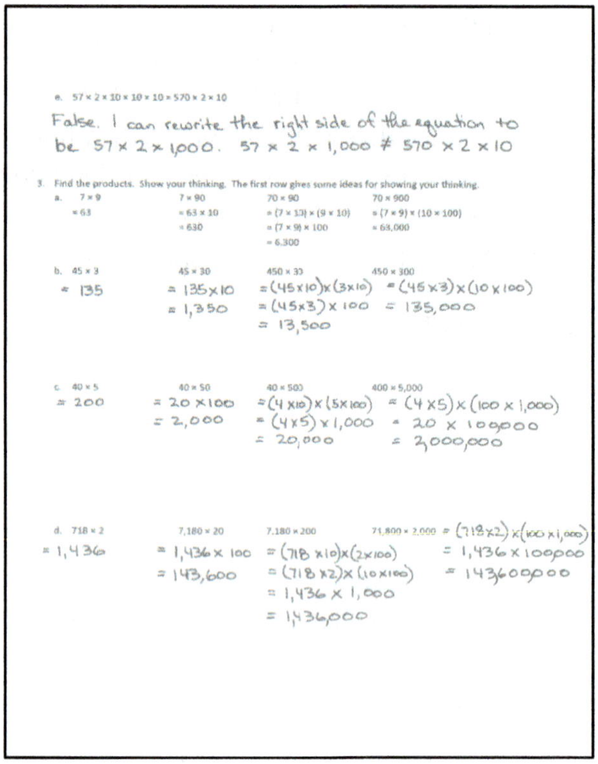

e. 57 × 2 × 10 × 10 × 10 = 570 × 2 × 10

False. I can rewrite the right side of the equation to be 57 × 2 × 1,000. 57 × 2 × 1,000 ≠ 570 × 2 × 10

3. Find the products. Show your thinking. The first row gives some ideas for showing your thinking.

a. 7 × 9 = 63; 7 × 90 = 63 × 10 = 630; 70 × 90 = (7 × 10) × (9 × 10) = (7 × 9) × 100 = 6,300; 70 × 900 = (7 × 9) × (10 × 100) = 63,000

b. 45 × 3 = 135; 45 × 30 = 135 × 10 = 1,350; 450 × 30 = (45 × 10) × (3 × 10) = (45 × 3) × 100 = 13,500; 450 × 300 = (45 × 3) × (10 × 100) = 135,000

c. 40 × 5 = 200; 40 × 50 = 20 × 100 = 2,000; 40 × 500 = (4 × 10) × (5 × 100) = (4 × 5) × 1,000 = 20,000; 400 × 5,000 = (4 × 5) × (100 × 1,000) = 20 × 100,000 = 2,000,000

d. 718 × 2 = 1,436; 7,180 × 20 = 1,436 × 100 = 143,600; 7,180 × 200 = (718 × 10) × (2 × 100) = (718 × 2) × (10 × 100) = 1,436 × 1,000 = 1,436,000; 71,800 × 2,000 = (718 × 2) × (100 × 1,000) = 1,436 × 100,000 = 143,600,000

## Student Debrief (10 minutes)

**Lesson Objective:** Multiply multi-digit whole numbers and multiples of 10 using place value patterns and the distributive and associative properties.

The Student Debrief is intended to invite reflection and active processing of the total lesson experience.

Invite students to review their solutions for the Problem Set. They should check work by comparing answers with a partner before going over answers as a class. Look for misconceptions or misunderstandings that can be addressed in the Debrief. Guide students in a conversation to debrief the Problem Set and process the lesson.

Any combination of the questions below may be used to lead the discussion.

- Take time to compare the various strategies used by students to find the products in Problem 3. Discuss how the parentheses that are used to show thinking directs us toward which part of the equation was grouped and, thus, which part of the expression is evaluated first.
- In Problem 3, for which problem was the distributive property most useful when solving? For which problems is the distributive property unnecessary?
- In Problem 2, was it necessary to solve each expression in order to compare the values? Why or why not? Lead the discussion toward the idea that the commutative, associative, and distributive properties allow us to make those comparisons without calculating.

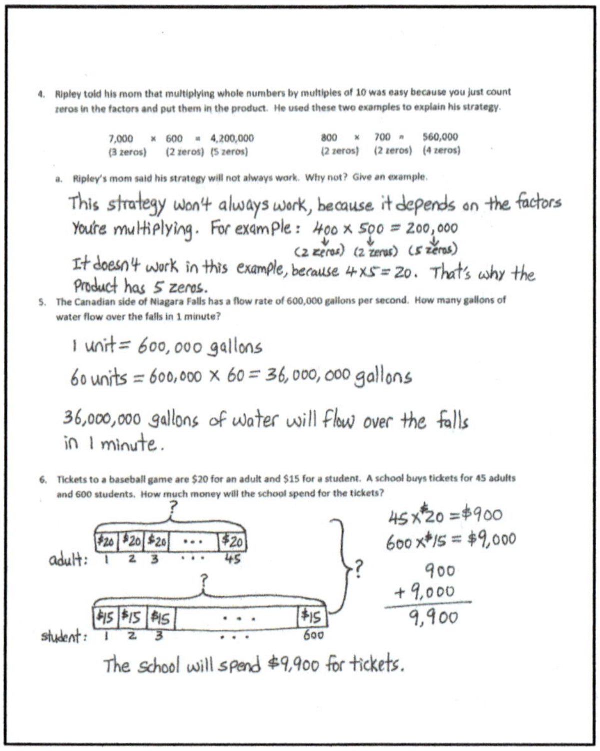

4. Ripley told his mom that multiplying whole numbers by multiples of 10 was easy because you just count zeros in the factors and put them in the product. He used these two examples to explain his strategy.

7,000 (3 zeros) × 600 (2 zeros) = 4,200,000 (5 zeros)    800 (2 zeros) × 700 (2 zeros) = 560,000 (4 zeros)

a. Ripley's mom said his strategy will not always work. Why not? Give an example.

This strategy won't always work, because it depends on the factors you're multiplying. For example: 400 (2 zeros) × 500 (2 zeros) = 200,000 (5 zeros). It doesn't work in this example, because 4 × 5 = 20. That's why the product has 5 zeros.

5. The Canadian side of Niagara Falls has a flow rate of 600,000 gallons per second. How many gallons of water flow over the falls in 1 minute?

1 unit = 600,000 gallons
60 units = 600,000 × 60 = 36,000,000 gallons

36,000,000 gallons of water will flow over the falls in 1 minute.

6. Tickets to a baseball game are $20 for an adult and $15 for a student. A school buys tickets for 45 adults and 600 students. How much money will the school spend for the tickets?



45 × $20 = $900
600 × $15 = $9,000
900 + 9,000 = 9,900

The school will spend $9,900 for tickets.

- Problem 4 raises one of the most common error patterns in multiplying by powers of 10. Take time to explore Ripley's error in thinking by allowing students to share their examples. Is there a pattern to the examples that we have shared? Any example involving 5 times an even number will produce such an example: 4 × 50; 50 × 60; 500 × 80; 2,000 × 50.
- How does understanding place value help you decompose large numbers to make them easier to multiply?
- About 36 million gallons of water leak from the New York City water supply every day. About how many gallons of water leak in one 30-day month? How can the patterns we discovered today about multiplying by 10s, 100s, and 1,000s help us solve this problem?

## Exit Ticket (3 minutes)

After the Student Debrief, instruct students to complete the Exit Ticket. A review of their work will help with assessing students' understanding of the concepts that were presented in today's lesson and planning more effectively for future lessons. The questions may be read aloud to the students.

Name ______________________________ Date ______________

1. Fill in the blanks using your knowledge of place value units and basic facts.

| | |
|---|---|
| a. 23 × 20<br><br>Think: 23 ones × 2 tens = __________ tens<br><br>23 × 20 = __________ | b. 230 × 20<br><br>Think: 23 tens × 2 tens = __________<br><br>230 × 20 = __________ |
| c. 41 × 4<br><br>41 ones × 4 ones = 164 __________<br><br>41 × 4 = __________ | d. 410 × 400<br><br>41 tens × 4 hundreds = 164 __________<br><br>410 × 400 = __________ |
| e. 3,310 × 300<br><br>______ tens × ______ hundreds = 993 ______<br><br>3,310 × 300 = __________ | f. 500 × 600<br><br>______ hundreds × _____ hundreds = 30 _____<br><br>500 × 600 = __________ |

2. Determine if these equations are true or false. Defend your answer using your knowledge of place value and the commutative, associative, and/or distributive properties.

   a. 6 tens = 2 tens × 3 tens

   b. 44 × 20 × 10 = 440 × 2

   c. 86 ones × 90 hundreds = 86 ones × 900 tens

   d. 64 × 8 × 100 = 640 × 8 × 10

e. $57 \times 2 \times 10 \times 10 \times 10 = 570 \times 2 \times 10$

3. Find the products. Show your thinking. The first row gives some ideas for showing your thinking.

a. $7 \times 9$
$= 63$

$7 \times 90$
$= 63 \times 10$
$= 630$

$70 \times 90$
$= (7 \times 10) \times (9 \times 10)$
$= (7 \times 9) \times 100$
$= 6{,}300$

$70 \times 900$
$= (7 \times 9) \times (10 \times 100)$
$= 63{,}000$

b. $45 \times 3$ $\quad$ $45 \times 30$ $\quad$ $450 \times 30$ $\quad$ $450 \times 300$

c. $40 \times 5$ $\quad$ $40 \times 50$ $\quad$ $40 \times 500$ $\quad$ $400 \times 5{,}000$

d. $718 \times 2$ $\quad$ $7{,}180 \times 20$ $\quad$ $7{,}180 \times 200$ $\quad$ $71{,}800 \times 2{,}000$

4. Ripley told his mom that multiplying whole numbers by multiples of 10 was easy because you just count zeros in the factors and put them in the product. He used these two examples to explain his strategy.

| 7,000 | × | 600 | = | 4,200,000 | | 800 | × | 700 | = | 560,000 |
|---|---|---|---|---|---|---|---|---|---|---|
| (3 zeros) | | (2 zeros) | | (5 zeros) | | (2 zeros) | | (2 zeros) | | (4 zeros) |

Ripley's mom said his strategy will not always work. Why not? Give an example.

5. The Canadian side of Niagara Falls has a flow rate of 600,000 gallons per second. How many gallons of water flow over the falls in 1 minute?

6. Tickets to a baseball game are $20 for an adult and $15 for a student. A school buys tickets for 45 adults and 600 students. How much money will the school spend for the tickets?

Name ______________________________ Date ________________

1. Find the products.

a. $1{,}900 \times 20$

b. $6{,}000 \times 50$

c. $250 \times 300$

2. Explain how knowing $50 \times 4 = 200$ helps you find $500 \times 400$.

Name ______________________________ Date ________________

1. Fill in the blanks using your knowledge of place value units and basic facts.

   a. $43 \times 30$

      Think: 43 ones × 3 tens = __________ tens

      $43 \times 30 =$ __________

   b. $430 \times 30$

      Think: 43 tens × 3 tens = __________ hundreds

      $430 \times 30 =$ __________

   c. $830 \times 20$

      Think: 83 tens × 2 tens = 166 __________

      $830 \times 20 =$ __________

   d. $4{,}400 \times 400$

      __________ hundreds × __________ hundreds = 176 __________

      $4{,}400 \times 400 =$ __________

   e. $80 \times 5{,}000$

      __________ tens × __________ thousands = 40 __________

      $80 \times 5{,}000 =$ __________

2. Determine if these equations are true or false. Defend your answer using your knowledge of place value and the commutative, associative, and/or distributive properties.

   a. 35 hundreds = 5 tens × 7 tens

   b. $770 \times 6 = 77 \times 6 \times 100$

   c. 50 tens × 4 hundreds = 40 tens × 5 hundreds

   d. $24 \times 10 \times 90 = 90 \times 2{,}400$





EUREKA MATH™

3. Find the products. Show your thinking. The first row gives some ideas for showing your thinking.

| | | | |
|---|---|---|---|
| a. 5 × 5<br>= 25 | 5 × 50<br>= 25 × 10<br>= 250 | 50 × 50<br>= (5 × 10) × (5 × 10)<br>= (5 × 5) × 100<br>= 2,500 | 50 × 500<br>= (5 × 5) × (10 × 100)<br>= 25,000 |
| b. 80 × 5 | 80 × 50 | 800 × 500 | 8,000 × 50 |
| c. 637 × 3 | 6,370 × 30 | 6,370 × 300 | 63,700 × 300 |

4. A concrete stepping-stone measures 20 square inches. What is the area of 30 such stones?

5. A number is 42,300 when multiplied by 10. Find the product of this number and 500.

| 1,000,000 | 100,000 | 10,000 | 1,000 | 100 | 10 | 1 | • | $\frac{1}{10}$ | $\frac{1}{100}$ | $\frac{1}{1,000}$ |
|---|---|---|---|---|---|---|---|---|---|---|
| Millions | Hundred Thousands | Ten Thousands | Thousands | Hundreds | Tens | Ones | • | Tenths | Hundredths | Thousandths |
| | | | | | | | • | | | |
| | | | | | | | • | | | |
| | | | | | | | • | | | |
| | | | | | | | • | | | |
| | | | | | | | • | | | |

millions to thousandths place value chart





EUREKA MATH™

# Lesson 2

## Objective: Estimate multi-digit products by rounding factors to a basic fact and using place value patterns.

### Suggested Lesson Structure

| | |
|---|---|
| Fluency Practice | (12 minutes) |
| Application Problem | (8 minutes) |
| Concept Development | (30 minutes) |
| Student Debrief | (10 minutes) |
| **Total Time** | **(60 minutes)** |

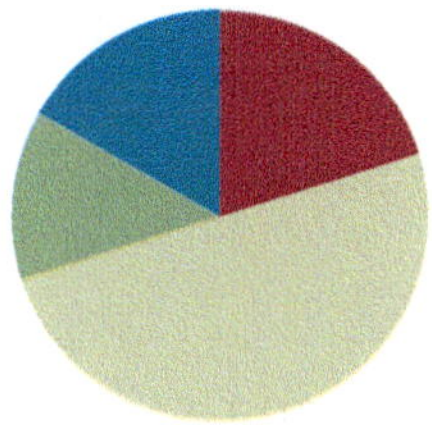

## Fluency Practice (12 minutes)

- Sprint: Multiply by 10, 100, and 1,000 **5.NBT.2** (8 minutes)
- Round to Different Place Values **5.NBT.4** (2 minutes)
- Multiply by Multiples of 10 **5.NBT.2** (2 minutes)

### Sprint: Multiply by 10, 100, and 1,000 (8 minutes)

Note: This review fluency activity helps preserve skills students learned and mastered in Module 1 and lays the groundwork for future concepts.

Materials: (S) Multiply by 10, 100, and 1,000 Sprint

### Round to Different Place Values (2 minutes)

Note: Practice with rounding to different place values in isolation helps students when they estimate to find products later in the module.

T: (Project 48,625.) Say the number.
S: 48,625.
T: (Write a vertical number line with two points and a midpoint.) Between which two ten thousands is 48,625?
S: 40,000 and 50,000.
T: (Write 40,000 at the bottom point and 50,000 at the top point.) What's the midpoint for 40,000 and 50,000?
S: 45,000.
T: (Write 45,000 at the midpoint.) Would 48,625 fall above or below 45,000?
S: Above.

T: (Write 48,625 ≈ ____.) What's 48,625 rounded to the nearest ten thousand?

S: 50,000.

Repeat the process for thousands, hundreds, and tens.

### Multiply by Multiples of 10 (2 minutes)

Note: This review fluency activity helps preserve skills students learned and mastered in Module 1 and lays the groundwork for future concepts.

Material: (S) Personal white board

T: (Write 31 × 10 = ____.) Say the multiplication sentence.

S: 31 × 10 = 310.

T: (Write 310 × 2 = ____ beside 31 × 10 = 310.) Say the multiplication sentence.

S: 310 × 2 = 620.

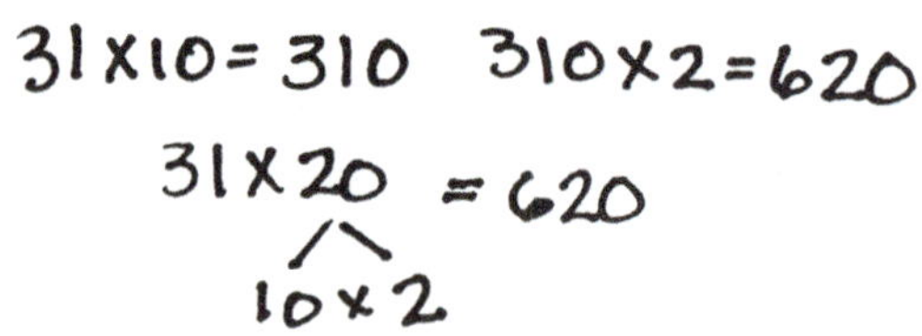


T: (Write 310 × 20 = ____ below 310 × 2 = 620.) Write 310 × 20 as a three-step multiplication sentence, taking out the ten.

S: 310 × 10 × 2 = 6,200.

T: Show your personal white board. (Check for accuracy.)

Direct students to solve using the same method for 23 × 40 and 32 × 30.

## Application Problem (8 minutes)

Jonas practices guitar 1 hour a day for 2 years. Bradley practices the guitar 2 hours a day more than Jonas. How many more minutes does Bradley practice the guitar than Jonas over the course of 2 years?

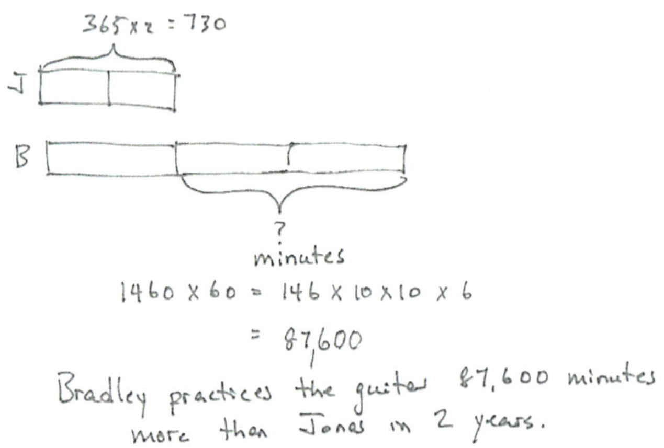


**NOTES ON MULTIPLE MEANS OF ACTION AND EXPRESSION:**

It may be helpful to offer a conversion table to students working below grade level and others, which includes the following:

- 1 hour = 60 minutes
- 1 year = 365 days

Note: The Application Problem is a multi-step word problem that asks students to convert units and multiply with multi-digit factors using their knowledge of the distributive and associative properties from Lesson 1. Allow students to share approaches with classmates.

## Concept Development (30 minutes)

**Problem 1**

Contextualize estimation using population of classroom and school.

T: How many students do we have in class? (Use class, school, and building numbers for the following that would yield a two-digit by two-digit estimation equation.)

S: 23.

T: Do all of the classes have exactly 23 students?

S: No.

T: There are 18 classes, but I'm not sure exactly how many students are in each class. What could I do to find a number that is close to the actual number of students in our school?

S: Estimate how many students are in each class.

T: Great idea. What number could help me make an estimate for the number of students in each class?

S: You could use the number in our class of 23.

T: True, but 23 is a little more difficult to multiply in my head. I'd like to use a number that I can multiply mentally. What could I round 23 students to so it is easier to multiply?

S: 20 students.

T: What could I round 18 classes to?

S: 20 classes.

T: How would I estimate the total number of students?

S: Multiply 20 by 20.

T: What would my estimate be? Explain your thinking.

S: 400. 2 times 2 is 4. Then you multiply 4 by 10 and 10.

T: (Write on the board (4 × 10) × 10 = 40 × 10 = 4 × 100.)

T: About 400 students. Estimates can help us understand a reasonable size of a product when we multiply the original numbers.

**Problems 2–4**

456 × 42 → 500 × 40 = 20,000

4,560 × 42 → 5,000 × 40 = 200,000

4,560 × 420 → 5,000 × 400 = 2,000,000

T: (Write on board 456 × 42 = ____.)

T: Suppose I don't need to know the exact product, just an estimate. How could I round the factors to estimate the product?

S: You could round to the nearest 10. → You'd get 460 × 40.

T: 460 × 40 is still pretty hard for me to do in my head. Could I round 456 to a different place value to make the product easier to find?

S: You could round to the hundreds place. → 500 × 40 is just like we did in the previous lesson!

T: 500 × 40 does sound pretty easy! What would my estimate be? Can you give me the multiplication sentence in unit form?

S: 5 hundreds × 4 tens equals 20 thousands.

T: (Write (5 × 100) × (4 × 10) = 20 × 1,000 = 20,000.) So, my product is about 20,000.

**Problems 5–7**

1,320 × 88

13,205 × 880

3,120 × 880

T: (Write on the board 1,320 × 88 = ____.) Round the factors to estimate the product.

S: (Estimate.)

T: Explain your thinking. (Accept any reasonable estimates of the factors. The most important thinking is how the properties are used to arrive at a product. Ask students to justify their choice of place value for rounding.)

S: I used 1,300 × 90, so I multiplied 13 × 9, then multiplied that by 1,000. This gave me 117,000. → I used 1,000 × 90 and got 90,000.

MP.2

T: Now, before you estimate 13,205 × 880, compare this to the problem we just did. What do you notice is different?

S: The factors are greater. → 13,205 is about 10 times as large as 1,320, and 880 is exactly 10 times as large as 88.

T: What do you think that will do to our estimate?

S: It should increase the product. → The product should be about 100 times as large as the first one.

T: Let's test that prediction. Round and find the estimated product. (Accept any reasonable estimate of the factors. The important thinking is the properties and the comparison of the relative sizes of the products.)

S: 13,205 → 10,000 and 880 → 900. So, 10,000 × 900 = (9 × 1) × 10,000 × 100 = 9,000,000.

T: Was our prediction correct?

S: Yes. 9 million is 100 times as large as 90,000.

Repeat the sequence for 3,120 × 880 and 31,200 × 880.

## Problem Set (10 minutes)

Students should do their personal best to complete the Problem Set within the allotted 10 minutes. For some classes, it may be appropriate to modify the assignment by specifying which problems they work on first. Some problems do not specify a method for solving. Students should solve these problems using the RDW approach used for Application Problems. (Please see "How to Implement *A Story of Units*" for more information on the Read–Draw–Write approach.)

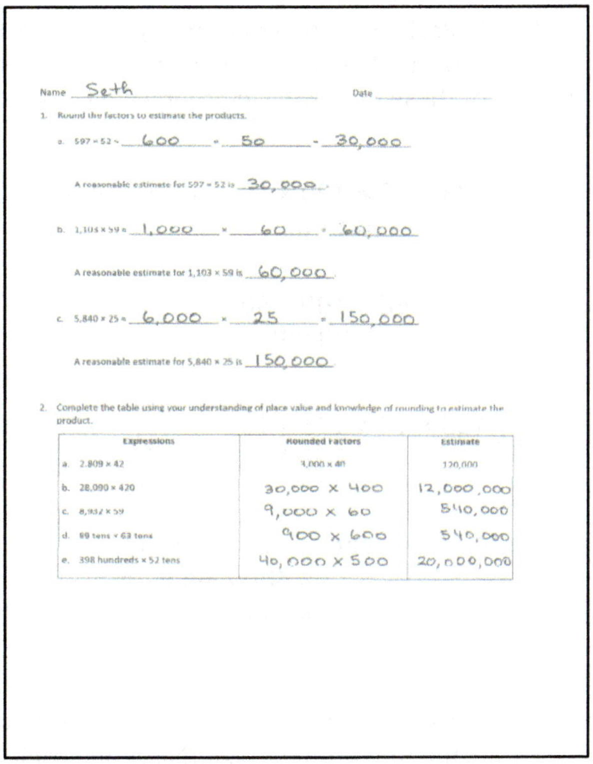

Name Seth Date

1. Round the factors to estimate the products.

a. 597 × 52 ≈ 600 × 50 = 30,000

A reasonable estimate for 597 × 52 is 30,000.

b. 1,103 × 59 ≈ 1,000 × 60 = 60,000

A reasonable estimate for 1,103 × 59 is 60,000.

c. 5,840 × 25 ≈ 6,000 × 25 = 150,000

A reasonable estimate for 5,840 × 25 is 150,000.

2. Complete the table using your understanding of place value and knowledge of rounding to estimate the product.

| Expressions | Rounded Factors | Estimate |
|---|---|---|
| a. 2,809 × 42 | 3,000 × 40 | 120,000 |
| b. 28,090 × 420 | 30,000 × 400 | 12,000,000 |
| c. 8,932 × 59 | 9,000 × 60 | 540,000 |
| d. 89 tens × 63 tens | 900 × 600 | 540,000 |
| e. 398 hundreds × 52 tens | 40,000 × 500 | 20,000,000 |

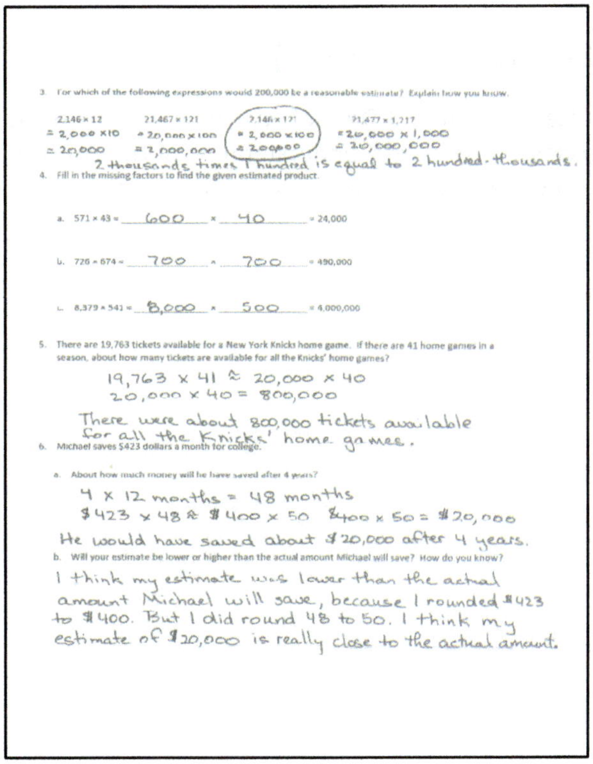

3. For which of the following expressions would 200,000 be a reasonable estimate? Explain how you know.

2,146 × 12 ≈ 2,000 × 10 ≈ 20,000

21,467 × 121 ≈ 20,000 × 100 ≈ 2,000,000

2,146 × 121 ≈ 2,000 × 100 ≈ 200,000 (circled)

21,477 × 1,217 ≈ 20,000 × 1,000 ≈ 20,000,000

2 thousands times 1 hundred is equal to 2 hundred-thousands.

4. Fill in the missing factors to find the given estimated product.

a. 571 × 43 ≈ 600 × 40 = 24,000

b. 726 × 674 ≈ 700 × 700 = 490,000

c. 8,379 × 541 ≈ 8,000 × 500 = 4,000,000

5. There are 19,763 tickets available for a New York Knicks home game. If there are 41 home games in a season, about how many tickets are available for all the Knicks' home games?

19,763 × 41 ≈ 20,000 × 40

20,000 × 40 = 800,000

There were about 800,000 tickets available for all the Knicks' home games.

6. Michael saves $423 dollars a month for college.

a. About how much money will he have saved after 4 years?

4 × 12 months = 48 months

$423 × 48 ≈ $400 × 50   $400 × 50 = $20,000

He would have saved about $20,000 after 4 years.

b. Will your estimate be lower or higher than the actual amount Michael will save? How do you know?

I think my estimate was lower than the actual amount Michael will save, because I rounded $423 to $400. But I did round 48 to 50. I think my estimate of $20,000 is really close to the actual amount.

## Student Debrief (10 minutes)

**Lesson Objective:** Estimate multi-digit products by rounding factors to a basic fact and using place value patterns.

The Student Debrief is intended to invite reflection and active processing of the total lesson experience.

Invite students to review their solutions for the Problem Set. They should check work by comparing answers with a partner before going over answers as a class. Look for misconceptions or misunderstandings that can be addressed in the Debrief. Guide students in a conversation to debrief the Problem Set and process the lesson.

Any combination of the questions below may be used to lead the discussion.

- Raise the idea of a different rounding strategy for Problem 1(c) using factors of 25 as "easy" mental factors. Ask students to consider the notion of rounding only one factor (e.g., 5,840 to 6,000). Multiply 6 × 25 = 150, and then multiply 150 × 1,000 to reach 150,000. What makes 25 an easy factor even though it is not a multiple of 10? Are there other numbers that students think of as easy like 25? Compare this to rounding both factors.

- In Problem 6, there are many ways to estimate the solution. Discuss the precision of each one. Which is the closest estimate? Does it matter in the context of this problem? Students may use any of these or may have other valid responses:

$$423 \times 12 \approx 400 \times 10 = 4{,}000 \Rightarrow 4{,}000 \times 4 = 16{,}000$$

$$423 \times 48 \approx 400 \times 50 = 20{,}000$$

$$423 \times 12 \approx 423 \times 10 = 4{,}230 \Rightarrow 4{,}230 \times 4 \approx 4{,}200 \times 4 = 16{,}800$$

$$423 \times 4 \text{ years} \approx 423 \times 5 \text{ years} \approx 400 \times 60 \text{ months} = 24{,}000$$

**NOTES ON MULTIPLE MEANS OF REPRESENTATION:**

Depending on the needs of English language learners, consider allowing students to discuss their responses to the Student Debrief in their first language or providing sentence frames or starters such as:

"I notice when I multiply 25 by any number that …."

"I chose to round to the ____ place because …."

"My estimate was close to the actual answer because …."

- Consider allowing students to generate other factors in Problem 4 that would round to produce the estimated product. Compare the problems to see how various powers of 10 multiplied by each other still yield a product in the thousands.

## Exit Ticket (3 minutes)

After the Student Debrief, instruct students to complete the Exit Ticket. A review of their work will help with assessing students' understanding of the concepts that were presented in today's lesson and planning more effectively for future lessons. The questions may be read aloud to the students.

# A

Number Correct: _______

Multiply by 10, 100, and 1,000

| | | |
|---|---|---|
| 1. | 9 × 10 = | |
| 2. | 9 × 100 = | |
| 3. | 9 × 1,000 = | |
| 4. | 8 × 10 = | |
| 5. | 80 × 10 = | |
| 6. | 80 × 100 = | |
| 7. | 80 × 1,000 = | |
| 8. | 7 × 10 = | |
| 9. | 70 × 10 = | |
| 10. | 700 × 10 = | |
| 11. | 700 × 100 = | |
| 12. | 700 × 1,000 = | |
| 13. | 2 × 10 = | |
| 14. | 30 × 10 = | |
| 15. | 32 × 10 = | |
| 16. | 4 × 10 = | |
| 17. | 50 × 10 = | |
| 18. | 54 × 10 = | |
| 19. | 37 × 10 = | |
| 20. | 84 × 10 = | |
| 21. | 84 × 100 = | |
| 22. | 84 × 1,000 = | |

| | | |
|---|---|---|
| 23. | 73 × 1,000 = | |
| 24. | 60 × 10 = | |
| 25. | 600 × 10 = | |
| 26. | 600 × 100 = | |
| 27. | 65 × 100 = | |
| 28. | 652 × 100 = | |
| 29. | 342 × 100 = | |
| 30. | 800 × 100 = | |
| 31. | 800 × 1,000 = | |
| 32. | 860 × 1,000 = | |
| 33. | 867 × 1,000 = | |
| 34. | 492 × 1,000 = | |
| 35. | 34 × 10 = | |
| 36. | 629 × 10 = | |
| 37. | 94 × 100 = | |
| 38. | 238 × 100 = | |
| 39. | 47 × 1,000 = | |
| 40. | 294 × 1,000 = | |
| 41. | 174 × 100 = | |
| 42. | 285 × 1,000 = | |
| 43. | 951 × 100 = | |
| 44. | 129 × 1,000 = | |

# B

Number Correct: _______

Improvement: _______

Multiply by 10, 100, and 1,000

| | | | | | |
|---|---|---|---|---|---|
| 1. | 8 × 10 = | | 23. | 37 × 1,000 = | |
| 2. | 8 × 100 = | | 24. | 50 × 10 = | |
| 3. | 8 × 1,000 = | | 25. | 500 × 10 = | |
| 4. | 7 × 10 = | | 26. | 500 × 100 = | |
| 5. | 70 × 10 = | | 27. | 56 × 100 = | |
| 6. | 70 × 100 = | | 28. | 562 × 100 = | |
| 7. | 70 × 1,000 = | | 29. | 432 × 100 = | |
| 8. | 6 × 10 = | | 30. | 700 × 100 = | |
| 9. | 60 × 10 = | | 31. | 700 × 1,000 = | |
| 10. | 600 × 10 = | | 32. | 760 × 1,000 = | |
| 11. | 600 × 100 = | | 33. | 765 × 1,000 = | |
| 12. | 600 × 1,000 = | | 34. | 942 × 1,000 = | |
| 13. | 3 × 10 = | | 35. | 74 × 10 = | |
| 14. | 20 × 10 = | | 36. | 269 × 10 = | |
| 15. | 23 × 10 = | | 37. | 49 × 100 = | |
| 16. | 5 × 10 = | | 38. | 328 × 100 = | |
| 17. | 40 × 10 = | | 39. | 37 × 1,000 = | |
| 18. | 45 × 10 = | | 40. | 924 × 1,000 = | |
| 19. | 73 × 10 = | | 41. | 147 × 100 = | |
| 20. | 48 × 10 = | | 42. | 825 × 1,000 = | |
| 21. | 48 × 100 = | | 43. | 651 × 100 = | |
| 22. | 48 × 1,000 = | | 44. | 192 × 1,000 = | |

Name ______________________________ Date ________________

1. Round the factors to estimate the products.

   a. 597 × 52 ≈ ____________ × ____________ = ____________

      A reasonable estimate for 597 × 52 is ____________.

   b. 1,103 × 59 ≈ ____________ × ____________ = ____________

      A reasonable estimate for 1,103 × 59 is ____________.

   c. 5,840 × 25 ≈ ____________ × ____________ = ____________

      A reasonable estimate for 5,840 × 25 is ____________.

2. Complete the table using your understanding of place value and knowledge of rounding to estimate the product.

| Expressions | Rounded Factors | Estimate |
|---|---|---|
| a. 2,809 × 42 | 3,000 × 40 | 120,000 |
| b. 28,090 × 420 | | |
| c. 8,932 × 59 | | |
| d. 89 tens × 63 tens | | |
| e. 398 hundreds × 52 tens | | |

3. For which of the following expressions would 200,000 be a reasonable estimate? Explain how you know.

   2,146 × 12     21,467 × 121     2,146 × 121     21,477 × 1,217

4. Fill in the missing factors to find the given estimated product.

   a. 571 × 43 ≈ ________________ × ________________ = 24,000

   b. 726 × 674 ≈ ________________ × ________________ = 490,000

   c. 8,379 × 541 ≈ ________________ × ________________ = 4,000,000

5. There are 19,763 tickets available for a New York Knicks home game. If there are 41 home games in a season, about how many tickets are available for all the Knicks' home games?

6. Michael saves $423 dollars a month for college.

   a. About how much money will he have saved after 4 years?

   b. Will your estimate be lower or higher than the actual amount Michael will save? How do you know?

Name ______________________________ Date ________________

Round the factors and estimate the products.

a. 656 × 106 ≈

b. 3,108 × 7,942 ≈

c. 425 × 9,311 ≈

d. 8,633 × 57,008 ≈

Name ______________________________ Date ________________

1. Round the factors to estimate the products.

   a. 697 × 82 ≈ ____________ × ____________ = ____________

      A reasonable estimate for 697 × 82 is ____________.

   b. 5,897 × 67 ≈ ____________ × ____________ = ____________

      A reasonable estimate for 5,897 × 67 is ____________.

   c. 8,840 × 45 ≈ ____________ × ____________ = ____________

      A reasonable estimate for 8,840 × 45 is ____________.

2. Complete the table using your understanding of place value and knowledge of rounding to estimate the product.

| Expressions | Rounded Factors | Estimate |
|---|---|---|
| a. 3,409 × 73 | 3,000 × 70 | 210,000 |
| b. 82,290 × 240 | | |
| c. 9,832 × 39 | | |
| d. 98 tens × 36 tens | | |
| e. 893 hundreds × 85 tens | | |

3. The estimated answer to a multiplication problem is 800,000. Which of the following expressions could result in this answer? Explain how you know.

   8,146 × 12    81,467 × 121    8,146 × 121    81,477 × 1,217

4. Fill in the blank with the missing estimate.

 a. 751 × 34 ≈ ________________ × ________________ = 24,000

 b. 627 × 674 ≈ ________________ × ________________ = 420,000

 c. 7,939 × 541 ≈ ________________ × ________________ = 4,000,000

5. In a single season, the New York Yankees sell an average of 42,362 tickets for each of their 81 home games. About how many tickets do they sell for an entire season of home games?

6. Raphael wants to buy a new car.

 a. He needs a down payment of $3,000. If he saves $340 each month, about how many months will it take him to save the down payment?

 b. His new car payment will be $288 each month for five years. What is the total of these payments?

## Topic B

# The Standard Algorithm for Multi-Digit Whole Number Multiplication

### 5.OA.1, 5.OA.2, 5.NBT.5

| | | |
|---|---|---|
| **Focus Standards:** | 5.OA.1 | Use parentheses, brackets, or braces in numerical expressions, and evaluate expressions with these symbols. |
| | 5.OA.2 | Write simple expressions that record calculations with numbers, and interpret numerical expressions without evaluating them. *For example, express the calculation "add 8 and 7, then multiply by 2" as 2 × (8 + 7). Recognize that 3 × (18932 + 921) is three times as large as 18932 + 921, without having to calculate the indicated sum or product.* |
| | 5.NBT.5 | Fluently multiply multi-digit whole numbers using the standard algorithm. |
| **Instructional Days:** | 7 | |
| **Coherence -Links from:** | G4–M3 | Multi-Digit Multiplication and Division |
| **-Links to:** | G6–M2 | Arithmetic Operations Including Division of Fractions |
| | G6–M4 | Expressions and Equations |

In Topic B, place value understanding moves toward understanding the distributive property by using area models to generate and record partial products (**5.OA.1**, **5.OA.2**), which are combined within the standard algorithm (**5.NBT.5**). Writing and interpreting numerical expressions in Lessons 1 and 2 and comparing those expressions using visual models, lay the necessary foundation for students to make connections between the distributive property, as depicted in area models, and the partial products within the standard multiplication algorithm. The algorithm is built over a period of days, increasing in complexity as the number of digits in both factors increases. Reasoning about zeros in the multiplier, along with considerations about the reasonableness of products, also provides opportunities to deepen understanding of the standard algorithm. Although word problems provide context throughout Topic B, the final lesson offers a concentration of multi-step problems that allows students to apply this new knowledge.

| A Teaching Sequence Toward Mastery of the Standard Algorithm for Multi-Digit Whole Number Multiplication | |
|---|---|
| **Objective 1:** | **Write and interpret numerical expressions, and compare expressions using a visual model. (Lesson 3)** |
| **Objective 2:** | **Convert numerical expressions into unit form as a mental strategy for multi-digit multiplication. (Lesson 4)** |
| **Objective 3:** | **Connect visual models and the distributive property to partial products of the standard algorithm without renaming. (Lesson 5)** |
| **Objective 4:** | **Connect area models and the distributive property to partial products of the standard algorithm with renaming. (Lessons 6–7)** |
| **Objective 5:** | **Fluently multiply multi-digit whole numbers using the standard algorithm and using estimation to check for reasonableness of the product. (Lesson 8)** |
| **Objective 6:** | **Fluently multiply multi-digit whole numbers using the standard algorithm to solve multi-step word problems. (Lesson 9)** |

# Lesson 3

## Objective: Write and interpret numerical expressions, and compare expressions using a visual model.

### Suggested Lesson Structure

| | |
|---|---|
| Fluency Practice | (12 minutes) |
| Application Problem | (7 minutes) |
| Concept Development | (31 minutes) |
| Student Debrief | (10 minutes) |
| **Total Time** | **(60 minutes)** |

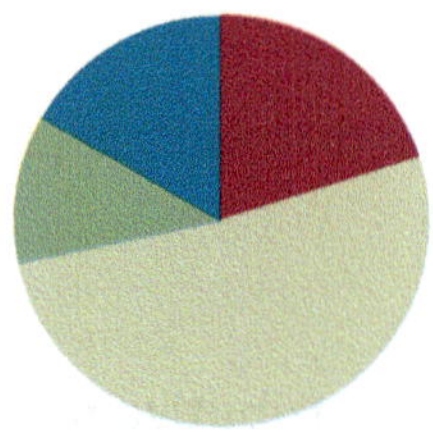

## Fluency Practice (12 minutes)

- Multiply by Multiples of 10 **5.NBT.2** (3 minutes)
- Estimate Products **5.NBT.6** (5 minutes)
- Decompose a Factor: The Distributive Property **3.OA.5** (4 minutes)

### Multiply by Multiples of 10 (3 minutes)

Note: This review fluency activity helps preserve skills students learned and mastered in Module 1 and lays the groundwork for future concepts.

Follow the same process and procedure as Lesson 2 for the following possible sequence: $21 \times 40$, $213 \times 30$, and $4{,}213 \times 20$.

### Estimate Products (5 minutes)

Materials: (S) Personal white board

T: (Write $421 \times 18 \approx$ ____ $\times$ ____ $=$ ____.) Round 421 to the nearest hundred.

S: 400.

T: (Write $421 \times 18 \approx 400 \times$ ____ $=$ ____.) Round 18 to the nearest ten.

S: 20.

T: (Write $421 \times 18 \approx 400 \times 20 =$ ____.) What's $400 \times 20$?

S: 8,000.

T: (Write $421 \times 18 \approx 400 \times 20 = 8{,}000$.)

T: (Write $323 \times 21 \approx$ ____ $\times$ ____ $=$ ____.) On your personal white board, write the multiplication sentence rounding each factor to arrive at a reasonable estimate of the product.

S: (Write $323 \times 21 \approx 300 \times 20 = 6{,}000$.)

Repeat the process and procedure for $1{,}950 \times 42$ and $2{,}480 \times 27$. Ask students to explain the reasoning behind their estimates.

### Decompose a Factor: The Distributive Property (4 minutes)

Note: Review of multiplication decomposition with low numbers prepares students for decomposing multiplication sentences with bigger numbers in the upcoming lessons. Students could be encouraged to generate their own decomposition to be used in the distribution (e.g., for the first, possible decompositions of 9 include 2 and 7 or 3 and 6). However, this will increase the time needed for this fluency activity.

Materials: (S) Personal white board

T: (Write 9 × 3 = ____.) Write the multiplication sentence.

S: (Write.)

T: (Write (5 × 3) + (____ × 3) = ____ below 9 × 3 = ____.) 9 is the same as 5 and what number?

S: 4.

T: (Write (5 × 3) + (4 × 3) = ____. Below it, write 15 + ____ = ____.) Fill in the blanks.

S: (Write 9 × 3 = 27. Below it, write (5 × 3) + (4 × 3) = 27. Below that line, write 15 + 12 = 27.)

9 × 3 = ____
(5 × 3) + (4 × 3) = ____
15 + 12 = ____

Repeat using the following possible sequence of 7 × 4, 8 × 2, and 9 × 6.

## Application Problem (7 minutes)

Robin is 11 years old. Her mother, Gwen, is 2 years more than 3 times Robin's age. How old is Gwen?

Note: This problem is simple enough that students can solve it prior to Lesson 3; however, in the Student Debrief, students are asked to return to the Application Problem and create a numerical expression to represent Gwen's age (i.e., (3 × 11) + 2). Accept any valid approach to solving the problem. The tape diagram is but one approach. Allow students to share.

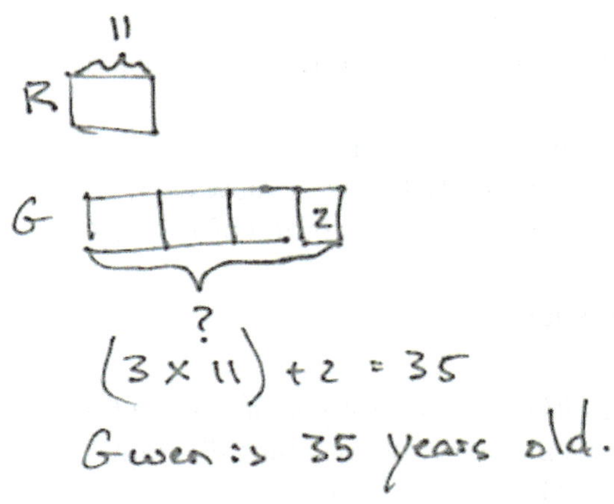

## Concept Development (31 minutes)

Materials: (S) Personal white board

Problems 1–3: From word form to numerical expressions and diagrams.

3 times the sum of 26 and 4.

6 times the difference between 60 and 51.

The sum of 2 twelves and 4 threes.

T: What expression describes the total value of these 3 equal units? Show a tape diagram.

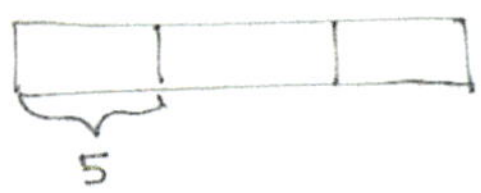


S: $3 \times 5$.

T: How about 3 times an unknown amount called *A*. Show a tape diagram and expression.

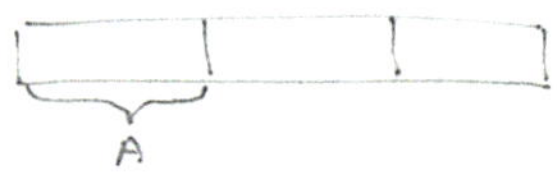


S: $3 \times A$.

T: 3 times the sum of 26 and 4? Show a tape diagram and expression.

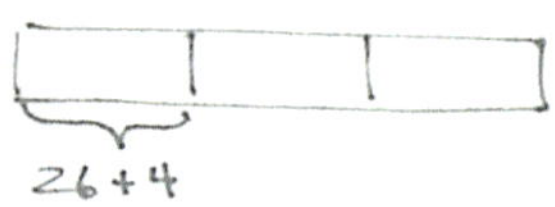


S: $3 \times (26 + 4)$ or $(26 + 4) \times 3$.

T: Why are **parentheses** necessary around 26 + 4? Talk to your partner.

S: We want 3 times as much as the total of 26 + 4. → If we don't put the parentheses, it doesn't show what we are counting. → We are counting the total of 26 and 4 three times.

T: Evaluate the expression.

S: 90.

T: (Write *6 times the difference between 60 and 51* on the board.) Work with a partner to show a tape diagram and expression to match these words.

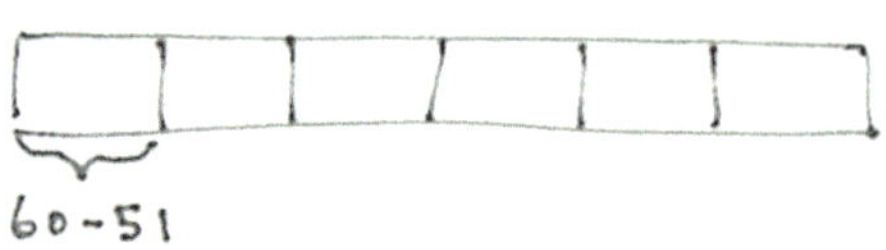


**NOTES ON MULTIPLE MEANS OF ENGAGEMENT:**

A review of relevant vocabulary may be in order for some students. Words such as *sum*, *product*, *difference*, and *quotient* might be reviewed, or a scaffold such as a word wall in the classroom might be appropriate.

**NOTES ON MULTIPLE MEANS OF REPRESENTATION:**

For some students, it may be more appropriate to begin with expressions in a more direct order such as, *the sum of 4 and 3 multiplied by 2* or *the difference between 14 and 6 times 5*.

**NOTES ON MULTIPLE MEANS OF REPRESENTATION:**

Some students may have difficulty understanding a number word like *twelves* as a noun—a unit to be counted. Substitute another more concrete noun like *apples* in the phrases, then transition to the noun *dozens* before using *twelves*. Use a concrete model of twelves like egg cartons to act out the problem.

S: 6 × (60 – 51) or (60 – 51) × 6.

T: You've offered two different expressions for these words: 6 × (60 – 51) and (60 – 51) × 6. Are these expressions equal? Why or why not?

S: Yes, they are equal. The two factors are just reversed.

T: What is the name of this property?

S: It's the commutative property.

T: Explain it in your own words to your partner.

S: (Share with partners.)

T: (Write *the sum of 2 twelves and 4 threes* on the board.) Represent this with a tape diagram and expression.

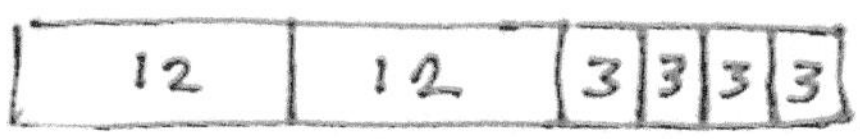


(2 × 12) + (4 × 3)

Repeat as necessary with examples such as *the sum of 2 nineteens and 8 nineteens* or *5 times the sum of 16 and 14.*

**Problems 4–6: From numerical expressions to word form.**

8 × (43 – 13)

(16 + 9) × 4

(20 × 3) + (5 × 3)

T: (Show 8 × (43 – 13) on the board.) Read this expression in words.

S: Eight times 43 minus 13.

T: Let me write down what I hear you saying. (Write 8 × 43 – 13.) It sounds like you are saying that we should multiply 8 and 43 and then subtract 13. Is that what you meant? Is this second expression equivalent to the one I wrote at first? Why or why not?

S: No. It's not the same. → You didn't write any parentheses. Without them, you will get a different answer because you won't subtract first. → We are supposed to subtract 13 from 43 and then multiply by 8.

T: Why can't we simply read every expression left to right and translate it?

S: We need to use words that tell which operation we should do first.

T: Let's name the two factors we are multiplying. Turn and talk.

S: 8 and the answer to 43 – 13. → We need to multiply the answer to the expression inside the parentheses by 8.

T: Since one of the factors is the answer to *this part* (make a circular motion around 43 – 13), what could we say to make sure we are talking about the answer to this subtraction problem? What do we call the answer to a subtraction problem?

S: The difference between 43 and 13.

T: What is happening to the difference of 43 and 13?

S: It's being multiplied by 8.

T: We can say and write, "8 times the difference of 43 and 13." Compare these words to the ones we said at first. Do they make sure we are multiplying the right numbers together? What other ways are there to say it?

S: Yes, the words clearly tell us what to multiply. → The product of 8 and the difference between 43 and 13. → 8 times as much as the difference between 43 and 13. → The difference of 43 and 13 multiplied 8 times.

Repeat the process with the following:

(16 + 9) × 4

Students should write *the sum of 16 and 9 times 4.* If students say *16 plus 9 times 4*, follow the sequence above to correct their thinking.

(20 × 3) + (5 × 3)

Students may write *the sum of 20 threes and 5 threes* or *the sum of 3 twenties and 3 fives,* or *the product of 20 and 3 plus the product of 5 and 3,* and so on. Similarly, discuss why *twenty times 3 plus 5 times 3* is unclear and imprecise.

**Problems 7–9: Comparison of expressions in word form and numerical form.**

| | |
|---|---|
| 9 × 13 | 8 thirteens |
| The sum of 10 and 9, doubled | (2 × 10) + (2 × 9) |
| 30 fifteens minus 1 fifteen | 29 × 15 |

T: Let's use < , > , or = to compare expressions. (Write 9 × 13 and 8 thirteens on the board.) Draw a tape diagram for each expression and compare them.

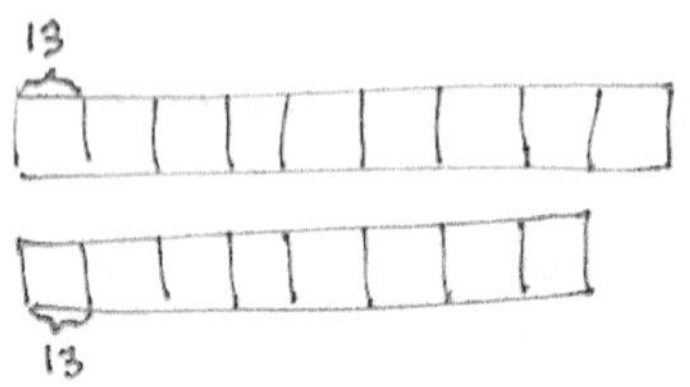


S: (Draw and write 9 × 13 > 8 thirteens.)

T: We don't even need to evaluate the solutions in order to compare them.

T: Now compare the next two expressions without evaluating, using diagrams.

S: They are equal because the sum of 10 and 9, doubled is (10 + 9) × 2. The expression on the right is the sum of 2 tens and 2 nines. There are 2 tens and 2 nines in each tape.

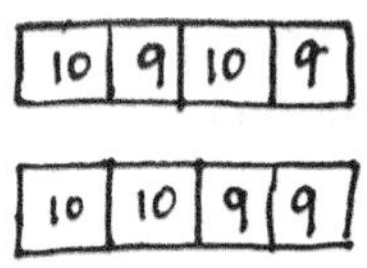


Repeat the process with the final example.

## Problem Set (10 minutes)

Students should do their personal best to complete the Problem Set within the allotted 10 minutes. For some classes, it may be appropriate to modify the assignment by specifying which problems they work on first. Some problems do not specify a method for solving. Students should solve these problems using the RDW approach used for Application Problems.

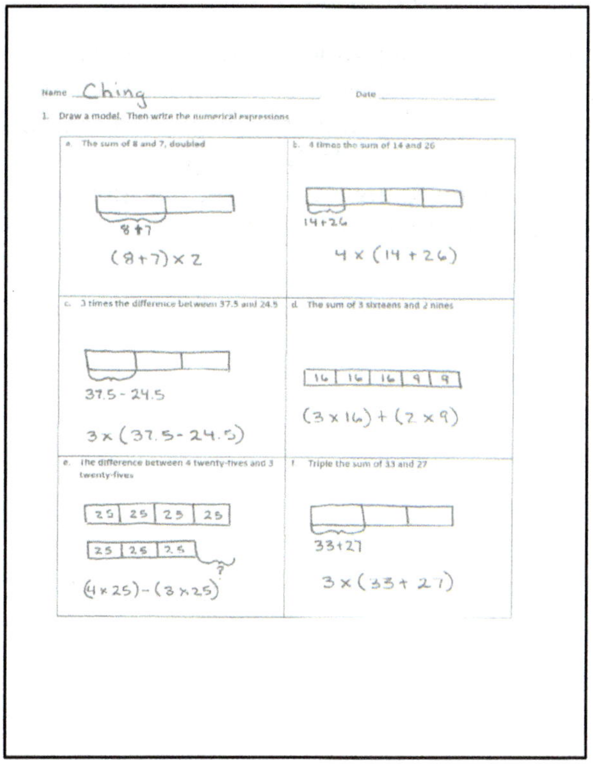

Name Ching Date

1. Draw a model. Then write the numerical expressions.

a. The sum of 8 and 7, doubled — 8+7 — $(8+7) \times 2$

b. 4 times the sum of 14 and 26 — 14+26 — $4 \times (14+26)$

c. 3 times the difference between 37.5 and 24.5 — 37.5 − 24.5 — $3 \times (37.5 - 24.5)$

d. The sum of 3 sixteens and 2 nines — 16 16 16 9 9 — $(3 \times 16) + (2 \times 9)$

e. The difference between 4 twenty-fives and 3 twenty-fives — 25 25 25 25 / 25 25 25 ? — $(4 \times 25) - (3 \times 25)$

f. Triple the sum of 33 and 27 — 33+27 — $3 \times (33+27)$

## Student Debrief (10 minutes)

**Lesson Objective:** Write and interpret numerical expressions, and compare expressions using a visual model.

The Student Debrief is intended to invite reflection and active processing of the total lesson experience.

Invite students to review their solutions for the Problem Set. They should check work by comparing answers with a partner before going over answers as a class. Look for misconceptions or misunderstandings that can be addressed in the Debrief. Guide students in a conversation to debrief the Problem Set and process the lesson.

Any combination of the questions below may be used to lead the discussion.

- Return to the Application Problem. Create a numerical expression to represent Gwen's age.
- In Problem 1(b) some of you wrote 4 × (14 + 26) and others wrote (14 + 26) × 4. Are both expressions acceptable? Explain.
- When evaluating the expression in Problem 2(a), a student got 85. Can you identify the error in thinking?
- Look at Problem 3(b). Talk in groups about how you know the expressions are not equal. How can you change the second expression to make it equivalent to 18 × 27?
- In Problem 5, be sure to point out that MeiLing's expression, while equivalent, does not accurately reflect what Mr. Huynh wrote on the board. As an extension, ask students to put the expressions that MeiLing and Angeline wrote into words.

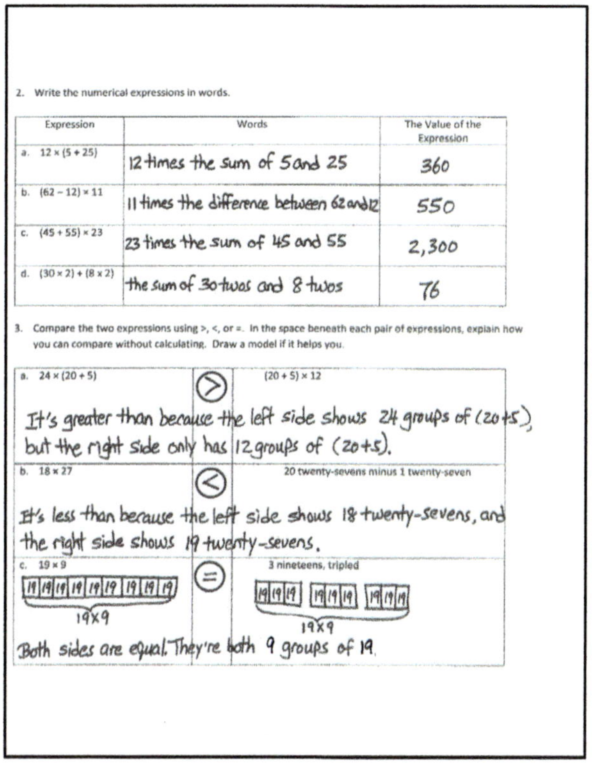

2. Write the numerical expressions in words.

| Expression | Words | The Value of the Expression |
|---|---|---|
| a. 12 × (5 + 25) | 12 times the sum of 5 and 25 | 360 |
| b. (62 − 12) × 11 | 11 times the difference between 62 and 12 | 550 |
| c. (45 + 55) × 23 | 23 times the sum of 45 and 55 | 2,300 |
| d. (30 × 2) + (8 × 2) | the sum of 30 twos and 8 twos | 76 |

3. Compare the two expressions using >, <, or =. In the space beneath each pair of expressions, explain how you can compare without calculating. Draw a model if it helps you.

a. 24 × (20 + 5) (>) (20 + 5) × 12
It's greater than because the left side shows 24 groups of (20+5), but the right side only has 12 groups of (20+5).

b. 18 × 27 (<) 20 twenty-sevens minus 1 twenty-seven
It's less than because the left side shows 18 twenty-sevens, and the right side shows 19 twenty-sevens.

c. 19 × 9 (=) 3 nineteens, tripled
19×9 — 19×9
Both sides are equal. They're both 9 groups of 19.

## Exit Ticket (3 minutes)

After the Student Debrief, instruct students to complete the Exit Ticket. A review of their work will help with assessing students' understanding of the concepts that were presented in today's lesson and planning more effectively for future lessons. The questions may be read aloud to the students.

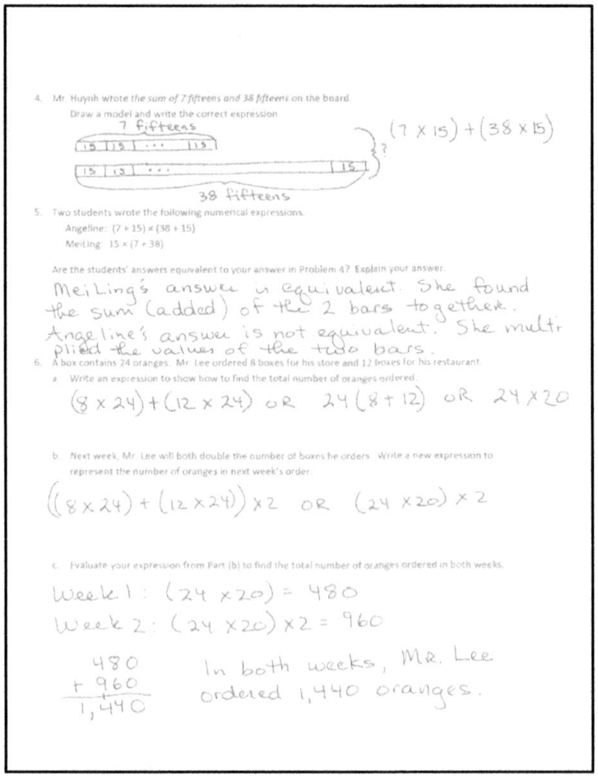

4. Mr. Huynh wrote *the sum of 7 fifteens and 38 fifteens* on the board.

   Draw a model and write the correct expression.

   7 fifteens

   38 fifteens

   (7 × 15) + (38 × 15)

5. Two students wrote the following numerical expressions.

   Angeline: (7 + 15) × (38 + 15)

   MeiLing: 15 × (7 + 38)

   Are the students' answers equivalent to your answer in Problem 4? Explain your answer.

   MeiLing's answer is equivalent. She found the sum (added) of the 2 bars together. Angeline's answer is not equivalent. She multiplied the values of the two bars.

6. A box contains 24 oranges. Mr. Lee ordered 8 boxes for his store and 12 boxes for his restaurant.

   a. Write an expression to show how to find the total number of oranges ordered.

   (8 × 24) + (12 × 24) OR 24(8 + 12) OR 24 × 20

   b. Next week, Mr. Lee will both double the number of boxes he orders. Write a new expression to represent the number of oranges in next week's order.

   ((8 × 24) + (12 × 24)) × 2 OR (24 × 20) × 2

   c. Evaluate your expression from Part (b) to find the total number of oranges ordered in both weeks.

   Week 1: (24 × 20) = 480

   Week 2: (24 × 20) × 2 = 960

   480 + 960 = 1,440

   In both weeks, Mr. Lee ordered 1,440 oranges.

Name ______________________________ Date ________________

1. Draw a model. Then, write the numerical expressions.

| | |
|---|---|
| a. The sum of 8 and 7, doubled | b. 4 times the sum of 14 and 26 |
| c. 3 times the difference between 37.5 and 24.5 | d. The sum of 3 sixteens and 2 nines |
| e. The difference between 4 twenty-fives and 3 twenty-fives | f. Triple the sum of 33 and 27 |

2. Write the numerical expressions in words. Then, solve.

| Expression | Words | The Value of the Expression |
|---|---|---|
| a. $12 \times (5 + 25)$ | | |
| b. $(62 - 12) \times 11$ | | |
| c. $(45 + 55) \times 23$ | | |
| d. $(30 \times 2) + (8 \times 2)$ | | |

3. Compare the two expressions using > , < , or = . In the space beneath each pair of expressions, explain how you can compare without calculating. Draw a model if it helps you.

| | | |
|---|---|---|
| a. $24 \times (20 + 5)$ | ◯ | $(20 + 5) \times 12$ |
| b. $18 \times 27$ | ◯ | 20 twenty-sevens minus 1 twenty-seven |
| c. $19 \times 9$ | ◯ | 3 nineteens, tripled |

4. Mr. Huynh wrote *the sum of 7 fifteens and 38 fifteens* on the board.

   Draw a model, and write the correct expression.

5. Two students wrote the following numerical expressions.

   Angeline: $(7 + 15) \times (38 + 15)$

   MeiLing: $15 \times (7 + 38)$

   Are the students' expressions equivalent to your answer in Problem 4? Explain your answer.

6. A box contains 24 oranges. Mr. Lee ordered 8 boxes for his store and 12 boxes for his restaurant.

   a. Write an expression to show how to find the total number of oranges ordered.

   b. Next week, Mr. Lee will double the number of boxes he orders. Write a new expression to represent the number of oranges in next week's order.

   c. Evaluate your expression from Part (b) to find the total number of oranges ordered in both weeks.

Name ______________________________ Date ______________

1. Draw a model. Then, write the numerical expressions.

| a. The difference between 8 forty-sevens and 7 forty-sevens | b. 6 times the sum of 12 and 8 |
|---|---|
| | |

2. Compare the two expressions using >, <, or =.

| 62 × (70 + 8) | ◯ | (70 + 8) × 26 |
|---|---|---|

Name ______________________________ Date ______________

1. Draw a model. Then, write the numerical expressions.

| a. The sum of 21 and 4, doubled | b. 5 times the sum of 7 and 23 |
|---|---|
| c. 2 times the difference between 49.5 and 37.5 | d. The sum of 3 fifteens and 4 twos |
| e. The difference between 9 thirty-sevens and 8 thirty-sevens | f. Triple the sum of 45 and 55 |

2. Write the numerical expressions in words. Then, solve.

| Expression | Words | The Value of the Expression |
|---|---|---|
| a. 10 × (2.5 + 13.5) | | |
| b. (98 – 78) × 11 | | |
| c. (71 + 29) × 26 | | |
| d. (50 × 2) + (15 × 2) | | |

3. Compare the two expressions using > , < , or = . In the space beneath each pair of expressions, explain how you can compare without calculating. Draw a model if it helps you.

| | | |
|---|---|---|
| a. 93 × (40 + 2) | ◯ | (40 + 2) × 39 |
| b. 61 × 25 | ◯ | 60 twenty-fives minus 1 twenty-five |

4. Larry claims that (14 + 12) × (8 + 12) and (14 × 12) + (8 × 12) are equivalent because they have the same digits and the same operations.

   a. Is Larry correct? Explain your thinking.

   b. Which expression is greater? How much greater?

# Lesson 4

## Objective: Convert numerical expressions into unit form as a mental strategy for multi-digit multiplication.

### Suggested Lesson Structure

| | |
|---|---|
| Fluency Practice | (12 minutes) |
| Application Problem | (6 minutes) |
| Concept Development | (32 minutes) |
| Student Debrief | (10 minutes) |
| **Total Time** | **(60 minutes)** |

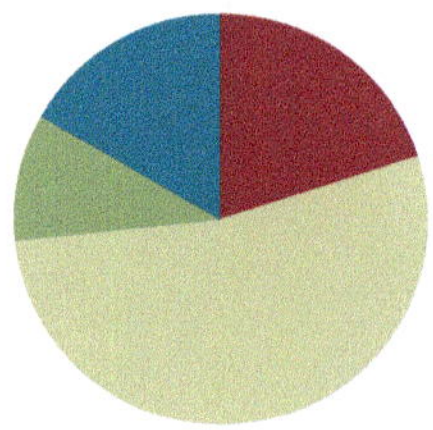

## Fluency Practice (12 minutes)

- Estimate Products **5.NBT.6** (4 minutes)
- Decompose Multiplication Sentences **3.OA.5** (4 minutes)
- Write the Value of the Expression **5.OA.1** (4 minutes)

### Estimate Products (4 minutes)

Materials: (S) Personal white board

T: (Write 409 × 21 ≈ ____ × ____ = ____.) On your personal white board, write the multiplication sentence rounding each factor to arrive at a reasonable estimate of the product.

S: (Write 409 × 21 ≈ 400 × 20 = 8,000.)

Repeat the process and procedure for 287 × 64, 3,875 × 92, and 6,130 × 37.

### Decompose Multiplication Sentences (4 minutes)

Materials: (S) Personal white board

T: (Write 12 × 3 = ____.) Write the multiplication sentence.

S: (Write.)

T: (Write (8 × 3) + (____ × 3) = ____. Below it, write 12 × 3 = ____.) 12 is the same as 8 and what number?

S: 4.

T: (Write (8 × 3) + (4 × 3) = ____. Below it, write 24 + ____ = ____.) Fill in the blanks.

S: (Write 12 × 3 = 36. Below it, they write (8 × 3) + (4 × 3) = 36. Below that line, they write 24 + 12 = 36.)

Repeat using the following possible sequence: 14 × 4, 13 × 3, and 15 × 6, changing the missing numbers that students need to fill in.

### Write the Value of the Expression (4 minutes)

Materials: (S) Personal white board

T: (On the board, write 11 × (15 + 5).) Write the expression as a single multiplication sentence without parentheses.

S: (Write 11 × 20 = 220.)

Repeat the process for (41 – 11) × 12, (75 + 25) × 38, and (20 × 2) + (6 × 2).

## Application Problem (6 minutes)

Jaxon earned $39 raking leaves. His brother, Dayawn, earned 7 times as much waiting on tables. Write a numerical expression to show Dayawn's earnings. How much money did Dayawn earn?

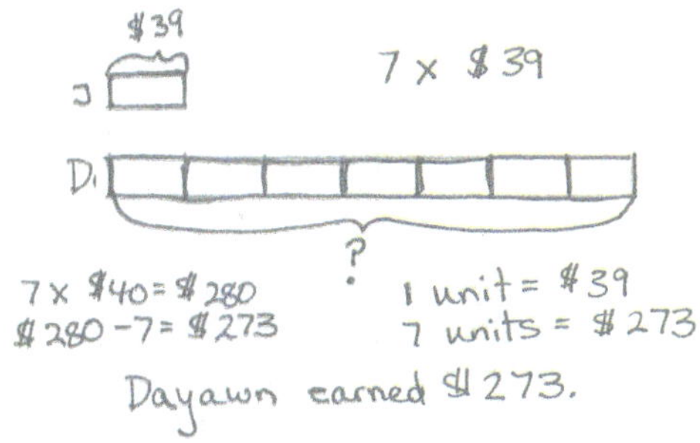


Note: This problem is simple enough that students can solve it using pencil and paper prior to this lesson. Allow students to share their approach to solving. However, in the Student Debrief, students are asked to return to the Application Problem and solve this problem again applying a new mental strategy to evaluate.

## Concept Development (32 minutes)

Materials: (S) Personal white board

**NOTES ON MULTIPLE MEANS OF ACTION AND EXPRESSION:**

This lesson requires students to work mentally with two-digit and three-digit numbers. If basic multiplication facts are not yet mastered, be prepared to adjust numbers in calculations to suit the learner's level. A good time to review mental math strategies is during Sprints and fluency activities. Spending time working on basic facts (with flash cards, computer games, etc.) may be necessary prior to this lesson.

**Problems 1–2**

8 × 31

8 × 29

T: (Show 8 × 31 on the board.) What does this expression mean when I designate 31 as the unit?

S: Add thirty-one 8 times. → 8 times as much as thirty-one.

T: What does it mean when I designate 8 as the unit?

S: Add eight 31 times. → 31 times as much as eight.

T: Does our choice of unit change the product of the two factors?

S: No.

T: Why not? What property allows for this?

S: The commutative property (any-order property) says that the order of the factors doesn't matter. The product will be the same.

T: Let's designate 8 as the unit. I've drawn diagrams of 8 × 31 and 8 × 30.

T: Use the diagrams to consider how 8 × 30 helps us to solve 8 × 31 when we designate eight as the unit (point to the diagram) and the other factor as the number of units, 31 and 30. (Run a finger down the length of each bar.) Turn and talk.

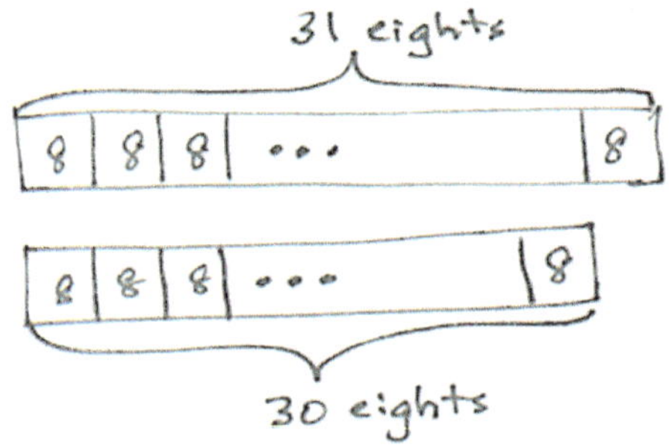


S: 31 eights is the same as 30 eights plus 1 eight.
→ 30 eights is 240 and one more eight makes 248.
→ 30 eights is easy, 240. 240 + 8 = 248.

T: How many more eights are in the first bar than in the second bar?

S: 1 more eight.

**NOTES ON MULTIPLE MEANS OF REPRESENTATION:**

Possibly challenge students to (a) solve the problem designating 31 as the unit and (b) think of other ways to decompose 31 units of 8.

T: Could we have decomposed 31 eights in another way? Turn and talk.

S: (Students share.)

T: Yes! 31 eights is also equivalent to 20 eights plus 11 eights. Would this way of decomposing 31 change the product of 8 × 31?

S: No. It would be the same because 20 eights is 160 and 11 eights is 88, which is the same as 160 + 88, which is 248.

T: Let's record our thinking. (Write 31 eights = 30 eights + 1 eight. 31 × 8 = (30 × 8) + (1 × 8).)

T: What is the value of 30 eights and 1 more eight? Say it in an addition sentence that corresponds to our last equation. (Point to (30 × 8) + (1 × 8).)

S: 240 + 8 = 248.

T: 31 times 8 is ...?

S: 248.

T: (Show 8 × 29 on the board.) What does this expression mean when we designate eight as the unit?

S: Add 8 twenty-nine times. → Add 8 over and over 29 times.

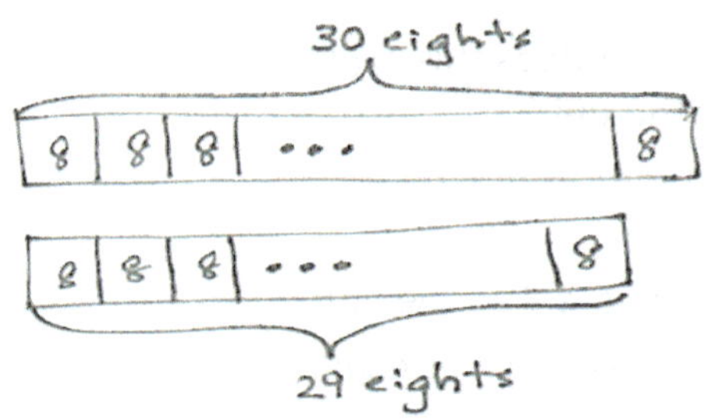


T: How does 8 × 30 helps us to solve 8 × 29? Turn and talk.

S: (Discuss.)

T: I heard Jackie say that 30 eights minus 1 eight is equal to 29 eights. (Write 30 eights – 1 eight = 29 eights. 29 × 8. (30 × 8) – (1 × 8) = 8 × 29.)

T: What is the value of 30 eights minus 1 eight?

S: 232.

T: Could we have decomposed 29 eights in another way to help us evaluate the expression mentally? Turn and talk.

S: (Share.)

T: Yes! 29 eights is also equivalent to 20 eights plus 9 eights. Would this way of decomposing 29 change the product of 8 × 29?

S: No.

T: Why not?

S: Because it is still 29 eights even though we found 20 eights then 9 eights. → 20 eights = 160 and 9 eights is 72. That's still 232.

**Problems 3–4**

49 × 20

20 × 51

T: (Write 49 × 20.) To solve this mentally using today's strategy, first determine which factor will be designated as the unit. Which is easier to work with: 49 twenties or 20 forty-nines? Turn and talk.

S: It is easier to think of 20 as the unit because then we can say 40 twenties and 9 twenties. → It's easier to think of twenty as the unit because it is 1 less than 50 twenties. (Students might also share why 49 is easier.)

T: Let's agree to designate 20 as the unit. Go ahead and find the value of the expression using today's unit form strategy. Use a tape diagram if you so choose.

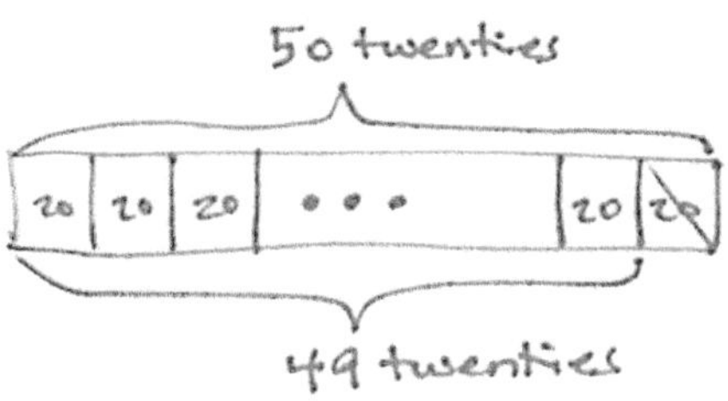


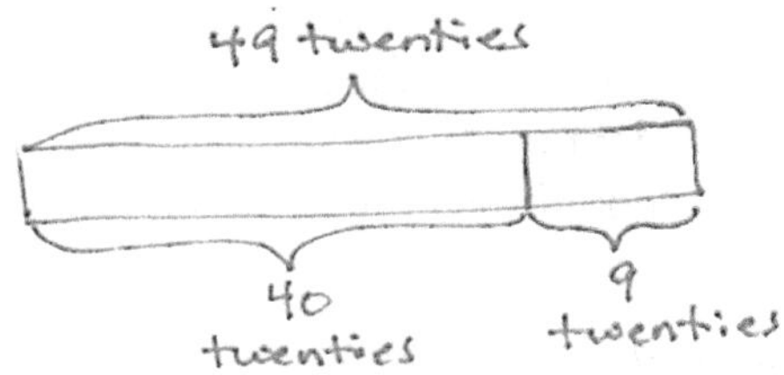


S: (Work and share.)

T: What is the value of 49 × 20?

S: 980.

T: Work with a partner to create an equivalent expression that you can use to help you solve 20 × 51 mentally. Write the equivalent expression and its value on your personal white board. As before, you may draw a tape diagram if you choose.

S: (Work and share.)

T: (Circulate and assess for understanding. Be receptive to any valid mental approach.)

**Problems 5–6**

101 × 12

12 × 98

T: Work independently to evaluate these two expressions mentally. (Write 12 × 98 and 12 × 101 on the board.) Compare your work with a neighbor when you're finished. Draw tape diagrams if you choose.

S: (Work.)

## Problem Set (10 minutes)

Students should do their personal best to complete the Problem Set within the allotted 10 minutes. For some classes, it may be appropriate to modify the assignment by specifying which problems they work on first. Some problems do not specify a method for solving. Students should solve these problems using the RDW approach used for Application Problems.

## Student Debrief (10 minutes)

**Lesson Objective:** Convert numerical expressions into unit form as a mental strategy for multi-digit multiplication.

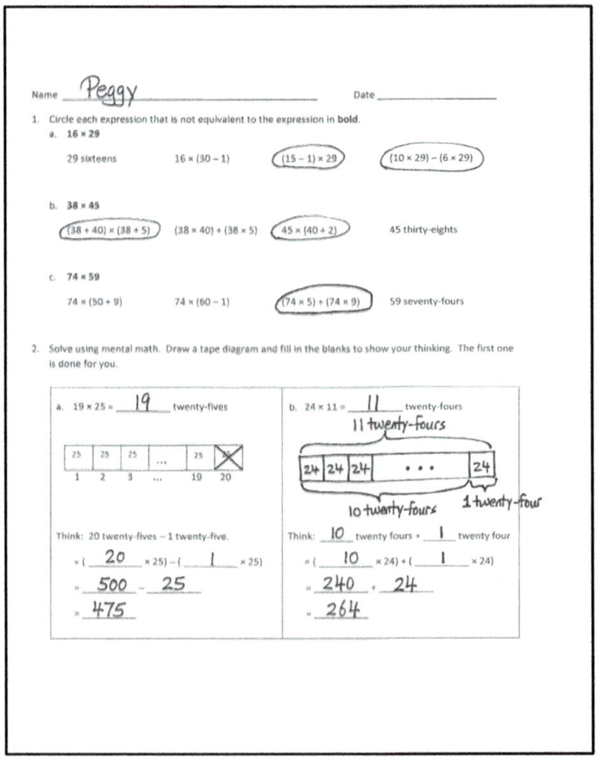

Name Peggy Date

1. Circle each expression that is not equivalent to the expression in **bold**.

a. **16 × 29**

29 sixteens  16 × (30 – 1)  (15 – 1) × 29  (10 × 29) – (6 × 29)

b. **38 × 45**

(38 + 40) × (38 + 5)  (38 × 40) + (38 × 5)  45 × (40 + 2)  45 thirty-eights

c. **74 × 59**

74 × (50 + 9)  74 × (60 – 1)  (74 × 5) + (74 × 9)  59 seventy-fours

2. Solve using mental math. Draw a tape diagram and fill in the blanks to show your thinking. The first one is done for you.

a. 19 × 25 = 19 twenty-fives

Think: 20 twenty-fives – 1 twenty-five.
= (20 × 25) – (1 × 25)
= 500 – 25
= 475

b. 24 × 11 = 11 twenty-fours

Think: 10 twenty fours + 1 twenty four
= (10 × 24) + (1 × 24)
= 240 + 24
= 264

The Student Debrief is intended to invite reflection and active processing of the total lesson experience.

Invite students to review their solutions for the Problem Set. They should check work by comparing answers with a partner before going over answers as a class. Look for misconceptions or misunderstandings that can be addressed in the Debrief. Guide students in a conversation to debrief the Problem Set and process the lesson.

Any combination of the questions below may be used to lead the discussion.

- What mental math strategy did you learn today? (Unit form.) Choose a problem in the Problem Set to support your answer.
- How did the Application Problem connect to today's lesson? Which factor did you decide to designate as the unit?
- In Problem 1(b) the first two possible expressions are very similar. How did you decide which one was not equivalent?
- Look at Problem 2. How did the think prompts help to guide you as you evaluated these expressions? Turn and talk.

- What was different about the think prompts in Problem 2 and Problem 3? (Problem 2 prompts give the units but not the number of units. Problem 3 prompts give the number of units but not the name of the units.)
- Explain to your partner how to solve Problem 5(a). (Some students may have thought 101 × 15 = (101 × 10) + (101 × 5), while others may see that 101 × 15 = (100 × 15) + (1 × 15). Both are acceptable.)

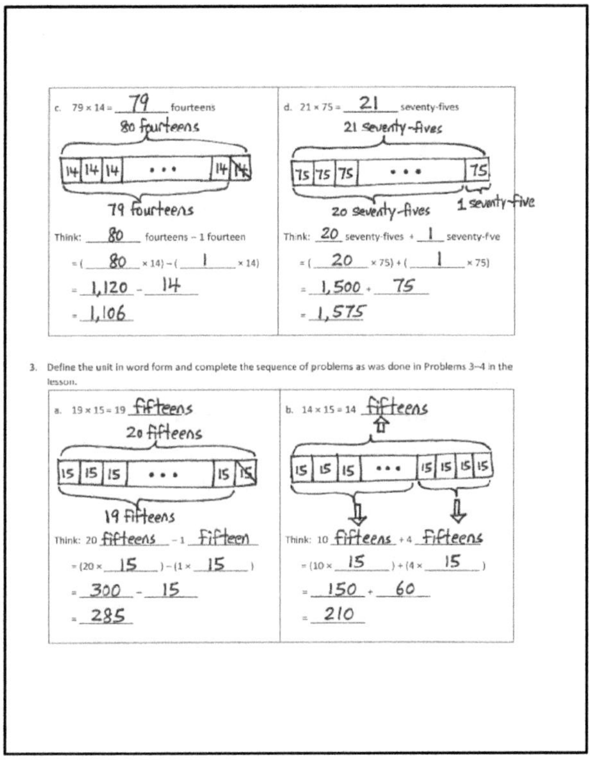

c. 79 × 14 = 79 fourteens
80 fourteens
79 fourteens
Think: 80 fourteens – 1 fourteen
= (80 × 14) – (1 × 14)
= 1,120 – 14
= 1,106

d. 21 × 75 = 21 seventy-fives
21 seventy-fives
20 seventy-fives 1 seventy-five
Think: 20 seventy-fives + 1 seventy-five
= (20 × 75) + (1 × 75)
= 1,500 + 75
= 1,575

3. Define the unit in word form and complete the sequence of problems as was done in Problems 3–4 in the lesson.

a. 19 × 15 = 19 fifteens
20 fifteens
19 fifteens
Think: 20 fifteens – 1 fifteen
= (20 × 15) – (1 × 15)
= 300 – 15
= 285

b. 14 × 15 = 14 fifteens
Think: 10 fifteens + 4 fifteens
= (10 × 15) + (4 × 15)
= 150 + 60
= 210

## Exit Ticket (3 minutes)

After the Student Debrief, instruct students to complete the Exit Ticket. A review of their work will help with assessing students' understanding of the concepts that were presented in today's lesson and planning more effectively for future lessons. The questions may be read aloud to the students.

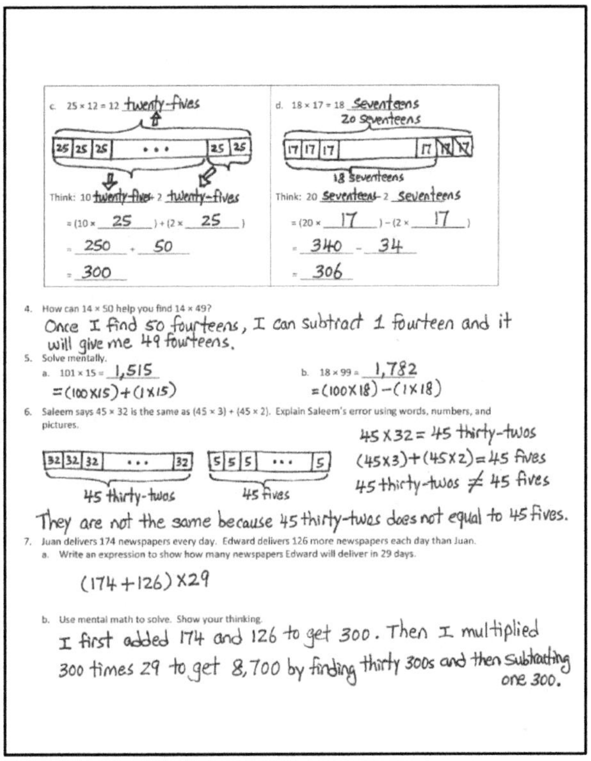

c. 25 × 12 = 12 twenty-fives
Think: 10 twenty-fives + 2 twenty-fives
= (10 × 25) + (2 × 25)
= 250 + 50
= 300

d. 18 × 17 = 18 seventeens
20 seventeens
18 seventeens
Think: 20 seventeens – 2 seventeens
= (20 × 17) – (2 × 17)
= 340 – 34
= 306

4. How can 14 × 50 help you find 14 × 49?
Once I find 50 fourteens, I can subtract 1 fourteen and it will give me 49 fourteens.

5. Solve mentally.
a. 101 × 15 = 1,515
=(100×15)+(1×15)
b. 18 × 99 = 1,782
=(100×18)–(1×18)

6. Saleem says 45 × 32 is the same as (45 × 3) + (45 × 2). Explain Saleem's error using words, numbers, and pictures.
45 thirty-twos
45 fives
45×32 = 45 thirty-twos
(45×3)+(45×2) = 45 fives
45 thirty-twos ≠ 45 fives
They are not the same because 45 thirty-twos does not equal to 45 fives.

7. Juan delivers 174 newspapers every day. Edward delivers 126 more newspapers each day than Juan.
a. Write an expression to show how many newspapers Edward will deliver in 29 days.
(174+126)×29
b. Use mental math to solve. Show your thinking.
I first added 174 and 126 to get 300. Then I multiplied 300 times 29 to get 8,700 by finding thirty 300s and then subtracting one 300.

Name ______________________________________ Date ____________________

1. Circle each expression that is not equivalent to the expression in **bold**.

   a. **16 × 29**

   29 sixteens      16 × (30 – 1)      (15 – 1) × 29      (10 × 29) – (6 × 29)

   b. **38 × 45**

   (38 + 40) × (38 + 5)      (38 × 40) + (38 × 5)      45 × (40 + 2)      45 thirty-eights

   c. **74 × 59**

   74 × (50 + 9)      74 × (60 – 1)      (74 × 5) + (74 × 9)      59 seventy-fours

2. Solve using mental math. Draw a tape diagram and fill in the blanks to show your thinking. The first one is partially done for you.

a. 19 × 25 = __________ twenty-fives

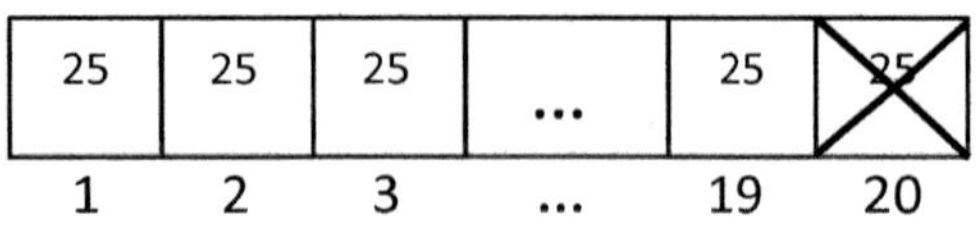


Think: 20 twenty-fives – 1 twenty-five.

= ( __________ × 25) – ( __________ × 25)

= __________ – __________

= __________

b. 24 × 11 = __________ twenty-fours

Think: _____ twenty fours + _____ twenty four

= ( __________ × 24) + ( __________ × 24)

= __________ + __________

= __________

| c. $79 \times 14 =$ ________ fourteens<br><br>Think: ________ fourteens – 1 fourteen<br>$= ($________$ \times 14) - ($________$ \times 14)$<br>= ________ – ________<br>= ________ | d. $21 \times 75 =$ ________ seventy-fives<br><br>Think: _____ seventy-fives + _____ seventy-five<br>$= ($________$ \times 75) + ($________$ \times 75)$<br>= ________ + ________<br>= ________ |
|---|---|

3. Define the unit in word form and complete the sequence of problems as was done in the lesson.

| a. $19 \times 15 = 19$ ________<br><br>Think: 20 ________ – 1 ________<br>$= (20 \times$ ________$) - (1 \times$ ________$)$<br>= ________ – ________<br>= ________ | b. $14 \times 15 = 14$ ________<br><br>Think: 10 ________ + 4 ________<br>$= (10 \times$ ________$) + (4 \times$ ________$)$<br>= ________ + ________<br>= ________ |
|---|---|

| c. $25 \times 12 = 12$ ________ | d. $18 \times 17 = 18$ ________ |
|---|---|
| Think: 10 ________ + 2 ________ | Think: 20 ________ − 2 ________ |
| = (10 × ________) + (2 × ________) | = (20 × ________) − (2 × ________) |
| = ________ + ________ | = ________ − ________ |
| = ________ | = ________ |

4. How can $14 \times 50$ help you find $14 \times 49$?

5. Solve mentally.

   a. $101 \times 15 =$ ________

   b. $18 \times 99 =$ ________

6. Saleem says $45 \times 32$ is the same as $(45 \times 3) + (45 \times 2)$. Explain Saleem's error using words, numbers, and/or pictures.

7. Juan delivers 174 newspapers every day. Edward delivers 126 more newspapers each day than Juan.

   a. Write an expression to show how many newspapers Edward will deliver in 29 days.

   b. Use mental math to solve. Show your thinking.

Name ______________________________ Date ________________

Solve using mental math. Draw a tape diagram and fill in the blanks to show your thinking.

| a. $49 \times 11 =$ __________ elevens | b. $25 \times 13 =$ __________ twenty-fives |
|---|---|
| Think: 50 elevens – 1 eleven<br>= (__________ × 11) – (__________ × 11)<br>= __________ – __________<br>= __________ | Think: _____ twenty-fives + _____ twenty-fives<br>= (__________ × 25) + (__________ × 25)<br>= __________ + __________<br>= __________ |

Name ______________________________ Date ______________

1. Circle each expression that is not equivalent to the expression in **bold**.

   a. **37 × 19**

   37 nineteens    (30 × 19) – (7 × 29)    37 × (20 – 1)    (40 – 2) × 19

   b. **26 × 35**

   35 twenty-sixes    (26 + 30) × (26 + 5)    (26 × 30) + (26 × 5)    35 × (20 + 60)

   c. **34 × 89**

   34 × (80 + 9)    (34 × 8) + (34 × 9)    34 × (90 – 1)    89 thirty-fours

2. Solve using mental math. Draw a tape diagram and fill in the blanks to show your thinking. The first one is partially done for you.

a. 19 × 50 = __________ fifties

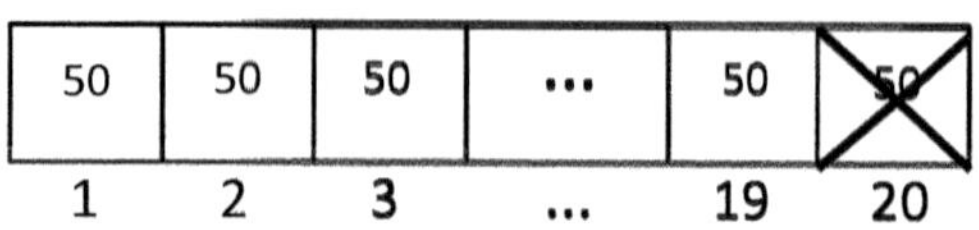


Think: 20 fifties – 1 fifty

= ( __________ × 50) – ( __________ × 50)

= __________ – __________

= __________

b. 11 × 26 = __________ twenty-sixes

Think: _____ twenty-sixes + _____ twenty-six

= ( __________ × 26) + ( __________ × 26)

= __________ + __________

= __________





EUREKA MATH

| c. 49 × 12 = __________ twelves<br><br>Think: _____ twelves – 1 twelve<br>= ( __________ × 12) – ( __________ × 12)<br>= __________ – __________<br>= __________ | d. 12 × 25 = __________ twenty-fives<br><br>Think: _____ twenty-fives + _____ twenty-fives<br>= ( __________ × 25) + ( __________ × 25)<br>= __________ + __________<br>= __________ |
|---|---|

3. Define the unit in word form and complete the sequence of problems as was done in the lesson.

| a. 29 × 12 = 29 __________<br><br>Think: 30 __________ – 1 __________<br>= (30 × __________ ) – (1 × __________ )<br>= __________ – __________<br>= __________ | b. 11 × 31 = 31 __________<br><br>Think: 30 __________ + 1 __________<br>= (30 × __________ ) + (1 × __________ )<br>= __________ + __________<br>= __________ |
|---|---|

| c. 19 × 11 = 19 __________ | d. 50 × 13 = 13 __________ |
|---|---|
| Think: 20 __________ – 1 __________ | Think: 10 __________ + 3 __________ |
| = (20 × __________ ) – (1 × __________ ) | = (10 × __________ ) + (3 × __________ ) |
| = __________ – __________ | = __________ + __________ |
| = __________ | = __________ |

4. How can 12 × 50 help you find 12 × 49?

5. Solve mentally.

   a. 16 × 99 = __________

   b. 20 × 101 = __________

6. Joy is helping her father to build a rectangular deck that measures 14 ft by 19 ft. Find the area of the deck using a mental strategy. Explain your thinking.

7. The Lason School turns 101 years old in June. In order to celebrate, they ask each of the 23 classes to collect 101 items and make a collage. How many total items will be in the collage? Use mental math to solve. Explain your thinking.

# Lesson 5

## Objective: Connect visual models and the distributive property to partial products of the standard algorithm without renaming.

### Suggested Lesson Structure

| | |
|---|---|
| Fluency Practice | (12 minutes) |
| Application Problem | (5 minutes) |
| Concept Development | (33 minutes) |
| Student Debrief | (10 minutes) |
| **Total Time** | **(60 minutes)** |

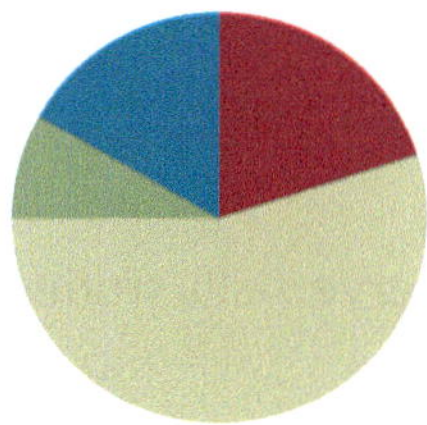

## Fluency Practice (12 minutes)

- Estimate Products **5.NBT.6** (4 minutes)
- Multiply Mentally **5.NBT.5** (4 minutes)
- Multiply by Multiples of 100 **5.NBT.2** (4 minutes)

### Estimate Products (4 minutes)

Materials: (S) Estimate Products Pattern Sheet

Note: This fluency activity helps bolster students' understanding of and automaticity with the estimation of products.

Distribute the Estimate Products pattern sheet, and give students two minutes to do as many problems as they can. Probe the room, correcting misunderstandings and encouraging students to use mental math strategies. When estimating, allow students flexibility when approximating factors. For example, when estimating the product of 23 × 42, a student may find that 25 × 40 *or* 20 × 40 are both logical approximations.

> **NOTES ON MULTIPLE MEANS OF REPRESENTATION:**
>
> Many students will arrive in Grade 5 with the requisite knowledge to make the connection between the tape diagram and the area model. The first problem set not only provides scaffolding for those lacking this knowledge, but it also makes clear the importance of the commutative property in interpreting multiplication equations when no context is present. That is, it is acceptable to name either factor as the unit being counted (multiplicand) and the number of times it is being counted (multiplier).

### Multiply Mentally (4 minutes)

Materials: (S) Personal white board

Note: This fluency activity helps bolster students' understanding of and automaticity with the distributive property of multiplication.

T: (Write 9 × 10 = ____.) Say the multiplication sentence.

S: 9 × 10 = 90. (Write 90 on the blank.)

T: (Write 9 × 9 = 90 – ____ below 9 × 10 = 90.) On your personal white board, write the number sentence, filling in the blank.

S: (Students write 9 × 9 = 90 – 9.)

T: What's 9 × 9?

S: 81.

Repeat the process and procedure for 9 × 100, 9 × 99, 15 × 10, 15 × 9, 29 × 100, and 29 × 99.

### Multiply by Multiples of 100 (4 minutes)

Materials: (S) Personal white board

Note: This review fluency activity helps preserve skills students learned and mastered in Module 1 and lays the groundwork for future concepts.

T: (Write 31 × 100 = ____.) Say the multiplication sentence with the answer.

S: 31 × 100 = 3,100.

T: (Write 3,100 × 2 = ____ below 31 × 100 = 3,100.) Say the multiplication sentence.

S: 3,100 × 2 = 6,200.

T: (Write 31 × 200 = ____ below 3,100 × 2 = 6,200.) Say 31 × 200 as a three-step multiplication sentence, taking out the hundred.

S: 31 × 100 × 2 = 6,200.

T: (Write 31 × 200 = 6,200.)

Direct students to solve using the same method for 24 × 300 and 34 × 200.

## Application Problem (5 minutes)

Aneisha is setting up a play space for her new puppy. She will be building a rectangular fence around part of her yard that measures 29 feet by 12 feet. How many square feet of play space will her new puppy have? If you have time, solve in more than one way.

① 30 twelves – 1 twelve = 29 twelves
(30 × 12) – (1 × 12) = 360 – 12 = 348

Aneisha's puppy will have 348 square feet of play space.

② 20 twelves + 9 twelves = 29 twelves
(20 × 12) + (9 × 12) = 240 + 108 = 348

Note: This problem is a bridge from Lesson 4's unit form mental math strategy. Students have significant practice in finding area and multiplying two digits by two digits in Grade 4.

## Concept Development (33 minutes)

**Problem 1: Represent units using first the tape diagram and then the area model.**

21 × 5

T: (Write on the board 21 × 5.) Can you solve mentally using the unit form strategy?

S: Think (20 × 5) plus 5 more. This is twenty 5s and 1 more 5. The product is 105.

T: Represent that thinking with a tape diagram.

S: (Draw.)

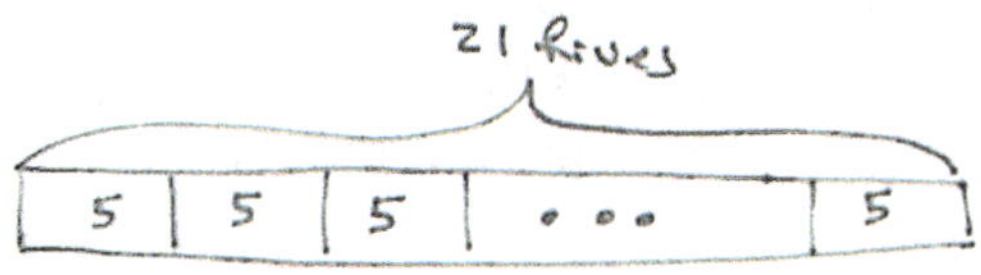


T: So you chose the factor 5 to be the unit. Can you imagine the area in each unit? (Draw the first image.)

T: Imagine that all 21 boxes are stacked vertically. (Draw second image.)

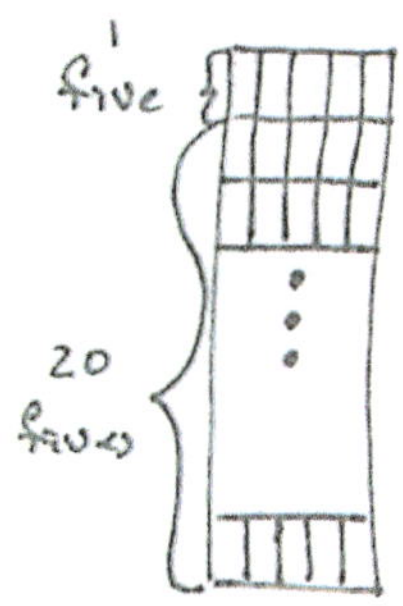


T: There are so many units in this drawing. Let's represent all the boxes using an area model like the type you used in fourth grade. (Draw third image.)

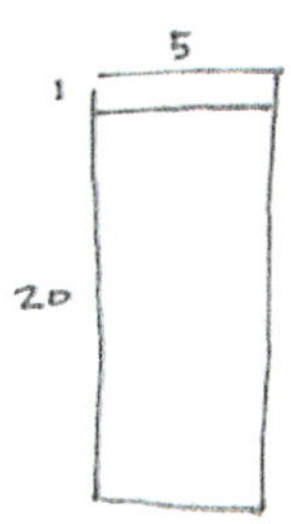


**NOTES ON MULTIPLE MEANS OF ACTION AND EXPRESSION:**

Today's lesson focuses on multiplication without renaming, which limits the digits and combination of digits that can be used when creating problems. Foster a student's curiosity about numbers and encourage them to notice and explore the patterns in today's problems. Have students share their observations and challenge them to create more multiplication problems that do not require renaming. Students might record their thoughts in a journal.

T: What values could you put in the area model? (How many units are in each part of the rectangle?)

S: 1 × 5 and 20 × 5. 1 five and 20 fives. 5 + 100 = 105.

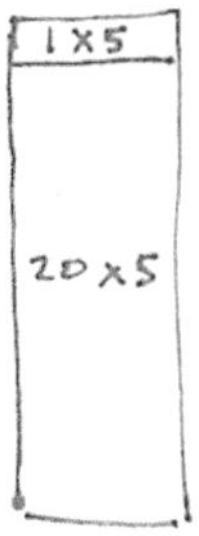


T: How are the area model and the tape diagram similar? How are they different?

S: They both show the same number of units. The tape diagram helps us think of 21 × 5 as (20 × 5) plus 5 more. The area model helped us show all the boxes in the tape diagram without having to draw every single one. It's easier to see (20 × 5) plus (1 × 5).

T: Can we turn this area model so that we count 5 groups of 21? What effect will turning the rectangle have on our area (product)? (Draw.)

S: Yes, we could have counted twenty-ones instead of fives by drawing lines horizontally across. We could count 5 twenty-ones or 21 fives, and it would be the same because of the commutative property. The area wouldn't change. 5 × 21 is the same as 21 × 5. 5 × 21 = 21 × 5.

**Problem 2: Products of two-digit and two-digit numbers using the area model and the standard algorithm.**

23 × 31

T: Now that we have discussed how the area model can show multiplication, let's connect it to a written method—the standard algorithm.

T: (Write 23 × 31 on the board.) Let's think about which factor we want to name as our unit. Which is easier to count, 31 twenty-threes or 23 thirty-ones? Turn and talk.

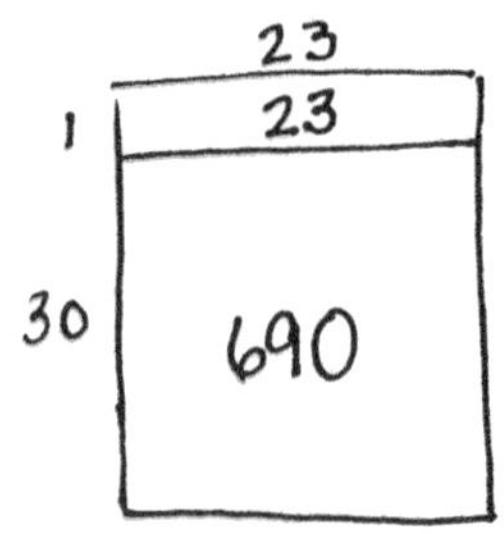


S: I think it's easier to count units of twenty-three because we can find 30 of them and then just add 1 more. → I think 20 twenty-threes plus 3 twenty-threes is easier.

T: Either works! Let's label the top with our unit of 23. (Label the top of area model.)

T: We showed units of five before. How can we show units of twenty-three now?

S: The 1 group of 23 is on top and the 30 group is on the bottom.

T: Does it matter how we split the rectangle? Does it change the area (product)?

S: No, it doesn't matter. The area will be the same.

S: (Talk and solve.)



EUREKA MATH™

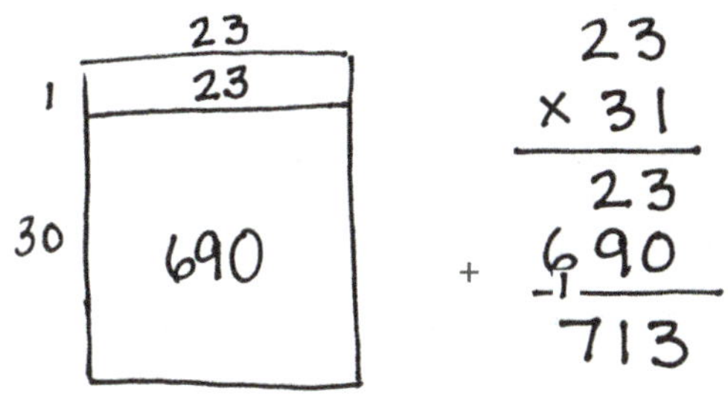


T: What's the product of 1 and 23?

S: 23.

T: What's the product of 30 and 23?

S: 690.

T: Now, add the partial products to find the total area of the rectangle.

S: (Add.)

T: What is 23 × 31?

S: 713.

T: Now, let's solve 23 × 31 using the standard algorithm. Show your neighbor how to set up this problem using the standard algorithm.

S: (Show and discuss.)

T: Work with your neighbor to solve using the standard algorithm.

S: (Solve.)

MP.7

T: Take a look at the area model and the standard algorithm. Compare them. What do you notice?

S: We added 1 unit of 23 to 30 units of 23. → In the area model we added two parts just like in the algorithm. → First, we wrote the value of 1 twenty-three. Then, we wrote the value of 30 twenty-threes.

T: Explain the connections between (30 × 23) + (1 × 23), the area model, and the algorithm.

S: (Explain the connections.)

**Problem 3: Products of two-digit and three-digit numbers.**

343 × 21

231 × 32

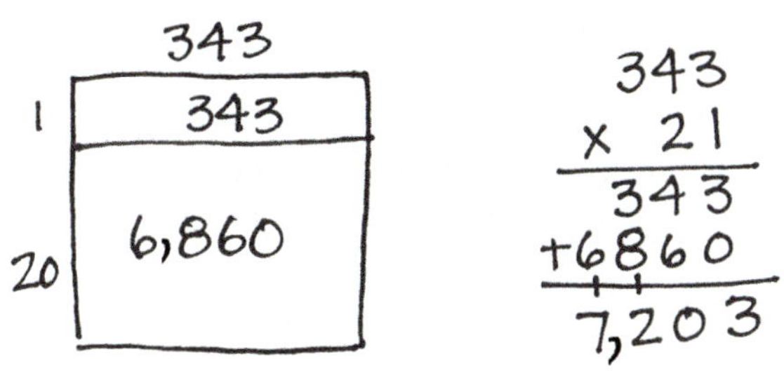


T: (Write 343 × 21 on the board.) What should we designate as the unit?

S: Three hundred forty-three.

T: Let's find the value of 21 units of 343. Draw an area model and solve. Then, solve with the algorithm. Compare. What do you notice?

S: (Draw and solve.)

T: Explain the connections between (20 × 343) + (1 × 343), the area model, and the algorithm.

S: (Explain connections.)

## Problem Set (10 minutes)

Students should do their personal best to complete the Problem Set within the allotted 10 minutes. For some classes, it may be appropriate to modify the assignment by specifying which problems they work on first. Some problems do not specify a method for solving. Students should solve these problems using the RDW approach used for Application Problems.

## Student Debrief (10 minutes)

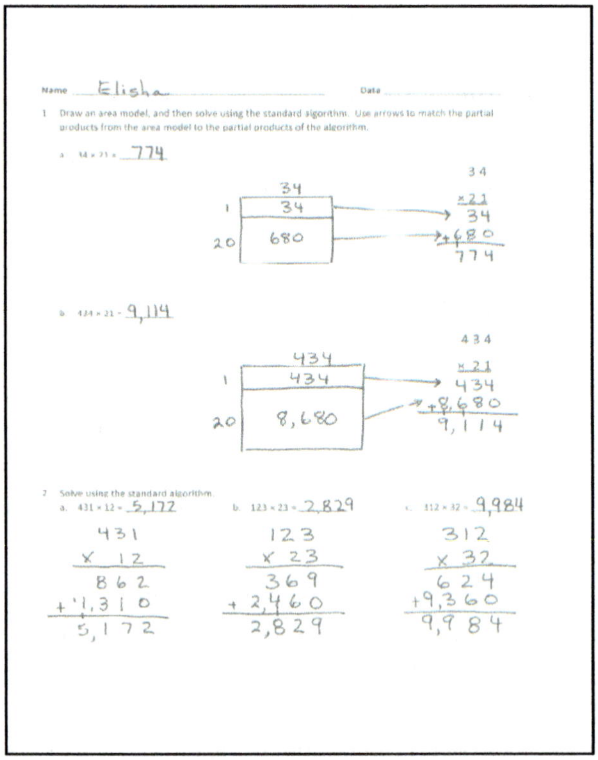


**Lesson Objective:** Connect visual models and the distributive property to partial products of the standard algorithm without renaming.

The Student Debrief is intended to invite reflection and active processing of the total lesson experience.

Invite students to review their solutions for the Problem Set. They should check work by comparing answers with a partner before going over answers as a class. Look for misconceptions or misunderstandings that can be addressed in the Debrief. Guide students in a conversation to debrief the Problem Set and process the lesson.

Any combination of the questions below may be used to lead the discussion.

- Look at the area models in Problems 1(a) and 1(b). What is the same about these two problems?
- How could you use Problem 1 to help you solve Problem 2?
- How is multiplying three digits by two digits different from multiplying two digits by two digits? How is it the same? (Make sure that students notice that the number of partial products is determined by the multiplier. Two-by-two-digit and two-by-three-digit multiplication still only has two partial products in the algorithm. The only real difference is that the unit being counted is a larger number.)
- What is different about Problem 4? (Decimal values.) Does using a decimal value like 12.1 as the unit being counted change the way you must think about the partial products? (Have students share their area models with the class and discuss.)

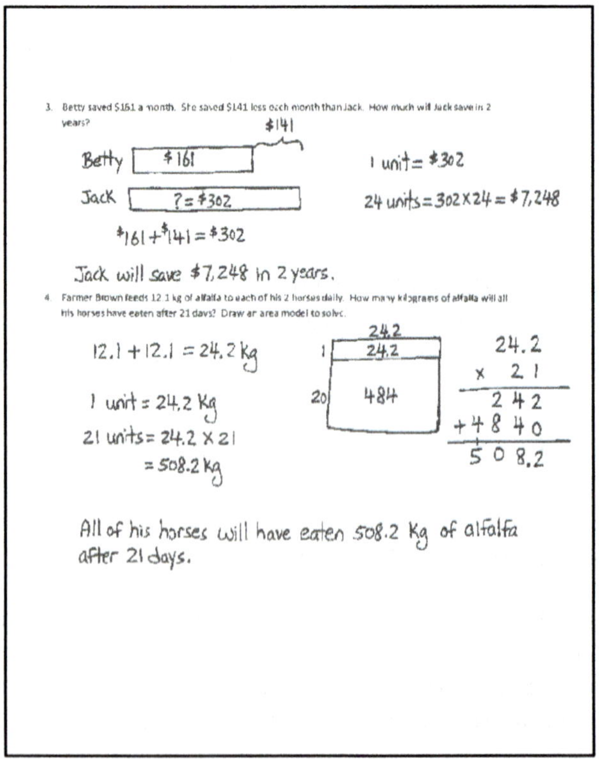


## Exit Ticket (3 minutes)

After the Student Debrief, instruct students to complete the Exit Ticket. A review of their work will help with assessing students' understanding of the concepts that were presented in today's lesson and planning more effectively for future lessons. The questions may be read aloud to the students.

Estimate and then multiply.

| | | | | | |
|---|---|---|---|---|---|
| 1 | 29 x 11 ≈ | | 23 | 801 x 31 ≈ | |
| 2 | 29 x 21 ≈ | | 24 | 803 x 31 ≈ | |
| 3 | 29 x 31 ≈ | | 25 | 703 x 31 ≈ | |
| 4 | 23 x 12 ≈ | | 26 | 43 x 34 ≈ | |
| 5 | 23 x 22 ≈ | | 27 | 53 x 34 ≈ | |
| 6 | 23 x 32 ≈ | | 28 | 53 x 31 ≈ | |
| 7 | 23 x 42 ≈ | | 29 | 53 x 51 ≈ | |
| 8 | 37 x 13 ≈ | | 30 | 93 x 31 ≈ | |
| 9 | 37 x 23 ≈ | | 31 | 913 x 31 ≈ | |
| 10 | 36 x 24 ≈ | | 32 | 73 x 31 ≈ | |
| 11 | 24 x 36 ≈ | | 33 | 723 x 31 ≈ | |
| 12 | 43 x 11 ≈ | | 34 | 78 x 34 ≈ | |
| 13 | 43 x 21 ≈ | | 35 | 798 x 34 ≈ | |
| 14 | 403 x 21 ≈ | | 36 | 62 x 33 ≈ | |
| 15 | 303 x 21 ≈ | | 37 | 642 x 33 ≈ | |
| 16 | 203 x 21 ≈ | | 38 | 374 x 64 ≈ | |
| 17 | 41 x 11 ≈ | | 39 | 64 x 374 ≈ | |
| 18 | 41 x 21 ≈ | | 40 | 740 x 36 ≈ | |
| 19 | 41 x 31 ≈ | | 41 | 750 x 36 ≈ | |
| 20 | 401 x 31 ≈ | | 42 | 65 x 680 ≈ | |
| 21 | 501 x 31 ≈ | | 43 | 849 x 84 ≈ | |
| 22 | 601 x 31 ≈ | | 44 | 85 x 849 ≈ | |

estimate products

Name ______________________________ Date ______________

1. Draw an area model, and then solve using the standard algorithm. Use arrows to match the partial products from the area model to the partial products of the algorithm.

   a. 34 × 21 = ________

$$\begin{array}{r} 3\,4 \\ \underline{\times\ 2\,1} \end{array}$$

   b. 434 × 21 = ________

$$\begin{array}{r} 4\,3\,4 \\ \underline{\times\ \ 2\,1} \end{array}$$

2. Solve using the standard algorithm.

   a. 431 × 12 = ________

   b. 123 × 23 = ________

   c. 312 × 32 = ________







EUREKA MATH™

3. Betty saves $161 a month. She saves $141 less each month than Jack. How much will Jack save in 2 years?

4. Farmer Brown feeds 12.1 kilograms of alfalfa to each of his 2 horses daily. How many kilograms of alfalfa will all his horses have eaten after 21 days? Draw an area model to solve.

Name ____________________________ Date ________________

Draw an area model, and then solve using the standard algorithm.

a. 21 × 23 = ________

$$\begin{array}{r} 2\,1 \\ \times\ \underline{2\,3} \end{array}$$

b. 143 × 12 = ________

$$\begin{array}{r} 1\,4\,3 \\ \times\ \underline{\ \ 1\,2} \end{array}$$

Name ______________________________ Date ______________

1. Draw an area model, and then solve using the standard algorithm. Use arrows to match the partial products from the area model to the partial products in the algorithm.

   a. $24 \times 21 =$ ________

$$\begin{array}{r} 24 \\ \times\ 21 \\ \hline \end{array}$$

   b. $242 \times 21 =$ ________

$$\begin{array}{r} 242 \\ \times\ 21 \\ \hline \end{array}$$

2. Solve using the standard algorithm.

   a. $314 \times 22 =$ ________ b. $413 \times 22 =$ ________ c. $213 \times 32 =$ ________

3. A young snake measures 0.23 meters long. During the course of his lifetime, he will grow to be 13 times his current length. What will his length be when he is full grown?

4. Zenin earns $142 per shift at his new job. During a pay period, he works 12 shifts. What would his pay be for that period?

# Lesson 6

## Objective: Connect area models and the distributive property to partial products of the standard algorithm with renaming.

### Suggested Lesson Structure

| | |
|---|---|
| Fluency Practice | (12 minutes) |
| Application Problem | (6 minutes) |
| Concept Development | (32 minutes) |
| Student Debrief | (10 minutes) |
| **Total Time** | **(60 minutes)** |

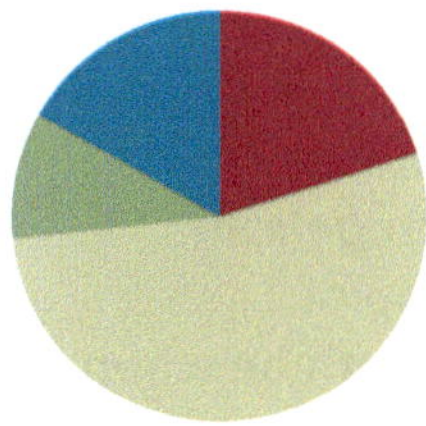

## Fluency Practice (12 minutes)

- Multiply Mentally **5.NBT.5** (4 minutes)
- Multiply by Multiples of 100 **5.NBT.2** (4 minutes)
- Multiply Using the Area Model **5.NBT.6** (4 minutes)

### Multiply Mentally (4 minutes)

Materials: (S) Mental Multiplication Pattern Sheet

Note: This fluency activity helps bolster students' understanding of, and automaticity with, the distributive property of multiplication.

Distribute the Mental Multiplication pattern sheet, and give students two minutes to do as many problems as they can. Probe the room, correcting misunderstandings and encouraging students to use mental math strategies.

### Multiply by Multiples of 100 (4 minutes)

Follow the same process and procedure as Lesson 5 for the following possible sequence: 21 × 400, 312 × 300, and 2,314 × 200.

### Multiply Using the Area Model (4 minutes)

Materials: (S) Personal white board

T: (Write 43 × 12 = ____.) Draw an area model on your personal white board to solve.

S: (Students draw area model.)

T: Fill in your area model and number sentence.

S: (Write 43 × 12 = 516.)

T: Solve using the algorithm.

S: (Solve.)

Repeat the procedure using the following possible sequence: 243 × 12 and 312 × 23.

## Application Problem (6 minutes)

Scientists are creating a material that may replace damaged cartilage in human joints. This *hydrogel* can stretch to 21 times its original length. If a strip of hydrogel measures 3.2 cm, what would its length be when stretched to capacity?

3.2 ≈ 3   3 × 21 = 63 So 67.2 is reasonable.

```
   32 tenths
 × 21
 ----
    32
 + 640
 -----
   672 tenths
```

672 tenths = 67.2 cm

The hydrogel's length when stretched would be 67.2 cm.

Note: This problem is designed to bridge from Lesson 5 where students are multiplying without renaming; however, it adds the twist of multiplying by a decimal. Students should be encouraged to estimate for a reasonable product prior to multiplying. The use of a tape diagram may be beneficial for some students.

(To show students a short video of the hydrogel in action, go to http://www.seas.harvard.edu/news-events/press-releases/tough-gel-stretches-to-21-times-its-length.)

## Concept Development (32 minutes)

Materials: (S) Personal white board

**Problem 1**

64 × 73

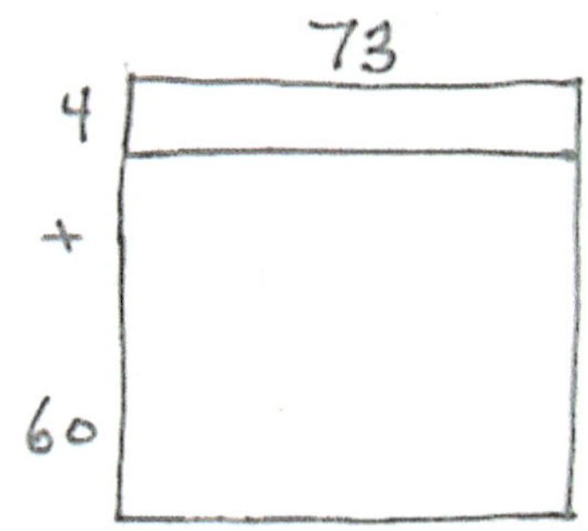


*Method 1: Area Model*

T: Please divide your personal white board into two sections. On one side, we'll solve with an area model, and on the other, we will connect it to the standard algorithm.

T: (Write 64 × 73 on the board.) Let's represent units of 73. Draw an area model with your partner and label the length as 73.

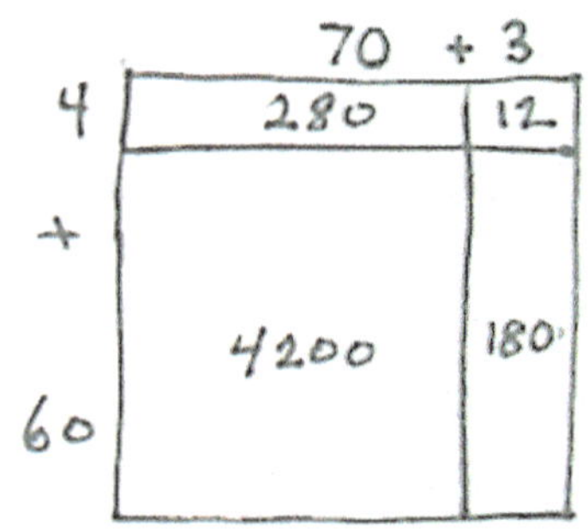


T: How many seventy-threes are we counting?

S: 64.

T: How can we decompose 64 to make our multiplication easier? Show this on your model.

S: Split it into 4 and 60. (Draw.)

T: $73 \times 4$ and $73 \times 60$ are both a bit more difficult to solve mentally. How could we decompose 73 to make finding these partial products easier to solve?

S: Split the length into 3 and 70.

T: Let's record that and begin solving. What's the product of 4 and 3?

S: 12.

T: (Continue recording the products in the area model.) Now, add each row's partial products to find the value of $64 \times 73$.

S: (Add.)

T: What is 64 groups of 73?

S: 4,672.

**NOTES ON MULTIPLE MEANS OF ENGAGEMENT AND EXPRESSION:**

Point to each portion of the area model as the solution is found. Gesture to clearly indicate the image or location that corresponds to the words.

Add variety to the way in which questions are asked. For example, 4 times 3 can also be expressed as 4 ones times 3 ones, 4 groups of 3, 4 copies of 3, or 4 threes. Students should be comfortable with the variety of language when multiplying.

*Method 2: Standard Algorithm*

T: Show your neighbor how to write $64 \times 73$ in order to solve using the standard algorithm.

T: First, we'll find the value of 4 units of seventy-three.

T: 4 times 3 ones equals?

S: 12 ones.

T: 12 ones equal 1 ten and how many ones?

S: 2 ones.

T: Watch how I record. (Write the 1 on the line under the tens place first, and the 2 in the ones place second.)

T: 4 times 7 tens equals?

S: 28 tens.

T: 28 tens plus 1 ten equals? (Point to the 1 placed on the line under the tens place.)

S: 29 tens.

T: I'll cross out the 1 ten and record 29 tens. 29 tens equal how many hundreds and how many tens?

S: 2 hundreds 9 tens.

T: What did we multiply to find this product? Find this product in your area model.

S: $4 \times 73$. It is the sum of the two products in the top row of the model.

T: Now, we'll find the value of 60 units of 73. What is 6 tens times 3 ones?

S: 18 tens.

T: How many hundreds can I make with 18 tens?

S: 1 hundred, 8 tens.

**NOTES ON MULTIPLE MEANS OF ENGAGEMENT:**

Point to the each digit and factor while carefully working through the recording process of the standard algorithm. Gesture to clearly indicate the image or location that corresponds to the words. Keep teacher-talk clear and concise.

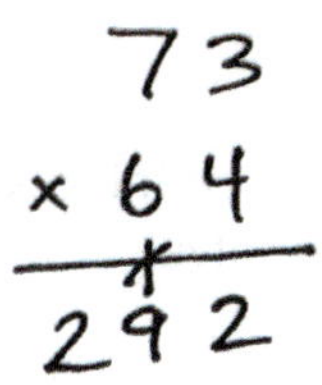


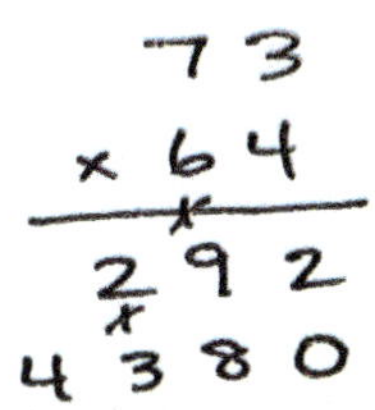

T: We'll record the hundred between the partial products. (Write a small 1 just below the 2 in 292 and the 8 in the tens place beneath the 9 in 292.)

T: What is 6 tens times 7 tens?

S: 42 hundreds.

T: 42 hundreds plus 1 hundred equals? (Point to the regrouped 1.)

S: 43 hundreds.

T: I'll cross out the 1 hundred and record 43 hundreds. 43 hundreds equals how many thousands and how many hundreds?

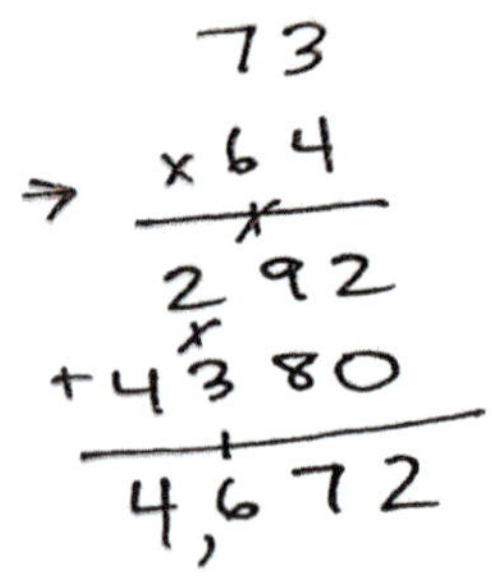

S: 4 thousands 3 hundreds.

T: What did we multiply to find this other product? Find it in your area model.

S: 60 × 73. It is the sum of the two products in the bottom row of the model.

T: Turn and tell your partner what the next step is.

T: I hear you saying that we should add these two products together.

T: Compare the area model with the algorithm. What do you notice?

S: Both of them have us multiply first and then add, and the answers are the same. → In the partial products, we had to add four sections of the rectangle that we combined into two products, and in the standard algorithm, there were only two the whole time. → The partial products method looks like the standard algorithm method, but the parts are decomposed.

**NOTES ON MULTIPLE MEANS OF ACTION AND EXPRESSION:**

Adjust numbers in calculations to suit learner's level. Some students may struggle with 7s, 8s, and 9s. Changing the digits to more comfortable factors allows students to focus more on the algorithm itself.

Throughout the multiplication work, students should continue to estimate products and then evaluate their calculation for reasonableness.

After having discussed the problem, have students complete the problem independently and check their work with a partner. Allowing students to generate other examples to calculate may also be fruitful.

**Problems 2–3**

814 × 39

624 × 82

T: (Write 814 × 39 on the board.) Partner A, use the standard algorithm to solve. Partner B, draw an area model to solve.

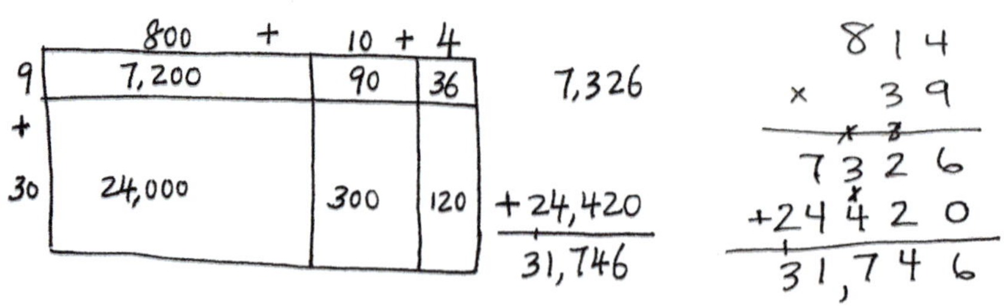

S: (Draw and solve.)

T: Compare your solutions.

T: (Post completed algorithm on board, for students to check.) Be sure you are recording your regrouped units correctly.

S: (Check.)

Have partners switch roles and complete the second problem in the same manner.

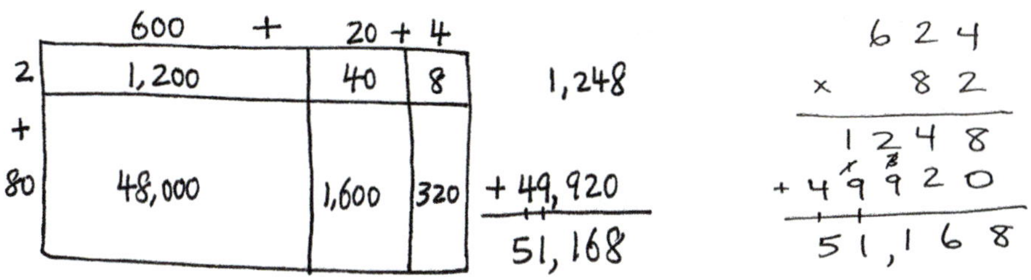


## Problem Set (10 minutes)

Students should do their personal best to complete the Problem Set within the allotted 10 minutes. For some classes, it may be appropriate to modify the assignment by specifying which problems they work on first. Some problems do not specify a method for solving. Students should solve these problems using the RDW approach used for Application Problems.

## Student Debrief (10 minutes)

**Lesson Objective:** Connect area models and the distributive property to partial products of the standard algorithm with renaming.

The Student Debrief is intended to invite reflection and active processing of the total lesson experience.

Invite students to review their solutions for the Problem Set. They should check work by comparing answers with a partner before going over answers as a class. Look for misconceptions or misunderstandings that can be addressed in the Debrief. Guide students in a conversation to debrief the Problem Set and process the lesson.

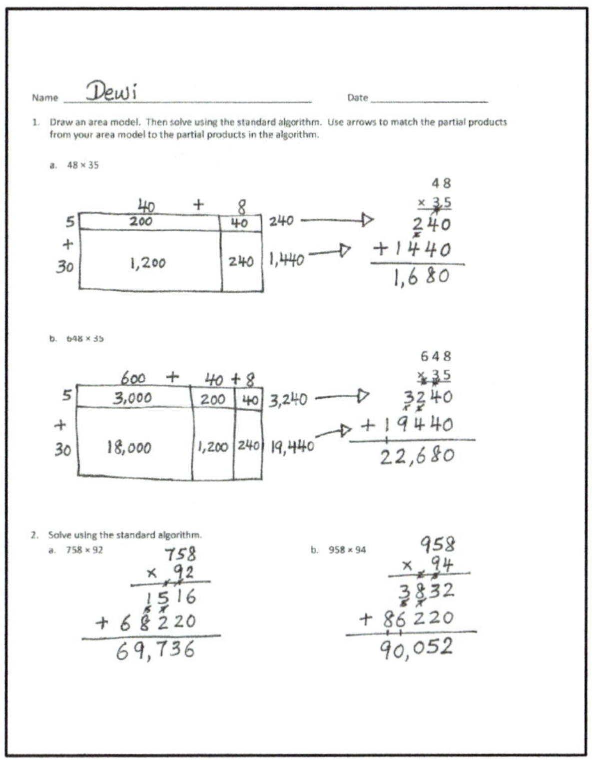


Any combination of the questions below may be used to lead the discussion.

- What pattern did you notice between Parts (a) and (b) of Problem 1? How did this slight difference in factors impact your final product?
- Explain to your partner how you recorded the regrouping in Problem 2(a). What were you thinking and what did you write as you multiplied 9 tens times 5 tens?

- Let's think about a problem like 23 × 45 and solve it with the algorithm. What is the first partial product that we would find? (3 × 45.) The second? (20 × 45.) Would this be the only order in which we could find the partial products? What else could we do? (Point out to students that it would also be appropriate to find 20 units of 45 and then 3 units of 45. It is simply a convention to find the smaller place value first. Use the area model to support this discussion.)
- What information did you need before you could find the cost of the carpet in Problem 3? (The area of the room.) How did you find that information? (Remind us how to find the area of a room.) Why is area measured in square units?
- Look at Problem 4. Discuss your thought process as you worked on solving this problem. There is more than one way to solve this problem. Work with your partner to show another way. How does your expression change? (Compare expressions that communicate students' thinking.)

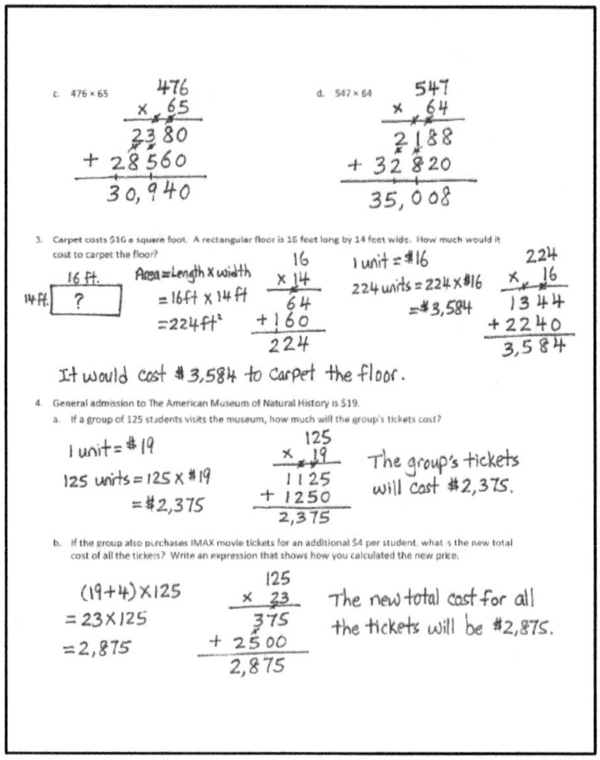
c. 476 × 65

d. 547 × 64

3. Carpet costs $16 a square foot. A rectangular floor is 16 feet long by 14 feet wide. How much would it cost to carpet the floor?

It would cost $3,584 to carpet the floor.

4. General admission to The American Museum of Natural History is $19.

a. If a group of 125 students visits the museum, how much will the group's tickets cost?

The group's tickets will cost $2,375.

b. If the group also purchases IMAX movie tickets for an additional $4 per student, what is the new total cost of all the tickets? Write an expression that shows how you calculated the new price.

The new total cost for all the tickets will be $2,875.

## Exit Ticket (3 minutes)

After the Student Debrief, instruct students to complete the Exit Ticket. A review of their work will help with assessing students' understanding of the concepts that were presented in today's lesson and planning more effectively for future lessons. The questions may be read aloud to the students.





EUREKA MATH™

Solve.

| | | | | | |
|---|---|---|---|---|---|
| 1 | 5 x 100 = | | 23 | 5000 - 50 = | |
| 2 | 500 - 5 = | | 24 | 50 x 99 = | |
| 3 | 5 x 99 = | | 25 | 80 x 100 = | |
| 4 | 3 x 100 = | | 26 | 80 x 99 = | |
| 5 | 300 - 3 = | | 27 | 60 x 100 = | |
| 6 | 3 x 99 = | | 28 | 60 x 99 = | |
| 7 | 2 x 100 = | | 29 | 11 x 100 = | |
| 8 | 200 - 2 = | | 30 | 1100 - 11 = | |
| 9 | 2 x 99 = | | 31 | 11 x 99 = | |
| 10 | 6 x 100 = | | 32 | 21 x 100 = | |
| 11 | 600 - 6 = | | 33 | 2100 - 21 = | |
| 12 | 6 x 99 = | | 34 | 21 x 99 = | |
| 13 | 4 x 100 = | | 35 | 31 x 100 = | |
| 14 | 4 x 99 = | | 36 | 31 x 99 = | |
| 15 | 7 x 100 = | | 37 | 71 x 100 = | |
| 16 | 7 x 99 = | | 38 | 71 x 99 = | |
| 17 | 9 x 100 = | | 39 | 42 x 100 = | |
| 18 | 9 x 99 = | | 40 | 42 x 99 = | |
| 19 | 8 x 100 = | | 41 | 53 x 99 = | |
| 20 | 8 x 99 = | | 42 | 64 x 99 = | |
| 21 | 5 x 100 = | | 43 | 75 x 99 = | |
| 22 | 50 x 100 = | | 44 | 97 x 99 = | |

mental multiplication

Name ______________________________________________ Date ______________________

1. Draw an area model. Then, solve using the standard algorithm. Use arrows to match the partial products from your area model to the partial products in the algorithm.

   a. 48 × 35

$$\begin{array}{r} 48 \\ \times\ 35 \\ \hline \end{array}$$

   b. 648 × 35

$$\begin{array}{r} 648 \\ \times\ 35 \\ \hline \end{array}$$





EUREKA MATH™

2. Solve using the standard algorithm.

   a. 758 × 92

   b. 958 × 94

   c. 476 × 65

   d. 547 × 64

3. Carpet costs $16 a square foot. A rectangular floor is 16 feet long by 14 feet wide. How much would it cost to carpet the floor?

4. General admission to The American Museum of Natural History is $19.
   a. If a group of 125 students visits the museum, how much will the group's tickets cost?

   b. If the group also purchases IMAX movie tickets for an additional $4 per student, what is the new total cost of all the tickets? Write an expression that shows how you calculated the new price.

Name ____________________________________________ Date ____________________

Draw an area model. Then, solve using the standard algorithm. Use arrows to match the partial products from your area model to the partial products in the algorithm.

a. 78 × 42

$$\begin{array}{r} 7\,8 \\ \times\ 4\,2 \\ \hline \end{array}$$

b. 783 × 42

$$\begin{array}{r} 7\,8\,3 \\ \times\ 4\,2 \\ \hline \end{array}$$

Name ______________________________ Date ________________

1. Draw an area model. Then, solve using the standard algorithm. Use arrows to match the partial products from your area model to the partial products in the algorithm.

   a. 27 × 36

$$\begin{array}{r} 27 \\ \times\ 36 \\ \hline \end{array}$$

   b. 527 × 36

$$\begin{array}{r} 527 \\ \times\ 36 \\ \hline \end{array}$$

2. Solve using the standard algorithm.

   a. $649 \times 53$

   b. $496 \times 53$

   c. $758 \times 46$

   d. $529 \times 48$

3. Each of the 25 students in Mr. McDonald's class sold 16 raffle tickets. If each ticket costs $15, how much money did Mr. McDonald's students raise?

4. Jayson buys a car and pays by installments. Each installment is $567 per month. After 48 months, Jayson owes $1,250. What was the total price of the vehicle?

# Lesson 7

## Objective: Connect area models and the distributive property to partial products of the standard algorithm with renaming.

### Suggested Lesson Structure

| | |
|---|---|
| ■ Fluency Practice | (12 minutes) |
| ■ Application Problem | (6 minutes) |
| ■ Concept Development | (32 minutes) |
| ■ Student Debrief | (10 minutes) |
| **Total Time** | **(60 minutes)** |

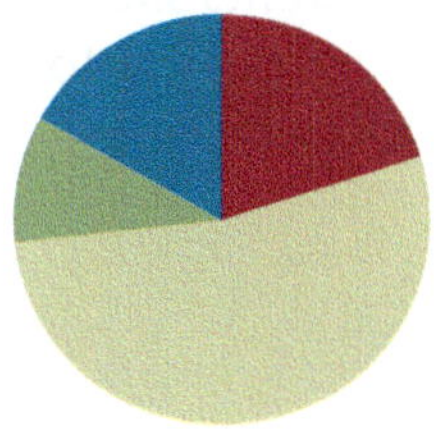

## Fluency Practice (12 minutes)

- Sprint: Multiply by Multiples of 10 and 100 **5.NBT.2** (8 minutes)
- Multiply Using the Area Model **5.NBT.6** (4 minutes)

### Sprint: Multiply by Multiples of 10 and 100 (8 minutes)

Materials: (S) Multiply by Multiples of 10 and 100 Sprint

Note: This review fluency activity helps preserve skills students learned and mastered in Module 1 and lays the groundwork for multiplying with three-digit factors in today's lesson.

### Multiply Using the Area Model (4 minutes)

Note: Since the area model will be used again in this lesson, a short review supports the solidity of the prior learning before adding on the complexity of factors with more digits.

Follow the same process and procedure as Lesson 6 using the following possible sequence: 24 × 15 and 824 × 15.

## Application Problem (6 minutes)

The length of a school bus is 12.6 meters. If 9 school buses park end-to-end with 2 meters between each one, what's the total length from the front of the first bus to the end of the last bus?

Note: This problem is designed to bridge to the current lesson with multi-digit multiplication while also reaching back to decimal multiplication work from Module 1. Students should be encouraged to estimate for a reasonable product prior to multiplying. Encourage students to use the most efficient method to solve this problem.

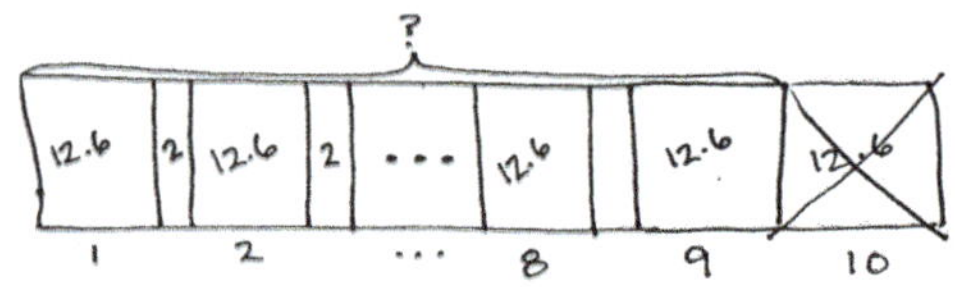


9 Buses : (12.6 × 10) − (12.6)
= 126 − 12.6
= 113.4 m

Space between:
8 × 2 = 16 m

113.4 + 16 = 129.4 m

The total length is 129.4 m.

**NOTES ON MULTIPLE MEANS OF REPRESENTATION:**

Some students may find it difficult to align digits in the standard algorithm. Consider offering graph paper as a scaffold to support them.

## Concept Development (32 minutes)

**Problem 1**

524 × 136

T: (Write 524 × 136.) Compare the problem on the board with the problems in the previous lesson. What do you notice?

S: In the previous lesson, we multiplied using only two-digit numbers as the number of units. → The problems yesterday had a two-digit number in them.

T: So, which one of these factors should we designate as our unit? Turn and talk.

S: I think it's easier to count 136 units of 524 than 524 units of 136. It seems like a lot less units to count that way. → I'm not sure which one to use as the unit. It seems like it won't really matter this time because they are both three-digit numbers. → I think we should count 136 units of 524 because then we just have to multiply by 100 and 30 and 6. These seem easier to me than multiplying by 500, 20, and 4. → I'm going to count 524 units of 136. I don't think multiplying by 500 then 20 then 4 will be any harder than the other way.

T: Very thoughtful conversations. Let's designate 524 as our unit. How will the area model for this problem be different than previous models?

S: There will be 3 columns and 3 rows. In the previous lesson, we only had 2 rows because we used the smaller number to tell the number of units. We used our larger numbers yesterday as our units.

T: Partner A, draw an area model to find the product. Partner B, solve using the standard algorithm.

S: (Work.)

T: What's the product of 524 × 136?

S: 71,264.

T: Compare your solutions by matching your partial products and final product.

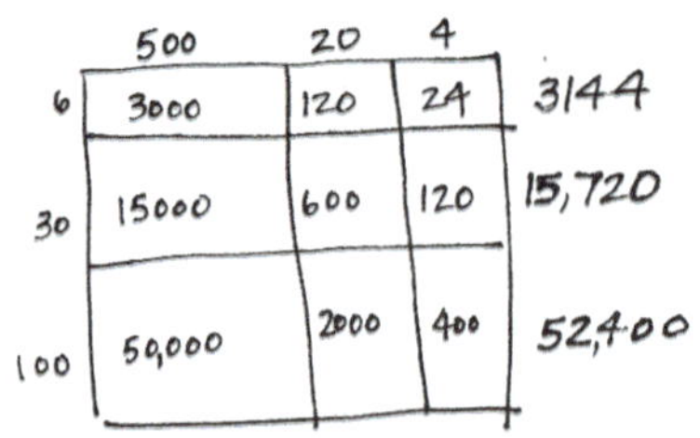


```
    524
  × 136
   3144
  15720
+ 52400
 71,264
```







EUREKA MATH™

**Problem 2**

4,519 × 326

T: What is different about this problem?

S: We have a four-digit number this time.

T: Which factor will be our unit? Is one more efficient to use than the other? Turn and talk.

S: (Discuss as in Problem 1.)

T: Does the presence of the fourth digit change anything about how we multiply? Why or why not?

S: We will have an extra column in the area model, but we just multiply the same way.

T: Before we solve this problem, let's estimate our product. Round the factors and make an estimate.

S: 5,000 × 300 = 1,500,000.

T: Now, solve this problem with your partner. Partner B should do the area model this time, and Partner A should use the algorithm. As you work, explain to your partner how you organized your thoughts to make this problem easier. (How did you decompose your factors?)

S: (Work and explain to partners.)

T: (Circulate and then review the answers. Return to the estimated product, and ask if the actual product is reasonable given the estimate.)

**NOTES ON MULTIPLE MEANS OF ACTION AND EXPRESSION:**

When multiplying multi-digit numbers, (especially those with three-digit multipliers) encourage students to remember which partial product they are finding. This will help to remind students about the zeros in the partial products. Ask, "Are we multiplying by ones, tens, or hundreds? When multiplying by a ten, what will the digit in the ones place be? When multiplying by hundreds, what will the digits in the ones and tens places always be?"

**Problem 3**

4,509 × 326. (Estimate the product first.)

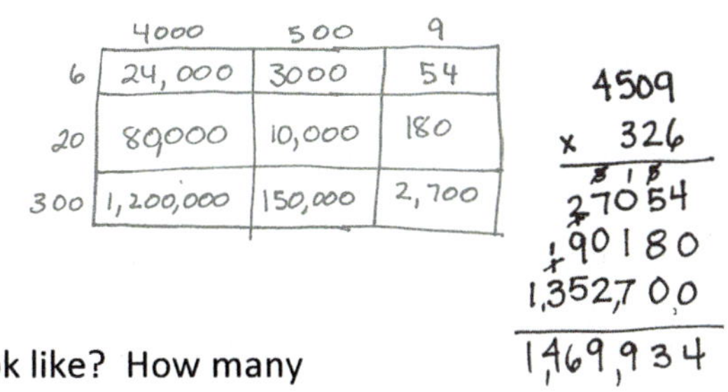


T: We will count 326 units of 4,509.

T: Compare 4,519 and 4,509. How are they different?

S: There's a zero in the tens place in 4,509.

T: What does 4,509 look like in expanded form?

S: 4,000 + 500 + 9.

T: Can you imagine what the length of our rectangle will look like? How many columns will we need to represent the total length?

S: We will need only three columns.

T: This is a four-digit number. Why only three columns?

S: The rectangle shows area. So, if we put a column in for the tens place, we would be drawing the rectangle bigger than it really is. → We are chopping the length of the rectangle into three parts—4,000, 500, and 9. That is the total length already. The width of the tens column would be zero, so it has no area.

T: Work with a partner to solve this problem. Partner A will use the area model, and Partner B will solve using the algorithm. Compare your work when you finish.

T: (Circulate and review answers. Have students assess the reasonableness of the product given the estimate.)

**Problem 4**

4,509 × 306. (Estimate the product first.)

T: This time we are counting 306 units of 4,509. How is this different from Problem 3?

S: It's going to be 20 units less of 4,509 than last time. → There is a zero in both factors this time.

T: Thinking about the expanded forms of the factors, imagine the area model. How will the length and width be decomposed? How will it compare to Problem 3?

S: Like Problem 3, there are only three columns in the length again even though it's a four-digit number. → The model doesn't need three rows because there's nothing in the tens place. We only need to show rows for hundreds and ones.

T: (Allow students time to solve with the model.) What two partial products do these two rows represent?

S: 6 × 4,509 and 300 × 4,509.

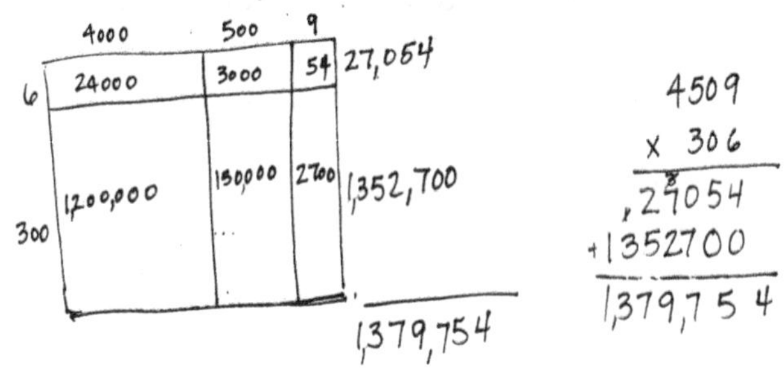


T: Let's record what we just drew with the algorithm. We'll begin with the first partial product 6 × 4,509. Find that partial product.

S: (Record first partial product.)

T: Now, let's record 300 × 4,509. When we multiply a number by 100, what happens to the value and position of each digit?

S: Each becomes 100 times as large and shifts two places to the left.

T: In the case of 4,509, when we multiply it by 300, what would need to be recorded in the ones and tens place after the digits shift?

S: Zeros would go in those places.

## Problem Set (10 minutes)

Students should do their personal best to complete the Problem Set within the allotted 10 minutes. For some classes, it may be appropriate to modify the assignment by specifying which problems they work on first. Some problems do not specify a method for solving. Students should solve these problems using the RDW approach used for Application Problems.

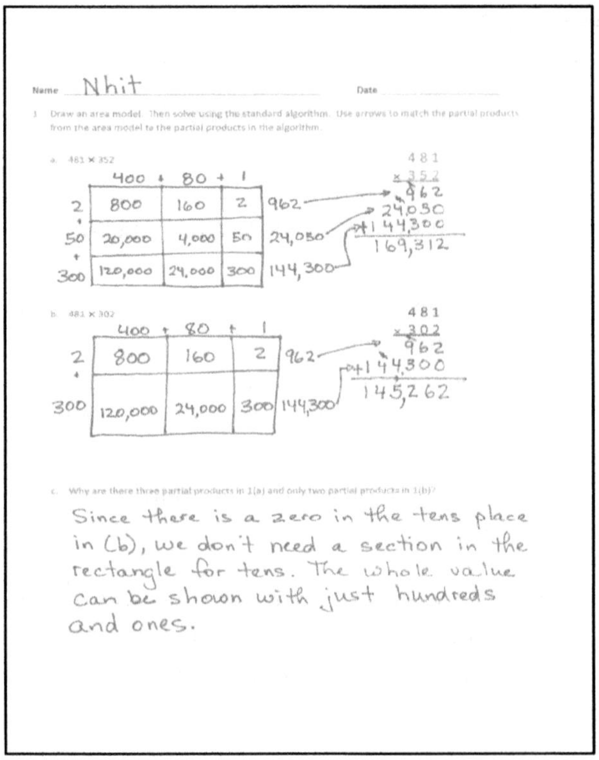

## Student Debrief (10 minutes)

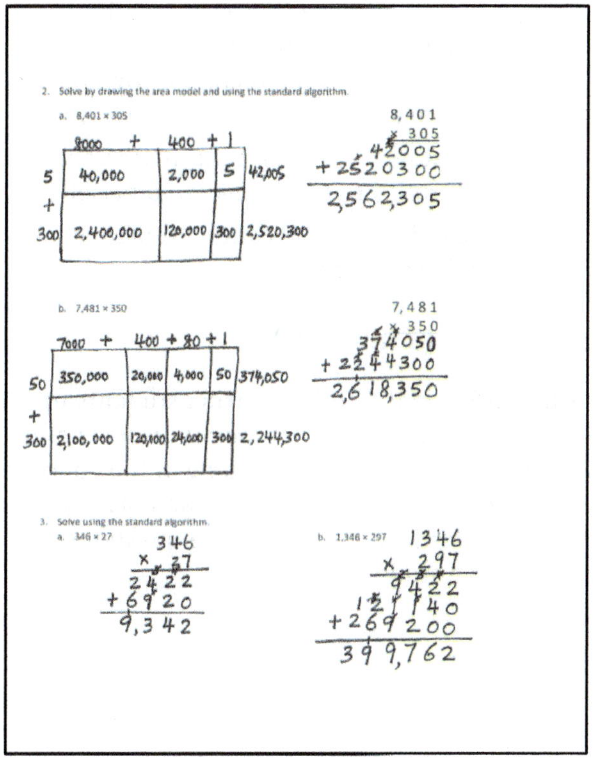


**Lesson Objective:** Connect area models and the distributive property to partial products of the standard algorithm with renaming.

The Student Debrief is intended to invite reflection and active processing of the total lesson experience.

Invite students to review their solutions for the Problem Set. They should check work by comparing answers with a partner before going over answers as a class. Look for misconceptions or misunderstandings that can be addressed in the Debrief. Guide students in a conversation to debrief the Problem Set and process the lesson.

Any combination of the questions below may be used to lead the discussion.

- Explain why a multiplication problem with a three-digit multiplier will not always have three partial products. Use Problems 1(a) and (b) as examples.
- How are the area models for Problems 2(a) and (b) alike, and how are they different?
- What pattern did you notice in Problem 3?
- Take time to discuss with students that the choice of decomposition in the area model and the order in which the partial products are found can be highly variable. Use a context, such as a rug or garden, to make the thinking even more concrete.
- It is important for students to understand that the standard algorithm's sequence of decomposition by place value unit is a convention. It is a useful convention as it allows us to make efficient use of multiples of ten, which make mental math easier. However, it is not a *rule*. Allow students to explore a multi-digit multiplication case like 52 × 35 by decomposing the area in many ways and comparing the results. A few examples are included below.

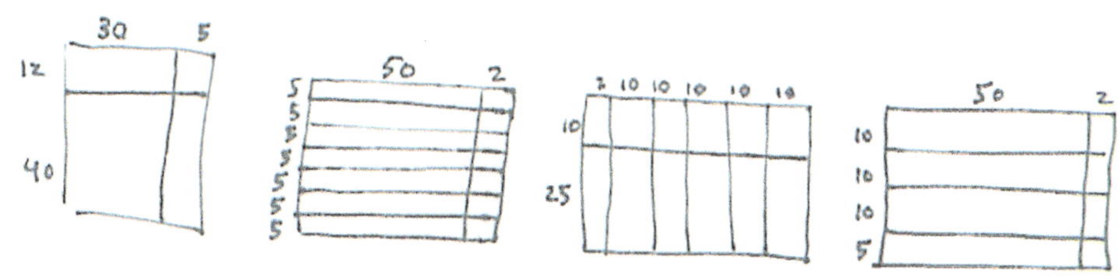

- Does it matter which factor goes on the top of the model or the algorithm? Why or why not? (The orientation of the rectangle does not change its area.)
- How many ways can you decompose the length? The width?
- What are you thinking about as you make these decisions on how to split the area into parts? (Mental math considerations, easier basic facts, etc.)

- Do any of these choices affect the size of the area (the product)? Why or why not? (The outer dimensions of the rectangle are unchanged regardless of the way in which it is partitioned.)
- What new (or significant) math vocabulary did we use today to communicate precisely?
- How did the Application Problem connect to today's lesson?

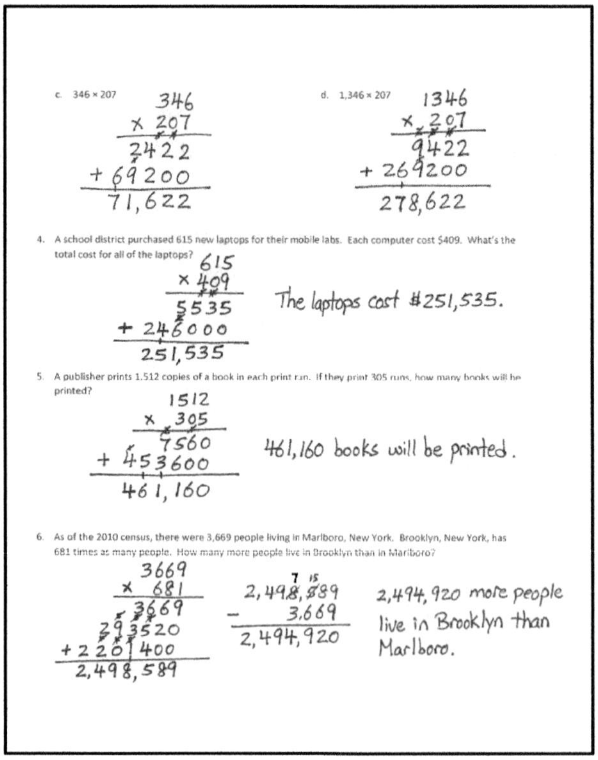
c. 346 × 207

346 × 207: 2422 + 69200 = 71,622

d. 1,346 × 207

1346 × 207: 9422 + 269200 = 278,622

4. A school district purchased 615 new laptops for their mobile labs. Each computer cost $409. What's the total cost for all of the laptops?

615 × 409: 5535 + 246000 = 251,535

The laptops cost $251,535.

5. A publisher prints 1,512 copies of a book in each print run. If they print 305 runs, how many books will be printed?

1512 × 305: 7560 + 453600 = 461,160

461,160 books will be printed.

6. As of the 2010 census, there were 3,669 people living in Marlboro, New York. Brooklyn, New York, has 681 times as many people. How many more people live in Brooklyn than in Marlboro?

3669 × 681: 3669 + 293520 + 2201400 = 2,498,589

2,498,589 − 3,669 = 2,494,920

2,494,920 more people live in Brooklyn than Marlboro.

## Exit Ticket (3 minutes)

After the Student Debrief, instruct students to complete the Exit Ticket. A review of their work will help with assessing students' understanding of the concepts that were presented in today's lesson and planning more effectively for future lessons. The questions may be read aloud to the students.

# A

Number Correct: _______

Multiply by Multiples of 10 and 100

| | | |
|---|---|---|
| 1. | 2 × 10 = | |
| 2. | 12 × 10 = | |
| 3. | 12 × 100 = | |
| 4. | 4 × 10 = | |
| 5. | 34 × 10 = | |
| 6. | 34 × 100 = | |
| 7. | 7 × 10 = | |
| 8. | 27 × 10 = | |
| 9. | 27 × 100 = | |
| 10. | 3 × 10 = | |
| 11. | 3 × 2 = | |
| 12. | 3 × 20 = | |
| 13. | 13 × 10 = | |
| 14. | 13 × 2 = | |
| 15. | 13 × 20 = | |
| 16. | 13 × 100 = | |
| 17. | 13 × 200 = | |
| 18. | 2 × 4 = | |
| 19. | 22 × 4 = | |
| 20. | 22 × 40 = | |
| 21. | 22 × 400 = | |
| 22. | 33 × 2 = | |

| | | |
|---|---|---|
| 23. | 33 × 20 = | |
| 24. | 33 × 200 = | |
| 25. | 24 × 10 = | |
| 26. | 24 × 20 = | |
| 27. | 24 × 100 = | |
| 28. | 24 × 200 = | |
| 29. | 23 × 30 = | |
| 30. | 23 × 300 = | |
| 31. | 71 × 2 = | |
| 32. | 71 × 20 = | |
| 33. | 14 × 2= | |
| 34. | 14 × 3 = | |
| 35. | 14 × 30 = | |
| 36. | 14 × 300 = | |
| 37. | 82 × 20 = | |
| 38. | 15 × 300 = | |
| 39. | 71 × 600 = | |
| 40. | 18 × 40 = | |
| 41. | 75 × 30 = | |
| 42. | 84 × 300 = | |
| 43. | 87 × 60 = | |
| 44. | 79 × 800 = | |

# B

Number Correct: _______

Improvement: _______

Multiply by Multiples of 10 and 100

| | | |
|---|---|---|
| 1. | 3 × 10 = | |
| 2. | 13 × 10 = | |
| 3. | 13 × 100 = | |
| 4. | 5 × 10 = | |
| 5. | 35 × 10 = | |
| 6. | 35 × 100 = | |
| 7. | 8 × 10 = | |
| 8. | 28 × 10 = | |
| 9. | 28 × 100 = | |
| 10. | 4 × 10 = | |
| 11. | 4 × 2 = | |
| 12. | 4 × 20 = | |
| 13. | 14 × 10 = | |
| 14. | 14 × 2 = | |
| 15. | 14 × 20 = | |
| 16. | 14 × 100 = | |
| 17. | 14 × 200 = | |
| 18. | 2 × 3 = | |
| 19. | 22 × 3 = | |
| 20. | 22 × 30 = | |
| 21. | 22 × 300 = | |
| 22. | 44 × 2 = | |

| | | |
|---|---|---|
| 23. | 44 × 20 = | |
| 24. | 44 × 200 = | |
| 25. | 42 × 10 = | |
| 26. | 42 × 20 = | |
| 27. | 42 × 100 = | |
| 28. | 42 × 200 = | |
| 29. | 32 × 30 = | |
| 30. | 32 × 300 = | |
| 31. | 81 × 2 = | |
| 32. | 81 × 20 = | |
| 33. | 13 × 3 = | |
| 34. | 13 × 4 = | |
| 35. | 13 × 40 = | |
| 36. | 13 × 400 = | |
| 37. | 72 × 30 = | |
| 38. | 15 × 300 = | |
| 39. | 81 × 600 = | |
| 40. | 16 × 40 = | |
| 41. | 65 × 30 = | |
| 42. | 48 × 300 = | |
| 43. | 89 × 60 = | |
| 44. | 76 × 800 = | |

Name ______________________________ Date ________________

1. Draw an area model. Then, solve using the standard algorithm. Use arrows to match the partial products from the area model to the partial products in the algorithm.

   a. 481 × 352

$$\begin{array}{r} 481 \\ \times\ 352 \\ \hline \end{array}$$

   b. 481 × 302

$$\begin{array}{r} 481 \\ \times\ 302 \\ \hline \end{array}$$

   c. Why are there three partial products in 1(a) and only two partial products in 1(b)?

2. Solve by drawing the area model and using the standard algorithm.

   a. 8,401 × 305

$$\begin{array}{r} 8,401 \\ \times \quad 305 \\ \hline \end{array}$$

   b. 7,481 × 350

$$\begin{array}{r} 7,481 \\ \times \quad 350 \\ \hline \end{array}$$

3. Solve using the standard algorithm.

   a. 346 × 27

   b. 1,346 × 297

c. $346 \times 207$

d. $1,346 \times 207$

4. A school district purchased 615 new laptops for their mobile labs. Each computer cost $409. What is the total cost for all of the laptops?

5. A publisher prints 1,512 copies of a book in each print run. If they print 305 runs, how many books will be printed?

6. As of the 2010 census, there were 3,669 people living in Marlboro, New York. Brooklyn, New York, has 681 times as many people. How many more people live in Brooklyn than in Marlboro?

Name ____________________ Date ____________

Draw an area model. Then, solve using the standard algorithm.

a. 642 × 257

$$\begin{array}{r} 642 \\ \times\ 257 \\ \hline \end{array}$$

b. 642 × 207

$$\begin{array}{r} 642 \\ \times\ 207 \\ \hline \end{array}$$

Name ______________________________ Date ______________

1. Draw an area model. Then, solve using the standard algorithm. Use arrows to match the partial products from your area model to the partial products in your algorithm.

a. 273 × 346

$$\begin{array}{r} 273 \\ \times\ 346 \\ \hline \end{array}$$

b. 273 × 306

$$\begin{array}{r} 273 \\ \times\ 306 \\ \hline \end{array}$$

c. Both Parts (a) and (b) have three-digit multipliers. Why are there three partial products in Part (a) and only two partial products in Part (b)?

2. Solve by drawing the area model and using the standard algorithm.

   a. 7,481 × 290

   b. 7,018 × 209

3. Solve using the standard algorithm.

   a. 426 × 357

   b. 1,426 × 357

c. 426 × 307

d. 1,426 × 307

4. The Hudson Valley Renegades Stadium holds a maximum of 4,505 people. During the height of their popularity, they sold out 219 consecutive games. How many tickets were sold during this time?

5. One Saturday at the farmer's market, each of the 94 vendors made $502 in profit. How much profit did all vendors make that Saturday?

# Lesson 8

## Objective: Fluently multiply multi-digit whole numbers using the standard algorithm and using estimation to check for reasonableness of the product.

### Suggested Lesson Structure

| | |
|---|---|
| Fluency Practice | (7 minutes) |
| Application Problem | (10 minutes) |
| Concept Development | (33 minutes) |
| Student Debrief | (10 minutes) |
| **Total Time** | **(60 minutes)** |

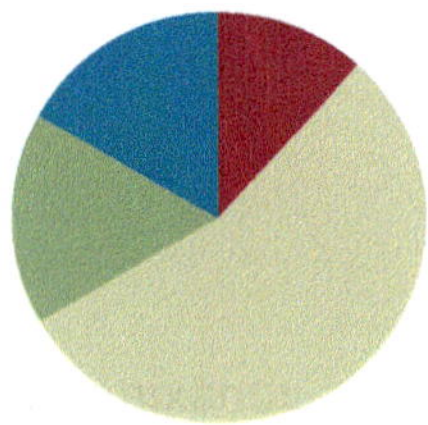

## Fluency Practice (7 minutes)

- State in Exponential Form Name **5.NBT.2** (3 minutes)
- Multiply Using the Area Model with a Zero in One Factor **5.NBT.6** (4 minutes)

### State in Exponential Form Name (3 minutes)

Materials: (S) Personal white board

Note: This fluency activity maintains earlier skills and encourages insights into the place value structure of multi-digit multiplication's partial products. A quick review of relevant vocabulary (*base, exponent, power*) may be in order.

T: (Write $10^2$ = _____.) Say the base.
S: 10.
T: Say the exponent.
S: 2.
T: Say 10 squared as a multiplication sentence starting with 10.
S: $10 \times 10 = 100$.
T: Say it as a number sentence without using a multiplication sentence.
S: 10 squared equals 100.

Repeat the process with $10^3$, $10^4$, $10^5$, and $10^7$.

### Multiply Using the Area Model with a Zero in One Factor (4 minutes)

Note: Students need additional practice when there is a zero in one of the factors. If deemed appropriate, students may be asked to share their observations about what they notice in these cases and then justify their thinking.

Follow the same process and procedure as Lesson 6, juxtaposing similar problems such as 342 × 251 and 342 × 201 whereby one factor has a zero.

## Application Problem (10 minutes)

Erin and Frannie entered a rug design contest. The rules stated that the rug's dimensions must be 32 inches × 45 inches and that they must be rectangular. They drew the following for their entries. Show at least three other designs they could have entered in the contest. Calculate the area of each section, and the total area of the rugs.

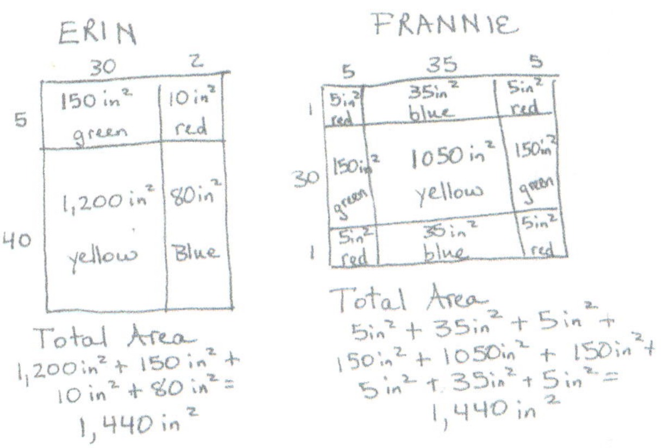


**NOTES ON MULTIPLE MEANS OF ACTION AND EXPRESSION:**

Students might be encouraged to actually produce the designs that they generate for this Application Problem. This offers opportunity for students not only to reinforce the notion that area can be partitioned into multiple partial products but also allows for a cross-curricular application of math concepts.

Note: This Application Problem echoes the Student Debrief discussion from Lesson 7. Accept any design whose partitions are accurate. Have students compare the total area of their designs to check.

## Concept Development (33 minutes)

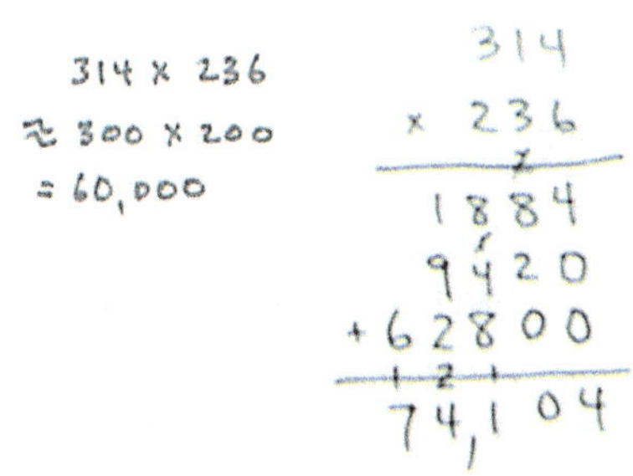


**Problem 1**

314 × 236

T: (Write 314 × 236 on the board.) Round each factor to estimate the product. Turn and talk.

S: 314 is closer to 3 hundreds than 4 hundreds on the number line.
→ 236 is closer to 2 hundreds than 3 hundreds on the number line.

T: Multiply your rounded factors to estimate the product. What is 300 times 200?

S: Hundreds times hundreds makes ten thousands. 3 × 2 is 6. So, we'll get 6 ten thousands, or 60,000.

T: Express 60,000 as a multiplication expression with an exponent.

S: $6 \times 10^4$.

T: I noticed that we rounded both of our factors *down* to the nearest hundred. Will our actual product be more than or less than our estimated product? Why? Tell a neighbor.

S: The answer should be more than 60,000. → Our actual factors are greater, therefore our actual product will be greater than 60,000.

T: Work with a partner to solve using the standard algorithm.

S: (Solve to find 74,104.)

T: Look back to our estimated product. Is our answer reasonable? Turn and talk.

S: Yes, it's greater like we thought it would be. Our answer makes sense.

**Problem 2**

1,882 × 296

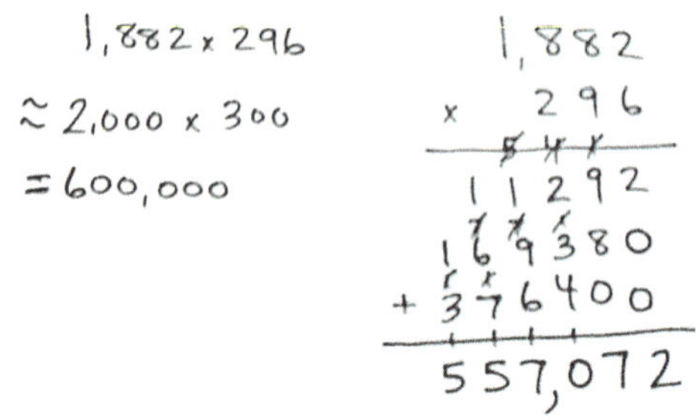


T: (Write 1,882 × 296 on the board.) Round each factor and estimate the product. Will the actual product be greater than or less than your estimate? Turn and talk.

S: 1,882 rounds to 2,000. 296 rounds up to 300. The estimated product is 600,000. We rounded both factors *up* this time. → Since our actual factors are less than 2,000 and 300, our actual product must be less than 600,000.

T: Work independently to solve 1,882 × 296.

S: (Solve.)

T: What is the product of 1,882 and 296?

S: 557,072.

T: Is our product reasonable considering our estimate? Turn and talk.

S: Yes, it is close to 600,000, but a bit less than our estimated product like we predicted it would be.

Possibly have students compare the estimates of Problems 1 and 2.

**NOTES ON MULTIPLE MEANS OF ACTION AND EXPRESSION:**

If students are not yet ready for independent work, have them work in pairs and talk as they estimate, solve, and check their solutions. These types of strategy-based discussions deepen understanding for students and allow them to see problems in different ways.

**Problem 3**

4,902 × 408

T: (Write 4,902 × 408 on the board.) Work independently to find an estimated product for this problem.

T: (Pause.) Let's read the estimated multiplication sentence without the product.

S: 4,902 × 408 is about as much as 5,000 × 400.

T: As I watched you work, I saw that some of you said our estimated product was 200,000, and others said 2,000,000. One is 10 times as much as the other. Analyze the error with your partner.

S: 5,000 × 400 is like (5 × 1,000) × (4 × 100). That's like (5 × 4) × 100,000, so 20 copies of 1 hundred thousand. That's 20 hundred thousands which is 2 million.

T: Simply counting the zeros in our factors is not an acceptable strategy. We should always be aware of our units and how many of those units we are counting.

T: Should our actual product be more or less than our estimated product? How do you know? Turn and talk.

4,902 × 408
≈ 5,000 × 400
= 2,000,000

$$\begin{array}{r} 4{,}902 \\ \times\ 408 \\ \hline 39216 \\ +\ 1960800 \\ \hline 2{,}000{,}016 \end{array}$$

S: We rounded one factor up and one factor down. Our actual product could be more or less. → We can't really tell yet, since we rounded 4,902 up and 408 down. Our actual product might be more or less than 2,000,000, but it should be close.

T: Work independently to solve 4,902 × 408.

S: (Solve to find 2,000,016.)

T: Is the actual product reasonable?

S: Yes.

## Problem Set (10 minutes)

Students should do their personal best to complete the Problem Set within the allotted 10 minutes. For some classes, it may be appropriate to modify the assignment by specifying which problems they work on first. Some problems do not specify a method for solving. Students should solve these problems using the RDW approach used for Application Problems.

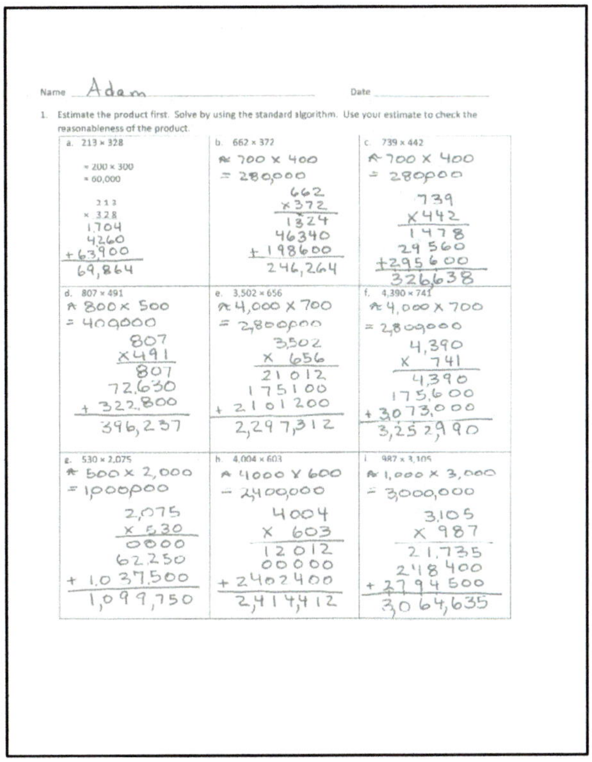
Name Adam Date
1. Estimate the product first. Solve by using the standard algorithm. Use your estimate to check the reasonableness of the product.

| a. 213 × 328 | b. 662 × 372 | c. 739 × 442 |
|---|---|---|
| ≈ 200 × 300 = 60,000; 69,864 | ≈ 700 × 400 = 280,000; 246,264 | ≈ 700 × 400 = 280,000; 326,638 |
| d. 807 × 491 | e. 3,502 × 656 | f. 4,390 × 741 |
| ≈ 800 × 500 = 400,000; 396,237 | ≈ 4,000 × 700 = 2,800,000; 2,297,312 | ≈ 4,000 × 700 = 2,800,000; 3,252,990 |
| g. 530 × 2,075 | h. 4,004 × 603 | i. 987 × 3,105 |
| ≈ 500 × 2,000 = 1,000,000; 1,099,750 | ≈ 4,000 × 600 = 2,400,000; 2,414,412 | ≈ 1,000 × 3,000 = 3,000,000; 3,064,635 |

## Student Debrief (10 minutes)

**Lesson Objective:** Fluently multiply multi-digit whole numbers using the standard algorithm and using estimation to check for reasonableness of the product.

The Student Debrief is intended to invite reflection and active processing of the total lesson experience.

Invite students to review their solutions for the Problem Set. They should check work by comparing answers with a partner before going over answers as a class. Look for misconceptions or misunderstandings that can be addressed in the Debrief. Guide students in a conversation to debrief the Problem Set and process the lesson.

Any combination of the questions below may be used to lead the discussion.

- What is the benefit of estimating before solving?
- Look at Problems 1 (b) and (c). What do you notice about the estimated products? Analyze why the estimates are the same, yet the products are so different. (You might point out the same issue in Problems 1 (e) and (f).)
- How could the cost of the chairs have been found using the unit form mental math strategy? (Students may have multiplied 355 × 200 and subtracted 355.)
- In Problem 4, Carmella estimated that she had 3,000 cards. How did she most likely round her factors?
- Would rounding the number of boxes of cards to 20 have been a better choice? Why or why not? (Students might consider that she is done collecting cards and will not need any more space. Others might argue that she is still collecting and could use more room for the future.)
- Do we always have to round to a multiple of 10, 100, or 1,000? Is there a number between 10 and 20 that would have been a better choice for Carmella?
- Can you identify a situation in a real-life example where overestimating would be most appropriate? Can you identify a situation in the real world where underestimation would be most appropriate? (For example, ordering food for a party where 73 people are invited. The answer, of course, depends on the circumstances, budget, the likelihood of the attendance of all who were invited, etc.)

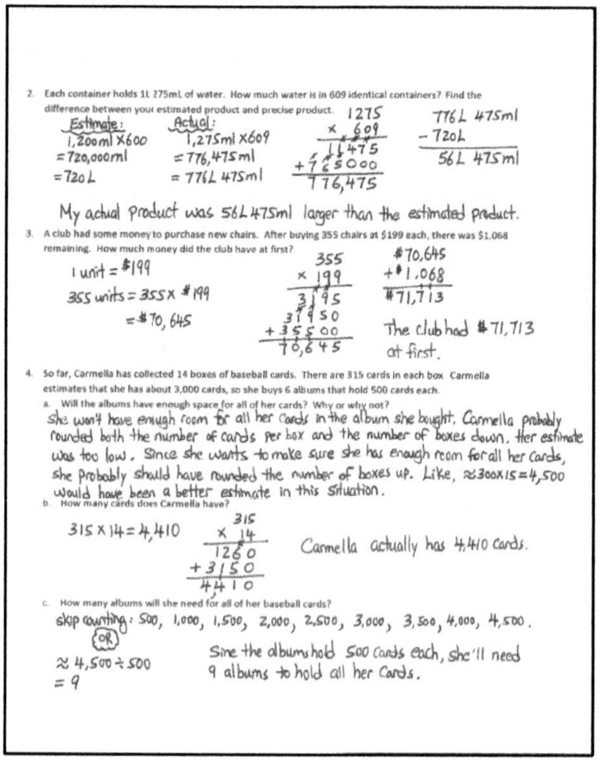
2. Each container holds 1 L 275 mL of water. How much water is in 609 identical containers? Find the difference between your estimated product and precise product.

Estimate: 1,200ml X 600 = 720,000ml = 720 L

Actual: 1,275ml X 609 = 776,475ml = 776L 475ml

1275 × 609 = 11475 + 765000 = 776,475

776L 475ml − 720L = 56L 475ml

My actual product was 56L 475ml larger than the estimated product.

3. A club had some money to purchase new chairs. After buying 355 chairs at $199 each, there was $1,068 remaining. How much money did the club have at first?

1 unit = $199

355 units = 355 X $199 = $70,645

355 × 199 = 3195 + 31950 + 35500 = 70,645

$70,645 + $1,068 = $71,713

The club had $71,713 at first.

4. So far, Carmella has collected 14 boxes of baseball cards. There are 315 cards in each box. Carmella estimates that she has about 3,000 cards, so she buys 6 albums that hold 500 cards each.

a. Will the albums have enough space for all of her cards? Why or why not?

She won't have enough room for all her cards in the album she bought. Carmella probably rounded both the number of cards per box and the number of boxes down. Her estimate was too low. Since she wants to make sure she has enough room for all her cards, she probably should have rounded the number of boxes up. Like, ≈300x15=4,500 would have been a better estimate in this situation.

b. How many cards does Carmella have?

315 x 14 = 4,410

315 × 14 = 1260 + 3150 = 4410

Carmella actually has 4,410 cards.

c. How many albums will she need for all of her baseball cards?

skip counting: 500, 1,000, 1,500, 2,000, 2,500, 3,000, 3,500, 4,000, 4,500.

OR ≈ 4,500 ÷ 500 = 9

Since the albums hold 500 cards each, she'll need 9 albums to hold all her cards.

## Exit Ticket (3 minutes)

After the Student Debrief, instruct students to complete the Exit Ticket. A review of their work will help with assessing students' understanding of the concepts that were presented in today's lesson and planning more effectively for future lessons. The questions may be read aloud to the students.

Name ______________________________ Date ______________

1. Estimate the product first. Solve by using the standard algorithm. Use your estimate to check the reasonableness of the product.

| | | |
|---|---|---|
| a. 213 × 328<br><br>≈ 200 × 300<br>= 60,000<br><br>  2 1 3<br>× 3 2 8 | b. 662 × 372 | c. 739 × 442 |
| d. 807 × 491 | e. 3,502 × 656 | f. 4,390 × 741 |
| g. 530 × 2,075 | h. 4,004 × 603 | i. 987 × 3,105 |

2. Each container holds 1 L 275 mL of water. How much water is in 609 identical containers? Find the difference between your estimated product and precise product.

3. A club had some money to purchase new chairs. After buying 355 chairs at $199 each, there was $1,068 remaining. How much money did the club have at first?

4. So far, Carmella has collected 14 boxes of baseball cards. There are 315 cards in each box. Carmella estimates that she has about 3,000 cards, so she buys 6 albums that hold 500 cards each.

   a. Will the albums have enough space for all of her cards? Why or why not?

   b. How many cards does Carmella have?

   c. How many albums will she need for all of her baseball cards?

Name ______________________________ Date ______________

Estimate the product first. Solve by using the standard algorithm. Use your estimate to check the reasonableness of the product.

a. 283 × 416

$$\begin{array}{r} 283 \\ \times\ 416 \\ \hline \end{array}$$

≈ __________ × __________

= __________

b. 2,803 × 406

$$\begin{array}{r} 2{,}803 \\ \times\ \ \ 406 \\ \hline \end{array}$$

≈ __________ × __________

= __________

Name ______________________________ Date ________________

1. Estimate the product first. Solve by using the standard algorithm. Use your estimate to check the reasonableness of the product.

| a. 312 × 149<br><br>≈ 300 × 100<br>= 30,000<br><br>312<br>× 149 | b. 743 × 295 | c. 428 × 637 |
|---|---|---|
| d. 691 × 305 | e. 4,208 × 606 | f. 3,068 × 523 |
| g. 430 × 3,064 | h. 3,007 × 502 | i. 254 × 6,104 |

2. When multiplying 1,729 times 308, Clayton got a product of 53,253. Without calculating, does his product seem reasonable? Explain your thinking.

3. A publisher prints 1,912 copies of a book in each print run. If they print 305 runs, the manager wants to know about how many books will be printed. What is a reasonable estimate?

# Lesson 9

## Objective: Fluently multiply multi-digit whole numbers using the standard algorithm to solve multi-step word problems.

### Suggested Lesson Structure

| | |
|---|---|
| Fluency Practice | (10 minutes) |
| Concept Development | (40 minutes) |
| Student Debrief | (10 minutes) |
| **Total Time** | **(60 minutes)** |

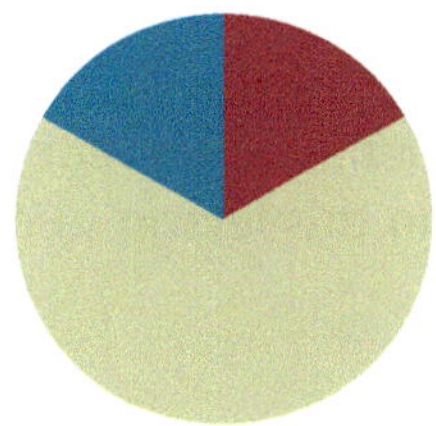

## Fluency Practice (10 minutes)

- Multiply and Divide by Exponents **5.NBT.2** (4 minutes)
- Estimate Products by Rounding **5.NBT.6** (6 minutes)

### Multiply and Divide by Exponents (4 minutes)

Materials: (T/S) Millions to thousandths place value chart (Lesson 1 Template) (S) Personal white board

Note: This review fluency activity encourages flexible thinking because of the inclusion of division. The notation of the exponent form is also important to revisit and reuse so that it becomes natural to students to think of powers of 10 written either as multiples of 10 or as exponents.

T: (Project place value chart from millions to thousandths.) Write 45 tenths as a decimal.
S: (Write 4 in the ones column and 5 in the tenths column.)
T: Say the decimal.
S: Four and five tenths.
T: Multiply it by $10^2$.
S: (Cross out 4.5 and write 450.)

Repeat the process and sequence for $0.4 \times 10^2$, $0.4 \div 10^2$, $3.895 \times 10^3$, and $5{,}472 \div 10^3$.

### Estimate Products by Rounding (6 minutes)

Materials: (S) Personal white board

Note: This fluency activity's focus is estimation, which will be used during this lesson.

T: (Write 412 × 231 ≈ ____ × ____.) Round both factors to the nearest hundred.
S: 400 × 200.

T: Write 412 × 231 ≈ 400 × 200. What is 400 × 200?

S: 80,000.

Repeat the process and procedure for 523 × 298 ≈ 500 × 300, 684 × 347, and 908 × 297.

## Concept Development (40 minutes)

Materials: (T/S) Problem Set

Note: This lesson omits the Application Problem component since the entire lesson is devoted to problem solving. Problems for this section are found in this lesson's Problem Set.

### Problem 1

An office space in New York City measures 48 feet by 56 feet. If it sells for $565 per square foot, what is the total cost of the office space?

T: We will work Problem 1 on your Problem Set together. (Project the problem on the board.) Let's read the word problem aloud.

S: (Read chorally.)

T: Now, let's re-read the problem sentence by sentence, and draw as we go.

S: (Read the first sentence.)

T: What do you see? Can you draw something?

S: (Draw.)

T: Read the next sentence. (Give students time to read.) What is the important information, and how can we show that in our drawing?

S: The office space sells for $565 for each square foot. We can draw a single square unit inside our rectangle to remind us. → We can write that 1 unit = $565.

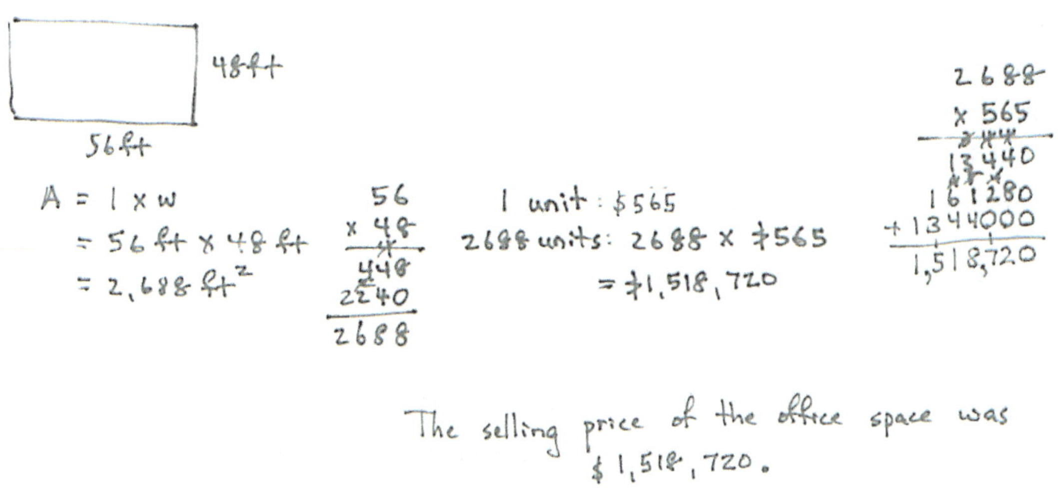




EUREKA MATH

T: How do we solve this problem? Turn and talk.

S: We have to multiply. We have to find the total square feet of the office space and then multiply by \$565. → We have to first find the area of the office space and then multiply by \$565.

T: What information are we given that would help us figure out the area?

S: We can multiply the length times the width.

S: (Solve to find 2,688 $ft^2$.)

T: Have we answered the question?

S: No. We need to multiply the area by the cost of one square foot, \$565, to find the total cost.

T: Solve and express your answer in a complete sentence.

S: (Work.) The cost of the office space is \$1,518,720.

**NOTES ON MULTIPLE MEANS OF REPRESENTATION:**

Guide students to select and practice using various models (e.g., tape diagram or area model) to represent the given information in each problem.

**Problem 2**

Gemma and Leah are both jewelry makers. Gemma made 106 beaded necklaces. Leah made 39 more necklaces than Gemma.

a. Each necklace they make has exactly 104 beads on it. How many beads did both girls use altogether while making their necklaces?

b. At a recent craft fair, Gemma sold each of her necklaces for \$14. Leah sold each of her necklaces for \$10 more. Who made more money at the craft fair? How much more?

MP.2

2)

106 + 39 = 145

a). (106 × 104) + (145 × 104)
= 11,024 + 15,080
= 26,104

Both girls used 26,104 beads altogether.

b). (\$24 × 145) – (\$14 × 106)
= \$3480 – \$1484
= \$1996

Leah made more money.
Leah made \$1996 more than Gemma

T: (Allow students to read the problem chorally, in pairs, or in silence.)

T: Can you draw something?

S: Yes.

T: What can you draw?

S: A tape for Gemma's necklaces and a second, longer bar for Leah's.

T: Go ahead and draw and label your tape diagrams. (Allow time for students to work.)

T: What is the question asking?

S: We have to find the total number of beads on all the necklaces.

T: What do we need to think about to solve this problem? What do you notice about it?

S: It is a multi-step problem. We need to know how many necklaces Leah made before we can find the total number of necklaces. Then, we need to find the number of beads.

T: Work together to complete the first steps by finding the total number of necklaces.

T: We haven't answered the question yet. Turn and talk to your partner about how we can finish solving Part (a).

S: We have to multiply to find the total beads for both girls. → Multiply Gemma's number of necklaces times 104 beads, multiply Leah's number of necklaces times 104, and then add them together. → Add Gemma and Leah's necklaces together, and then multiply by 104.

T: Use an expression to show your strategy for solving.

S: (106 × 104) + (145 × 104) or (106 + 145) × 104.

T: Solve the problem with your partner, and make a statement to answer the question.

S: Gemma and Leah used 26,104 beads altogether.

T: Let's read Part (b) together.

S: (Read.)

T: Who made more money? Without calculating, can we answer this question? Turn and talk.

S: Leah made more necklaces than Gemma, and she charged more per necklace. Therefore, it makes sense that Leah made more money than Gemma.

T: Find out how much more money Leah made.

S: (Work.)

S: Leah made $1,996 more than Gemma.

T: Complete Problems 3, 4, 5, and 6 of the Problem Set independently or in pairs.

**NOTES ON MULTIPLE MEANS OF ACTIONS AND EXPRESSION:**

Vary the grouping size in the classroom. Smaller groups support English language learners to navigate the language of word problems and allow students to find full proficiency of the mathematics first, without the obstacles of vocabulary.

## Problem Set (10 minutes)

Students should do their personal best to complete the Problem Set within the allotted 10 minutes. For some classes, it may be appropriate to modify the assignment by specifying which problems they work on first. Some problems do not specify a method for solving. Students should solve these problems using the RDW approach used for Application Problems.

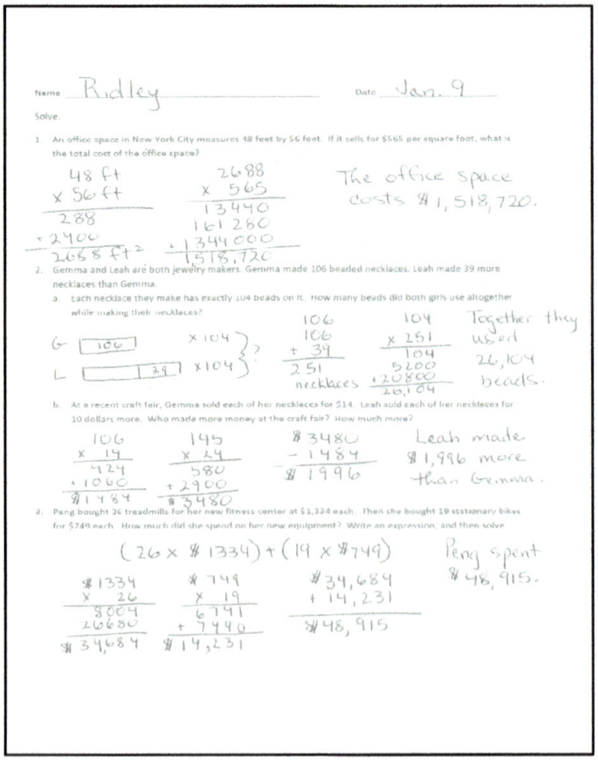

## Student Debrief (10 minutes)

**Lesson Objective:** Fluently multiply multi-digit whole numbers using the standard algorithm to solve multi-step word problems.

The Student Debrief is intended to invite reflection and active processing of the total lesson experience.

Invite students to review their solutions for the Problem Set. They should check work by comparing answers with a partner before going over answers as a class. Look for misconceptions or misunderstandings that can be addressed in the Debrief. Guide students in a conversation to debrief the Problem Set and process the lesson.

Any combination of the questions below may be used to lead the discussion.

- Share and explain to your partner the numerical expressions you wrote to help you solve Problems 3 and 5.
- Explain how Problems 3 and 5 could both be solved in more than one way.
- What type of problem are Problem 1 and Problem 5? How are these two problems different from the others? (Problems 1 and 5 are measurement problems.)

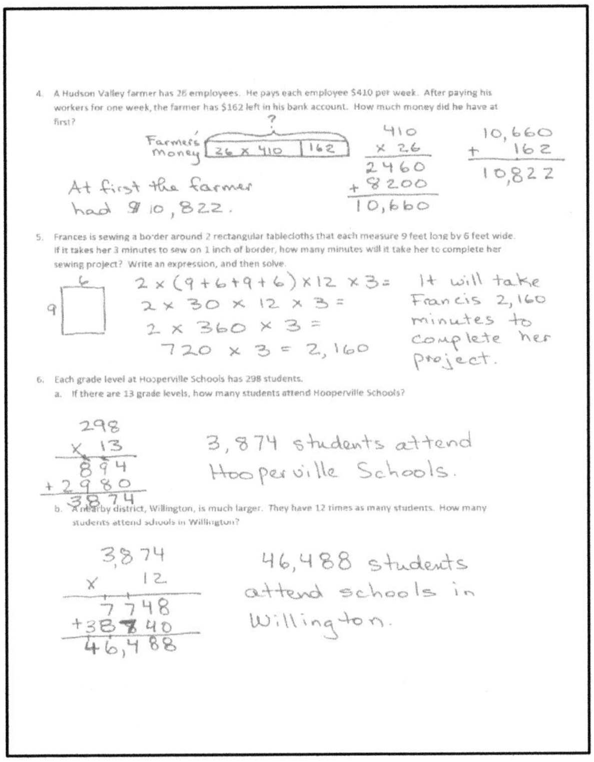

4. A Hudson Valley farmer has 26 employees. He pays each employee $410 per week. After paying his workers for one week, the farmer has $162 left in his bank account. How much money did he have at first?

Farmer's Money: 26 × 410 | 162 ; ?

410 × 26 = 2460 + 8200 = 10,660

10,660 + 162 = 10,822

At first the farmer had $10,822.

5. Frances is sewing a border around 2 rectangular tablecloths that each measure 9 feet long by 6 feet wide. If it takes her 3 minutes to sew on 1 inch of border, how many minutes will it take her to complete her sewing project? Write an expression, and then solve.

2 × (9 + 6 + 9 + 6) × 12 × 3 =
2 × 30 × 12 × 3 =
2 × 360 × 3 =
720 × 3 = 2,160

It will take Francis 2,160 minutes to complete her project.

6. Each grade level at Hooperville Schools has 298 students.

a. If there are 13 grade levels, how many students attend Hooperville Schools?

298 × 13 = 894 + 2980 = 3874

3,874 students attend Hooperville Schools.

b. A nearby district, Willington, is much larger. They have 12 times as many students. How many students attend schools in Willington?

3,874 × 12 = 7748 + 38740 = 46,488

46,488 students attend schools in Willington.

## Exit Ticket (3 minutes)

After the Student Debrief, instruct students to complete the Exit Ticket. A review of their work will help with assessing students' understanding of the concepts that were presented in today's lesson and planning more effectively for future lessons. The questions may be read aloud to the students.

Name ______________________________ Date ______________

Solve.

1. An office space in New York City measures 48 feet by 56 feet. If it sells for $565 per square foot, what is the total cost of the office space?

2. Gemma and Leah are both jewelry makers. Gemma made 106 beaded necklaces. Leah made 39 more necklaces than Gemma.

   a. Each necklace they make has exactly 104 beads on it. How many beads did both girls use altogether while making their necklaces?

   b. At a recent craft fair, Gemma sold each of her necklaces for $14. Leah sold each of her necklaces for $10 more. Who made more money at the craft fair? How much more?

3. Peng bought 26 treadmills for her new fitness center at $1,334 each. Then, she bought 19 stationary bikes for $749 each. How much did she spend on her new equipment? Write an expression, and then solve.

4. A Hudson Valley farmer has 26 employees. He pays each employee $410 per week. After paying his workers for one week, the farmer has $162 left in his bank account. How much money did he have at first?

5. Frances is sewing a border around 2 rectangular tablecloths that each measure 9 feet long by 6 feet wide. If it takes her 3 minutes to sew on 1 inch of border, how many minutes will it take her to complete her sewing project? Write an expression, and then solve.

6. Each grade level at Hooperville Schools has 298 students.

   a. If there are 13 grade levels, how many students attend Hooperville Schools?

   b. A nearby district, Willington, is much larger. They have 12 times as many students. How many students attend schools in Willington?

Name ______________________________ Date ______________

Solve.

Juwad picked 30 bags of apples on Monday and sold them at his fruit stand for $3.45 each. The following week he picked and sold 26 bags.

a. How much money did Juwad earn in the first week?

b. How much money did he earn in the second week?

c. How much did Juwad earn selling bags of apples these two weeks?

d. **Extension:** Each bag Juwad picked holds 15 apples. How many apples did he pick in two weeks? Write an expression to represent this problem.

Name ______________________________ Date ______________

Solve.

1. Jeffery bought 203 sheets of stickers. Each sheet has a dozen stickers. He gave away 907 stickers to his family and friends on Valentine's Day. How many stickers does Jeffery have remaining?

2. During the 2011 season, a quarterback passed for 302 yards per game. He played in all 16 regular season games that year.

   a. For how many total yards did the quarterback pass?

   b. If he matches this passing total for each of the next 13 seasons, how many yards will he pass for in his career?

3. Bao saved $179 a month. He saved $145 less than Ada each month. How much would Ada save in three and a half years?

4. Mrs. Williams is knitting a blanket for her newborn granddaughter. The blanket is 2.25 meters long and 1.8 meters wide. What is the area of the blanket? Write the answer in centimeters.

5. Use the chart to solve.

**Soccer Field Dimensions**

| | FIFA Regulation (in yards) | New York State High Schools (in yards) |
|---|---|---|
| **Minimum Length** | 110 | 100 |
| **Maximum Length** | 120 | 120 |
| **Minimum Width** | 70 | 55 |
| **Maximum Width** | 80 | 80 |

a. Write an expression to find the difference in the maximum area and minimum area of a NYS high school soccer field. Then, evaluate your expression.

b. Would a field with a width of 75 yards and an area of 7,500 square yards be within FIFA regulation? Why or why not?

c. It costs $26 to fertilize, water, mow, and maintain each square yard of a full size FIFA field (with maximum dimensions) before each game. How much will it cost to prepare the field for next week's match?

# Topic C
# Decimal Multi-Digit Multiplication

**5.NBT.7,** 5.OA.1, 5.OA.2, 5.NBT.1

| | | |
|---|---|---|
| **Focus Standard:** | 5.NBT.7 | Add, subtract, multiply, and divide decimals to hundredths, using concrete models or drawings and strategies based on place value, properties of operations, and/or the relationship between addition and subtraction; relate the strategy to a written method and explain the reasoning used. |
| **Instructional Days:** | 3 | |
| **Coherence -Links from:** | G5–M1 | Place Value and Decimal Fractions |
| **-Links to:** | G5–M4 | Multiplication and Division of Fractions and Decimal Fractions |
| | G6–M2 | Arithmetic Operations Including Division of Fractions |

Throughout Topic C, students make connections between what they know of whole number multiplication to its parallel role in multiplication with decimals by using place value to reason and make estimations about products (**5.NBT.7**). Knowledge of multiplicative patterns from Grade 4 experiences, as well as those provided in Grade 5 Module 1, provide support for converting decimal multiplication to whole number multiplication. Students reason about how products of such converted cases must be adjusted through division, giving rise to explanations about how the decimal must be placed.

**A Teaching Sequence Toward Mastery of Decimal Multi-Digit Multiplication**

**Objective 1:** **Multiply decimal fractions with tenths by multi-digit whole numbers using place value understanding to record partial products.**
**(Lesson 10)**

**Objective 2:** **Multiply decimal fractions by multi-digit whole numbers through conversion to a whole number problem and reasoning about the placement of the decimal.**
**(Lesson 11)**

**Objective 3:** **Reason about the product of a whole number and a decimal with hundredths using place value understanding and estimation.**
**(Lesson 12)**

# Lesson 10

## Objective: Multiply decimal fractions with tenths by multi-digit whole numbers using place value understanding to record partial products.

### Suggested Lesson Structure

| | |
|---|---|
| Fluency Practice | (12 minutes) |
| Application Problem | (6 minutes) |
| Concept Development | (32 minutes) |
| Student Debrief | (10 minutes) |
| **Total Time** | **(60 minutes)** |

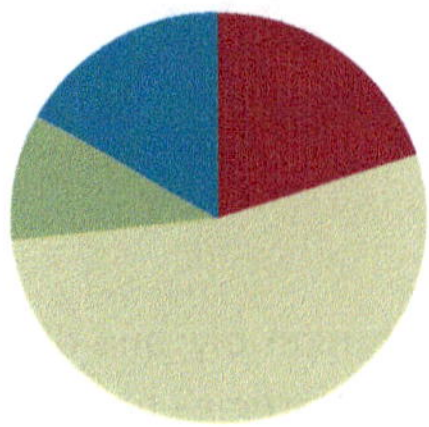

## Fluency Practice (12 minutes)

- Multiply then Divide by the Same Number **5.NBT.2** (6 minutes)
- Decompose Decimals **5.NBT.3** (6 minutes)

### Multiply then Divide by the Same Number (6 minutes)

Note: This fluency activity reviews what happens when any number or expression is divided and then multiplied by the same number in preparation for today's lesson.

T: 3 × 2 is ...?

S: 6.

T: 3 × 2 × 10 ÷ 10 is ...?

S: 6.

T: 5 × 0.3 is ...?

S: 1.5.

T: 5 × 0.3 × 10 ÷ 10 is ...?

S: 1.5.

T: (Continue the sequence with 3 × 2.5 and 2 × 3.4.)

T: Why are the products the same when we multiply by 10 and then divide by 10?

S: You are undoing what you did when you multiplied by 10. → We're moving over one place to the left on the place value chart and then back to the right again. → Because it's just like multiplying by 1.





EUREKA MATH™

### Decompose Decimals (6 minutes)

Materials: (S) Personal white board

Note: This fluency activity reviews decimal place value concepts and emphasizes part–whole decomposition through the use of the number bond.

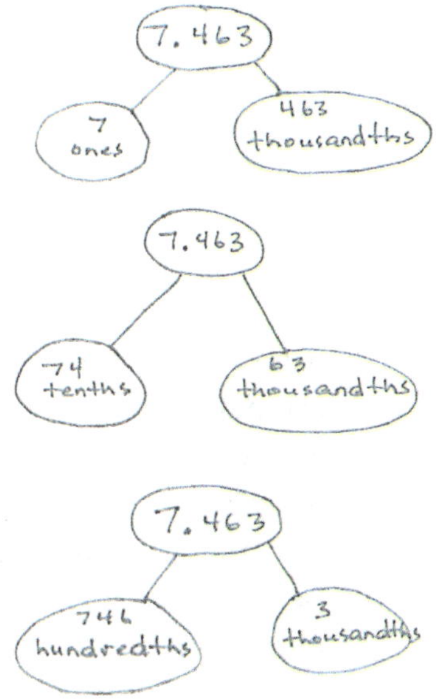


T: (Project 7.463.) Say the number.

S: 7 and 463 thousandths.

T: Represent this number in a two-part number bond with ones as one part and thousandths as the other part (pictured to the right).

S: (Draw.)

T: Represent it again with tenths and thousandths.

T: Represent it again with hundredths and thousandths.

Follow the same process for 8.972 and 6.849.

## Application Problem (6 minutes)

MP.2

The fifth-grade craft club is making aprons to sell. Each apron takes 1.25 yards of fabric that costs \$3 per yard and 4.5 yards of trim that costs \$2 per yard. What does it cost the club to make one apron? If the club wants to make \$1.75 profit on each apron, how much should they charge per apron?

Note: This problem requires students to not only use their Module 1 knowledge of decimal by single-digit multiplier, but also asks them to reason about a start unknown problem type.

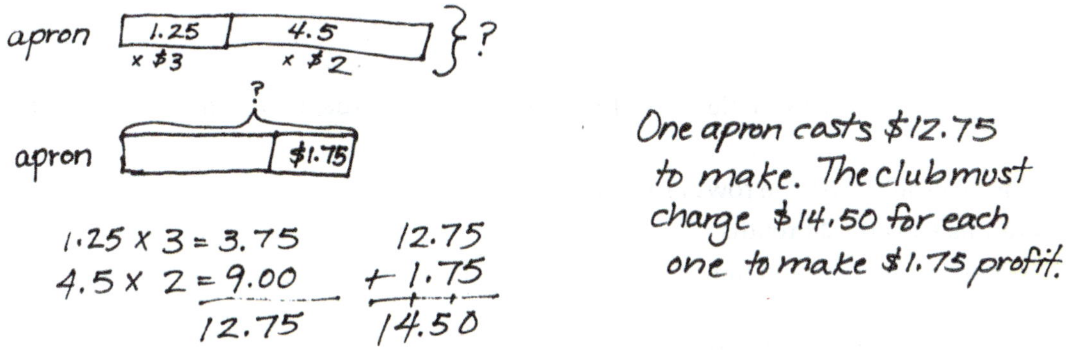

## Concept Development (32 minutes)

Materials: (S) Personal white board

**Problems 1–3**

43 × 2.4

3.5 × 42

15.6 × 73

T: (Write 43 × 2.4 on the board.) Round the factors to estimate the product.

S: (Show.) 40 × 2 = 80.

T: Predict whether our estimate is greater than or less than the actual product.

S: Less than because both factors were rounded to numbers less than the actual factors. → Our actual answer might be about 90.

T: We have 43 units of 2.4. I'd like to rename 2.4 using only tenths. How many tenths would that be?

S: 24 tenths.

T: Decompose those 24 tenths into expanded form along the length of our rectangle. Let's write tenths out to the right to remind us of the unit. (Demonstrate.)

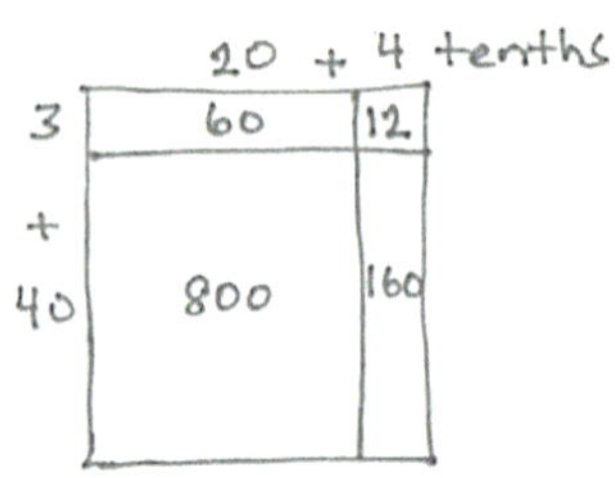


S: (Draw.)

T: Our rectangle's width is 43 whole units. Decompose 43 into expanded form along the width.

S: (Draw.)

T: What partial products do the rows represent?

S: 3 × 24 tenths and 40 × 24 tenths.

T: Find the partial products and the final product.

S: (Multiply the cells and add the rows.)

T: We found that we have 1,032 of what unit?

S: Tenths.

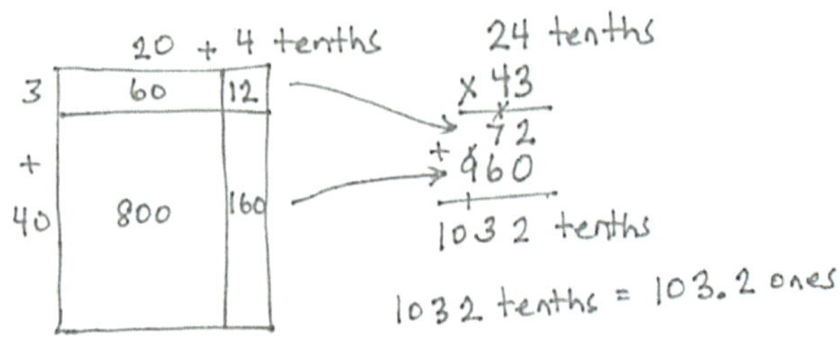


> **NOTES ON MULTIPLE MEANS OF REPRESENTATION:**
>
> The decimal multiplication in this and following lessons builds on the concept of whole number multiplication in earlier module lessons and the single-digit decimal multiplication from Module 1. It is important for students to note that, because multiplication is commutative, multiplication sentences may be notated in any order. In this part of the module, the decimal factor will be designated as the unit (*multiplicand*—the *what* that is being multiplied), while the whole number will be treated as the multiplier (the *how many copies* number). This interpretation allows students to build on the *repeated addition* concept of multiplying whole numbers, which has formed the basis of the area model as students understand it. This makes the distributive property and partial products of the algorithm a direct parallel to whole number work.

T: Write 1,032 tenths in standard form.

S: 103.2.

T: Compare this to our estimate. Is our product reasonable?

S: Our estimate was 80, and our exact product is 103.2. Our product is reasonable.

T: Let's solve this same problem using the algorithm. (Write 24 tenths × 43 on the board as shown on previous page.) When we find the product, we have to remember that we copied tenths. Solve this problem, and then share with your partner.

S: (Work and share.)

T: Look back at your area model. Find these partial products in your algorithm. Turn and talk.

S: 72 is the first row in the area model and the first row in the algorithm. → I see 72 tenths in both of them. → I see 960 tenths in the second row of both.

T: We've found 1,032 tenths using a second strategy. Let's write it in standard form.

S: 103.2.

It's important to have students recognize that the area model drawn using whole number values would be 10 times as wide as the model that would be drawn using tenths.

T: We don't have to do this process in such a long way. Here is a simplifying shortcut for multiplying by 1. We can first multiply one of the factors by 10 and then divide the product by 10.

Students demonstrate this with the algorithm by multiplying by 10 and then dividing by 10. "It's like multiplying by 1. 2 times 3 times 10 divided by 10 is 6. See, it's the same idea, just with bigger numbers."

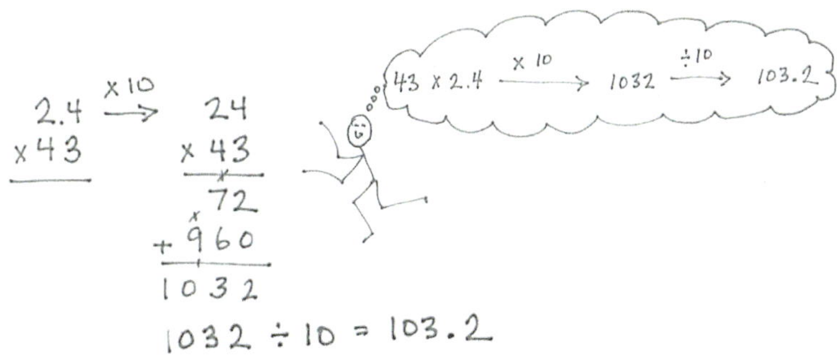


T: Solve 3.5 × 42. Round the factors, and estimate the product.

S: 4 × 40 = 160.

MP.4

T: Naming 3.5 using tenths, draw an area model to show 3.5 × 42. Check your work with your partner. Remember to compare your final product with your estimate to see if your answer is reasonable.

S: (Work.)

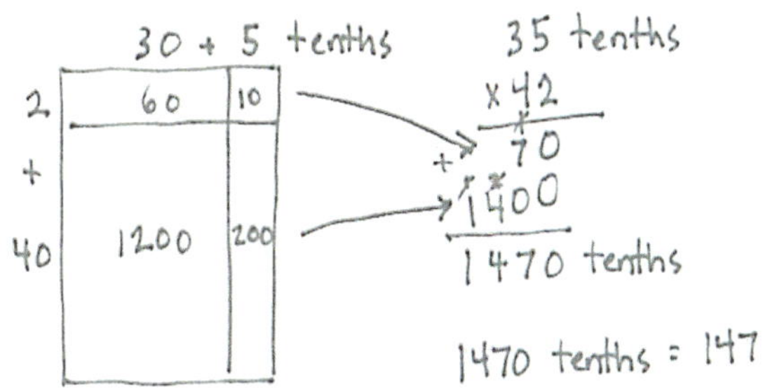

T: Partner A, confirm this product by naming 3.5 in tenths and using the standard algorithm to solve. Partner B, confirm this product by first multiplying 3.5 by 10 and then multiplying by 42. Then, divide the product by 10.

35 (tenths)
× 42
70
+ 1400
1470 (tenths) = 147.0

T: How are these two ways of thinking different? Turn and talk to your partner.

S: In the first way, we thought of 3.5 as 35 tenths. After we multiplied by 42, we still had tenths. → In the second way, we first multiplied 3.5 times 10 and then multiplied by 42. Then, we used the final product to divide by 10. → Multiplying by 10 and then dividing by 10 doesn't change the value of the answer because we are really just multiplying by 1.

3.5 × 42 —×10→ 35 × 42
70
+ 1400
1470 —÷10→ 147.0

T: How are these two ways of thinking similar? Turn and talk to your partner.

S: In both cases, we needed to think about the original units of the first factor. → In both cases, we had the same partial products. → In both cases, the multiplication process was exactly the same. After we adjusted the product, the answer to both was the same.

Repeat the sequence for 15.6 × 73. Have students compare this problem to the others in the set, making sure to elicit from them that the presence of the third column in the area model does not change the thinking behind the area model, nor does it affect the partial products. Also, encourage students to think about multiplying the decimal factor by 10 and then adjusting the product through division by 10.

## Problem Set (10 minutes)

Students should do their personal best to complete the Problem Set within the allotted 10 minutes. For some classes, it may be appropriate to modify the assignment by specifying which problems they work on first. Some problems do not specify a method for solving. Students should solve these problems using the RDW approach used for Application Problems.

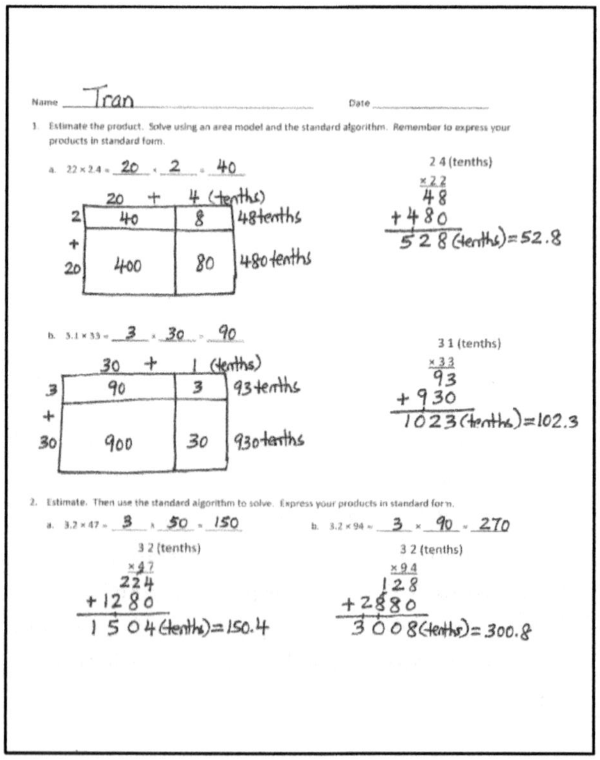
Name Tran Date

1. Estimate the product. Solve using an area model and the standard algorithm. Remember to express your products in standard form.

a. 22 × 2.4 ≈ 20 × 2 = 40

24 (tenths) × 22 = 48 + 480 = 528 (tenths) = 52.8

b. 3.1 × 33 ≈ 3 × 30 = 90

31 (tenths) × 33 = 93 + 930 = 1023 (tenths) = 102.3

2. Estimate. Then use the standard algorithm to solve. Express your products in standard form.

a. 3.2 × 47 ≈ 3 × 50 = 150

32 (tenths) × 47 = 224 + 1280 = 1504 (tenths) = 150.4

b. 3.2 × 94 ≈ 3 × 90 = 270

32 (tenths) × 94 = 128 + 2880 = 3008 (tenths) = 300.8

## Student Debrief (10 minutes)

**Lesson Objective:** Multiply decimal fractions with tenths by multi-digit whole numbers using place value understanding to record partial products.

The Student Debrief is intended to invite reflection and active processing of the total lesson experience.

Invite students to review their solutions for the Problem Set. They should check work by comparing answers with a partner before going over answers as a class. Look for misconceptions or misunderstandings that can be addressed in the Debrief. Guide students in a conversation to debrief the Problem Set and process the lesson.

Any combination of the questions below may be used to lead the discussion.

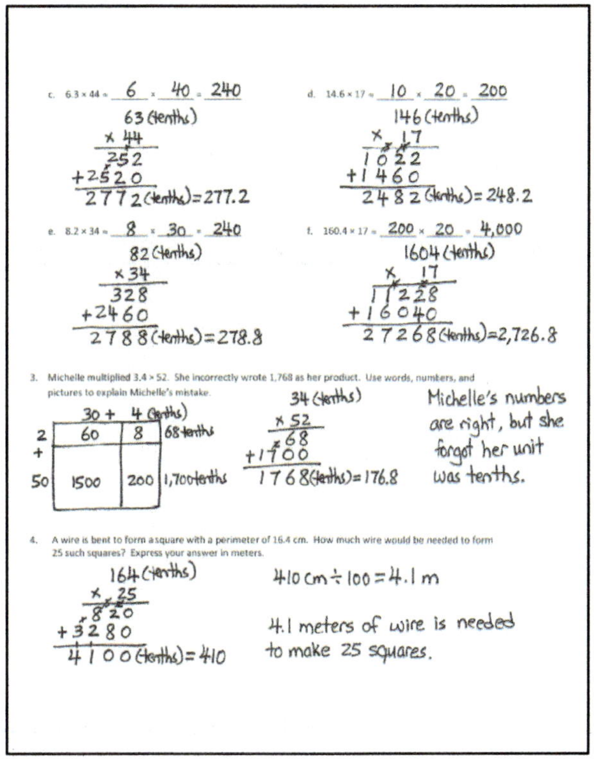
c. 6.3 × 44 ≈ 6 × 40 = 240
63 (tenths) × 44 = 252 + 2520 = 2772 (tenths) = 277.2

d. 14.6 × 17 ≈ 10 × 20 = 200
146 (tenths) × 17 = 1022 + 1460 = 2482 (tenths) = 248.2

e. 8.2 × 34 ≈ 8 × 30 = 240
82 (tenths) × 34 = 328 + 2460 = 2788 (tenths) = 278.8

f. 160.4 × 17 ≈ 200 × 20 = 4,000
1604 (tenths) × 17 = 11228 + 16040 = 27268 (tenths) = 2,726.8

3. Michelle multiplied 3.4 × 52. She incorrectly wrote 1,768 as her product. Use words, numbers, and pictures to explain Michelle's mistake.

30 + 4 (tenths); 2: 60, 8 — 68 tenths; 50: 1500, 200 — 1,700 tenths
34 (tenths) × 52 = 68 + 1700 = 1768 (tenths) = 176.8
Michelle's numbers are right, but she forgot her unit was tenths.

4. A wire is bent to form a square with a perimeter of 16.4 cm. How much wire would be needed to form 25 such squares? Express your answer in meters.

164 (tenths) × 25 = 820 + 3280 = 4100 (tenths) = 410
410 cm ÷ 100 = 4.1 m
4.1 meters of wire is needed to make 25 squares.

- Discuss Michelle's error in Problem 3 by allowing students to share their representations and explanations. Some students may explain her error by saying that she should have said 1,768 *tenths*. Others may offer that she should have written her answer in standard form as 176.8. Either explanation's premise is that Michelle did not consider the unit of her final product.
- How does being fluent in whole number multi-digit multiplication help you multiply decimals? (Focus student attention on the notion that the algorithm is exactly the same, but different units must be considered when multiplying decimals.)
- Extend student reasoning about decimal multiplication by offering a case, such as 0.3 × 42. Ask students how they would draw an area model and/or record this case vertically. Point out that the convention is to write the numeral with the most digits as the *top* number in the algorithm but that this is not expressly necessary. Ask students to discuss how putting the single-digit numeral (3 tenths) as the top number affects the recording of partial products? (It does not. The process is the same. The order is different.)

## Exit Ticket (3 minutes)

After the Student Debrief, instruct students to complete the Exit Ticket. A review of their work will help with assessing students' understanding of the concepts that were presented in today's lesson and planning more effectively for future lessons. The questions may be read aloud to the students.

Name ______________________________ Date ________________

1. Estimate the product. Solve using an area model and the standard algorithm. Remember to express your products in standard form.

   a. 22 × 2.4 ≈ _______ × _______ = _______

   2 4 (tenths)
   × 2 2

   b. 3.1 × 33 _______ × _______ = _______

   3 1 (tenths)
   × 3 3

2. Estimate. Then, use the standard algorithm to solve. Express your products in standard form.

   a. 3.2 × 47 ≈ _______ × _______ = _______

   3 2 (tenths)
   × 4 7

   b. 3.2 × 94 ≈ _______ × _______ = _______

   3 2 (tenths)
   × 9 4

c. 6.3 × 44 ≈ _______ × _______ = _______

d. 14.6 × 17 ≈ _______ × _______ = _______

e. 8.2 × 34 ≈ _______ × _______ = _______

f. 160.4 × 17 ≈ _______ × _______ = _______

3. Michelle multiplied 3.4 × 52. She incorrectly wrote 1,768 as her product. Use words, numbers, and/or pictures to explain Michelle's mistake.

4. A wire is bent to form a square with a perimeter of 16.4 cm. How much wire would be needed to form 25 such squares? Express your answer in meters.

Name ____________________ Date ____________

1. Estimate the product. Solve using an area model and the standard algorithm. Remember to express your products in standard form.

   a. 33.2 × 21 ≈ _______ × _______ = _______

   b. 1.7 × 55 ≈ _______ × _______ = _______

2. If the product of 485 × 35 is 16,975, what is the product of 485 × 3.5? How do you know?

Name ______________________________ Date ______________

1. Estimate the product. Solve using an area model and the standard algorithm. Remember to express your products in standard form.

   a. 53 × 1.2 ≈ _______ × _______ = _______

   1 2 (tenths)
   × 5 3

   b. 2.1 × 82 ≈ _______ × _______ = _______

   2 1 (tenths)
   × 8 2

2. Estimate. Then, use the standard algorithm to solve. Express your products in standard form.

   a. 4.2 × 34 ≈ _______ × _______ = _______

   4 2 (tenths)
   × 3 4

   b. 65 × 5.8 ≈ _______ × _______ = _______

   5 8 (tenths)
   × 6 5

c. $3.3 \times 16 \approx$ _______ $\times$ _______ = _______

d. $15.6 \times 17 \approx$ _______ $\times$ _______ = _______

e. $73 \times 2.4 \approx$ _______ $\times$ _______ = _______

f. $193.5 \times 57 \approx$ _______ $\times$ _______ = _______

3. Mr. Jansen is building an ice rink in his backyard that will measure 8.4 meters by 22 meters. What is the area of the rink?

4. Rachel runs 3.2 miles each weekday and 1.5 miles each day of the weekend. How many miles will she have run in 6 weeks?

# Lesson 11

## Objective: Multiply decimal fractions by multi-digit whole numbers through conversion to a whole number problem and reasoning about the placement of the decimal.

### Suggested Lesson Structure

| | |
|---|---|
| ■ Fluency Practice | (12 minutes) |
| ■ Application Problem | (6 minutes) |
| ■ Concept Development | (32 minutes) |
| ■ Student Debrief | (10 minutes) |
| **Total Time** | **(60 minutes)** |

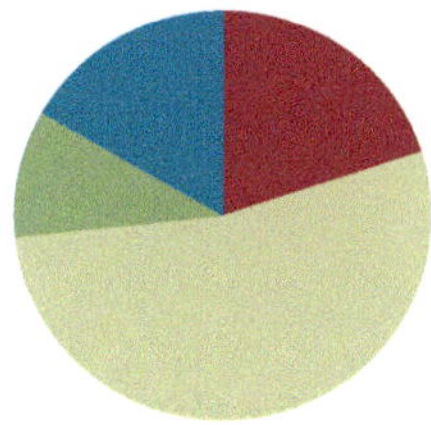

## Fluency Practice (12 minutes)

- Sprint: Multiply Decimals **5.NBT.2** (8 minutes)
- Multiply then Divide by the Same Number **5.NBT.2** (4 minutes)

### Sprint: Multiply Decimals (8 minutes)

Materials: (S) Multiply Decimals Sprint

Note: This fluency activity provides single-digit multiplication practice with decimals. This provides practice with computation required during Concept Development.

### Multiply then Divide by the Same Number (4 minutes)

Note: This fluency activity reviews what happens when any number or expression is divided and then multiplied by the same number in preparation for today's lesson.

T: 3 × 4.1 is ...?
S: 12.3.
T: 12.3 × 10 ÷ 10 is ...?
S: 12.3.
T: 3 × 4.1 × 1 is ...?
S: 12.3.
T: (Repeat with 3 × 2.4.)
T: 3 × 4 × 17.6 ÷ 17.6 is ...?
S: 12.

## Application Problem (6 minutes)

Mr. Mohr wants to build a rectangular patio using concrete tiles that are 12 square inches. The patio will measure 13.5 feet by 43 feet. What is the area of the patio? How many concrete tiles will he need to complete the patio?

```
   135 (tenths)
 ×  43
 -----
   405
+ 5400
 -----
 5,805 (tenths) = 580.5
```

The patio's area is 580.5 $ft^2$ so Mr Mohr will need to buy 581 tiles because each tile is 1 $ft^2$.

Note: This Application Problem asks students to use the decimal multiplication concepts from Lesson 10. Additionally, students must demonstrate understanding of area and use that understanding to reason with respect to the number of tiles needed in the second question. This problem involves a decimal factor of tenths. Use this problem as a springboard for today's lesson, which extends to multiplication of decimal factors of hundredths.

## Concept Development (32 minutes)

Materials: (S) Personal white board

**Problems 1–3**

7.38 × 41

8.26 × 128

82.51 × 63

T: (Write 7.38 × 41.) Compare this problem to our Application Problem.

MP.7

S: It's still multiplication of a decimal by a whole number. → The decimal in the Application Problem was tenths. This is hundredths.

T: Estimate this product.

S: 7 × 40 ÷ 280.

T: Predict whether our estimate is greater than or less than the actual product.

S: The estimate is less than because both factors were rounded to numbers less than the actual factors. → Our actual answer will be more than 280, but it will still be in the hundreds.

**NOTES ON MULTIPLE MEANS OF ACTION AND EXPRESSION:**

The compensation strategy of multiplying a decimal number by a multiple of 10 and then dividing the product by the same multiple of 10 may require some time for students to internalize. The following scaffolds may be appropriate:

- Encourage students to draw the *think* bubble next to their work, or encourage them to label the units.
- Encourage students who are struggling with the standard algorithm to use the area model. The area model provides support by calculating all of the partial products of the problem.

T: We have 41 units of 7.38. I'd like to rename 7.38 using only hundredths. How many hundredths would that be? How do you know?

S: 738 hundredths because 7 is 700 hundredths plus another 38 hundredths equals 738 hundredths. → 7 and 38 hundredths times 100 equals 738 ones.

T: Let's use an area model to find the actual product of this expression. Decompose those 738 hundredths into expanded form along the length of our rectangle. Write *hundredths* out to the right to remind us that we've named 7.38 as hundredths. (Demonstrate.)

S: (Draw area model.)

T: Our rectangle's width is 41 whole units. Decompose 41 into expanded form along the width.

S: (Draw area model.)

T: What two partial products do these rows represent?

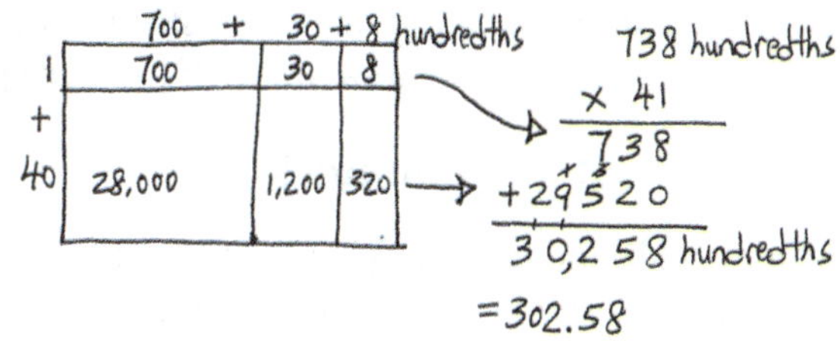


S: 1 × 738 hundredths and 40 × 738 hundredths.

T: Find the partial products and the final product.

S: (Multiply the cells, and add the rows.)

T: We found that we have 30,258 of what unit?

S: Hundredths.

T: We need to write this in standard form. How can our estimate help us convert our product back to wholes and hundredths?

MP.8

S: The estimate told us that our answer was in the hundreds, not the ten thousands or the thousands. → 30,258 is about 100 times as large as our estimate said the real answer should be, so we need to divide by 100 to make the answer make sense.

**NOTES ON MULTIPLE MEANS OF REPRESENTATION:**

Students may discover the pattern that the number of decimal digits in the factors equals the number of decimal digits in the product. While this can be a useful observation, keep students focused on the reason for the pattern. "We multiplied a factor by a power of 10; therefore, we must divide the product by the same power of 10 to adjust it."

T: What is 30,258 hundredths written in standard form?

S: 302.58.

T: Let's solve this same problem using the algorithm. Yesterday, we rewrote our first factor as a whole number with the unit name to the right. (Write 738 hundredths × 41 on the board as shown.) Today, let's think about the units without removing the decimal from our first factor. We see 7.38, but we think 738 hundredths. Multiply 738 × 41 and find the product. Look back at your area model to confirm the partial products in your algorithm.

S: (Work.)

T: This product is 100 times as large as the product of our original problem. What should we do to adjust this product so that it answers our original problem of 7.38 × 41?

S: We should divide by 100.

T: Let me record what I hear you saying. (Write on board as shown.) So, is our adjusted product of 302.58 reasonable given our estimate?

S: Yes.

Work with the other two problems in this set according to what is best for students. Continue with other examples, if necessary. Encourage students who struggle with the algorithm to use the area model. Allow students to forego the area model if they are proficient with the algorithm.

### Problem Set (10 minutes)

Students should do their personal best to complete the Problem Set within the allotted 10 minutes. For some classes, it may be appropriate to modify the assignment by specifying which problems they work on first. Some problems do not specify a method for solving. Students should solve these problems using the RDW approach used for Application Problems.

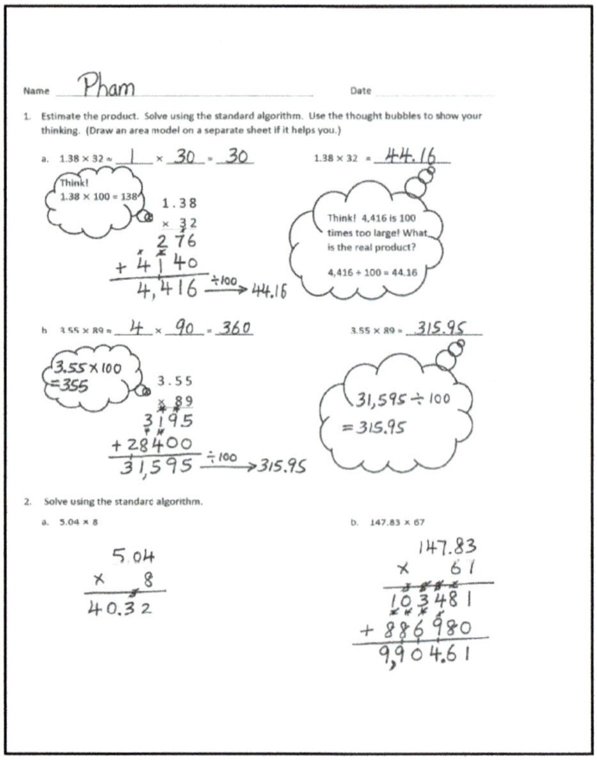


## Student Debrief (10 minutes)

**Lesson Objective:** Multiply decimal fractions by multi-digit whole numbers through conversion to a whole number problem and reasoning about the placement of the decimal.

The Student Debrief is intended to invite reflection and active processing of the total lesson experience.

Invite students to review their solutions for the Problem Set. They should check work by comparing answers with a partner before going over answers as a class. Look for misconceptions or misunderstandings that can be addressed in the Debrief. Guide students in a conversation to debrief the Problem Set and process the lesson.

Any combination of the questions below may be used to lead the discussion.

- For Problem 1, what did you write in your think bubbles? Compare what you wrote with a partner.
- What strategies did you use to solve Problem 2(d)? Does the decimal number affect the process for solving? Why or why not? (It is important to note with students that, while convention dictates the number with more digits is put *on top* in the algorithm, this is not strictly necessary.)

- What is the relationship between the relative size of the factors in the whole number problems and the factors in the decimal problems? What is the relationship between the products? (Problem 3 provides an opportunity for students to reason about the compensation strategy without the burden of the actual multiplication. One factor in the whole number problem is 100 times as large as the corresponding decimal factor. This results in products that share the same digits but are one-hundredth the size. Refer to the second UDL box in the lesson.)

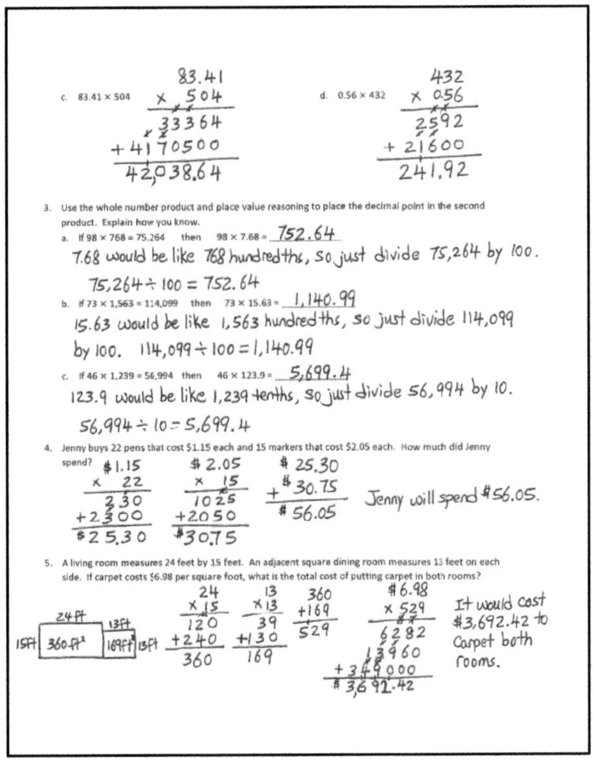
c. 83.41 × 504 = 42,038.64

d. 0.56 × 432 = 241.92

3. Use the whole number product and place value reasoning to place the decimal point in the second product. Explain how you know.

a. If 98 × 768 = 75,264 then 98 × 7.68 = 752.64

7.68 would be like 768 hundredths, so just divide 75,264 by 100. 75,264 ÷ 100 = 752.64

b. If 73 × 1,563 = 114,099 then 73 × 15.63 = 1,140.99

15.63 would be like 1,563 hundredths, so just divide 114,099 by 100. 114,099 ÷ 100 = 1,140.99

c. If 46 × 1,239 = 56,994 then 46 × 123.9 = 5,699.4

123.9 would be like 1,239 tenths, so just divide 56,994 by 10. 56,994 ÷ 10 = 5,699.4

4. Jenny buys 22 pens that cost \$1.15 each and 15 markers that cost \$2.05 each. How much did Jenny spend?

Jenny will spend \$56.05.

5. A living room measures 24 feet by 15 feet. An adjacent square dining room measures 13 feet on each side. If carpet costs \$6.98 per square foot, what is the total cost of putting carpet in both rooms?

It would cost \$3,692.42 to carpet both rooms.

## Exit Ticket (3 minutes)

After the Student Debrief, instruct students to complete the Exit Ticket. A review of their work will help with assessing students' understanding of the concepts that were presented in today's lesson and planning more effectively for future lessons. The questions may be read aloud to the students

# A

Number Correct: _______

Multiply Decimals

| | | |
|---|---|---|
| 1. | 3 × 3 = | |
| 2. | 0.3 × 3 = | |
| 3. | 0.03 × 3 = | |
| 4. | 3 × 2 = | |
| 5. | 0.3 × 2 = | |
| 6. | 0.03 × 2 = | |
| 7. | 2 × 2 = | |
| 8. | 0.2 × 2 = | |
| 9. | 0.02 × 2 = | |
| 10. | 5 × 3 = | |
| 11. | 0.5 × 3 = | |
| 12. | 0.05 × 3 = | |
| 13. | 0.04 × 3 = | |
| 14. | 0.4 × 3 = | |
| 15. | 4 × 3 = | |
| 16. | 5 × 5 = | |
| 17. | 0.5 × 5 = | |
| 18. | 0.05 × 5 = | |
| 19. | 7 × 4 = | |
| 20. | 0.7 × 4 = | |
| 21. | 0.07 × 4 = | |
| 22. | 0.9 × 4 = | |

| | | |
|---|---|---|
| 23. | 8 × 5 = | |
| 24. | 0.8 × 5 = | |
| 25. | 0.08 × 5 = | |
| 26. | 0.06 × 5 = | |
| 27. | 0.06 × 3 = | |
| 28. | 0.6 × 5 = | |
| 29. | 0.06 × 2 = | |
| 30. | 0.06 × 7 = | |
| 31. | 0.9 × 6 = | |
| 32. | 0.06 × 9 = | |
| 33. | 0.09 × 9 = | |
| 34. | 0.8 × 8 = | |
| 35. | 0.07 × 7 = | |
| 36. | 0.6 × 6 = | |
| 37. | 0.05 × 5 = | |
| 38. | 0.6 × 8 = | |
| 39. | 0.07 × 9 = | |
| 40. | 0.8 × 3 = | |
| 41. | 0.09 × 6 = | |
| 42. | 0.5 × 7 = | |
| 43. | 0.12 × 4 = | |
| 44. | 0.12 × 9 = | |

# B

Number Correct: _______

Improvement: _______

Multiply Decimals

| | | |
|---|---|---|
| 1. | 2 × 2 = | |
| 2. | 0.2 × 2 = | |
| 3. | 0.02 × 2 = | |
| 4. | 4 × 2 = | |
| 5. | 0.4 × 2 = | |
| 6. | 0.04 × 2 = | |
| 7. | 3 × 3 = | |
| 8. | 0.3 × 3 = | |
| 9. | 0.03 × 3 = | |
| 10. | 4 × 3 = | |
| 11. | 0.4 × 3 = | |
| 12. | 0.04 × 3 = | |
| 13. | 0.05 × 3 = | |
| 14. | 0.5 × 3 = | |
| 15. | 5 × 3 = | |
| 16. | 4 × 4 = | |
| 17. | 0.4 × 4 = | |
| 18. | 0.04 × 4 = | |
| 19. | 8 × 4 = | |
| 20. | 0.8 × 4 = | |
| 21. | 0.08 × 4 = | |
| 22. | 0.6 × 4 = | |

| | | |
|---|---|---|
| 23. | 6 × 5 = | |
| 24. | 0.6 × 5 = | |
| 25. | 0.06 × 5 = | |
| 26. | 0.08 × 5 = | |
| 27. | 0.08 × 3 = | |
| 28. | 0.8 × 5 = | |
| 29. | 0.08 × 2 = | |
| 30. | 0.08 × 7 = | |
| 31. | 0.9 × 8 = | |
| 32. | 0.08 × 9 = | |
| 33. | 0.9 × 9 = | |
| 34. | 0.08 × 8 = | |
| 35. | 0.7 × 7 = | |
| 36. | 0.06 × 6 = | |
| 37. | 0.5 × 5 = | |
| 38. | 0.06 × 8 = | |
| 39. | 0.7 × 9 = | |
| 40. | 0.08 × 3 = | |
| 41. | 0.9 × 6 = | |
| 42. | 0.05 × 7 = | |
| 43. | 0.12 × 6 = | |
| 44. | 0.12 × 8 = | |

Name ______________________________ Date ______________

1. Estimate the product. Solve using the standard algorithm. Use the thought bubbles to show your thinking. (Draw an area model on a separate sheet if it helps you.)

   a. $1.38 \times 32 \approx$ _______ $\times$ _______ $=$ _______ 1.38 × 32 = ____________

   Think!
   $1.38 \times 100 = 138$

   $$\begin{array}{r} 1.38 \\ \times\ \ 32 \\ \hline \end{array}$$

   Think! 4,416 is 100 times too large! What is the real product?

   $4{,}416 \div 100 = 44.16$

   b. $3.55 \times 89 \approx$ _______ $\times$ _______ $=$ _______ 3.55 × 89 = ____________

   $$\begin{array}{r} 3.55 \\ \times\ \ 89 \\ \hline \end{array}$$

2. Solve using the standard algorithm.

a. $5.04 \times 8$

b. $147.83 \times 67$

c. $83.41 \times 504$

d. $0.56 \times 432$

3. Use the whole number product and place value reasoning to place the decimal point in the second product. Explain how you know.

a. If $98 \times 768 = 75{,}264$ then $98 \times 7.68 =$ _______________

b. If $73 \times 1{,}563 = 114{,}099$ then $73 \times 15.63 =$ _______________

c. If $46 \times 1{,}239 = 56{,}994$ then $46 \times 123.9 =$ _______________

4. Jenny buys 22 pens that cost $1.15 each and 15 markers that cost $2.05 each. How much did Jenny spend?

5. A living room measures 24 feet by 15 feet. An adjacent square dining room measures 13 feet on each side. If carpet costs $6.98 per square foot, what is the total cost of putting carpet in both rooms?

Name ______________________________ Date ________________

Use estimation and place value reasoning to find the unknown product. Explain how you know.

1. If $647 \times 63 = 40{,}761$ then $6.47 \times 63 =$ ____________

2. Solve using the standard algorithm.

   a. $6.13 \times 14$

   b. $104.35 \times 34$

Name ______________________________ Date ______________

1. Estimate the product. Solve using the standard algorithm. Use the thought bubbles to show your thinking. (Draw an area model on a separate sheet if it helps you.)

   a. 2.42 × 12 ≈ ______ × ______ = ______ 2.42 × 12 = ____________

   Think!
   2.42 × 100 = 242

   $$\begin{array}{r} 2.42 \\ \times\ 12 \\ \hline \end{array}$$

   Think! 2,904 is 100 times too large! What is the real product?

   2,904 ÷ 100 = 29.04

   b. 4.13 × 37 ≈ ______ × ______ = ______ 4.13 × 37 = ____________

   $$\begin{array}{r} 4.13 \\ \times\ 37 \\ \hline \end{array}$$





EUREKA MATH

2. Solve using the standard algorithm.

   a. 2.03 × 13

   b. 53.16 × 34

   c. 371.23 × 53

   d. 1.57 × 432

3. Use the whole number product and place value reasoning to place the decimal point in the second product. Explain how you know.

   a. If 36 × 134 = 4,824 then 36 × 1.34 = ____________

   b. If 84 × 2,674 = 224,616 then 84 × 26.74 = ____________

   c. 19 × 3,211 = 61,009 then 321.1 × 19 = ____________

4. A slice of pizza costs $1.57. How much will 27 slices cost?

5. A spool of ribbon holds 6.75 meters. A craft club buys 21 spools.

   a. What is the total cost if the ribbon sells for $2 per meter?

   b. If the club uses 76.54 meters to complete a project, how much ribbon will be left?

# Lesson 12

## Objective: Reason about the product of a whole number and a decimal with hundredths using place value understanding and estimation.

### Suggested Lesson Structure

| | |
|---|---|
| Fluency Practice | (10 minutes) |
| Application Problem | (7 minutes) |
| Concept Development | (33 minutes) |
| Student Debrief | (10 minutes) |
| **Total Time** | **(60 minutes)** |

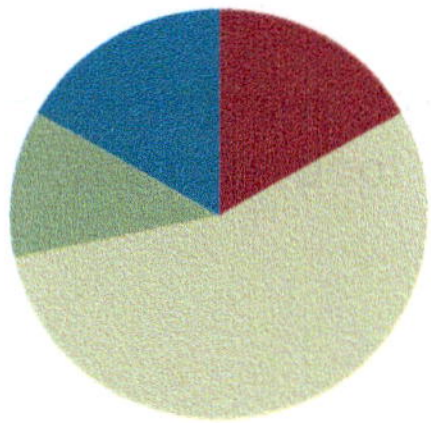

## Fluency Practice (10 minutes)

- Unit Conversions **5.MD.1** (5 minutes)
- State the Decimal **5.NBT.3** (5 minutes)

### Unit Conversions (5 minutes)

Materials: (S) Personal white board

Note: Reviewing this fluency activity builds a foundation for upcoming Topic D lessons on measurement problem solving.

T: (Write 12 in = ____ ft.) 12 inches is the same as how many feet?

S: 1 foot.

Repeat the process for possible sequence: 24 in, 36 in, 48 in, and 120 in.

T: (Write 1 ft = ____ in.) 1 foot is the same as how many inches?

S: 12 inches.

Repeat the process and procedure for 2 ft, 2.5 ft, 3 ft, 3.5 ft, 4 ft, 4.5 ft, 9 ft, 9.5 ft, 27 ft, and 27.5 ft.

### State the Decimal (5 minutes)

Note: This fluency activity reviews Module 1's concepts.

T: Say the number as you would write it. 8 tenths.

S: Zero point eight.

Repeat process using the following possible sequence: 9 tenths, 10 tenths, 11 tenths, 19 tenths, 20 tenths, 30 tenths, 35 tenths, 45 tenths, 85 tenths, 83 tenths, 63 tenths, and 47 tenths.

T: Say the number as you would write it. 8 hundredths.

S: Zero point zero eight.

Repeat the process for the following possible sequence: 9 hundredths, 10 hundredths, 20 hundredths, 30 hundredths, 90 hundredths, 95 hundredths, 99 hundredths, 199 hundredths, 299 hundredths, 357 hundredths, and 463 hundredths.

## Application Problem (7 minutes)

MP.2 MP.4

Thirty-two cyclists make a seven-day trip. Each cyclist requires 8.33 kilograms of food for the entire trip. If each cyclist wants to eat an equal amount of food each day, how many kilograms of food will the group be carrying at the end of Day 5?

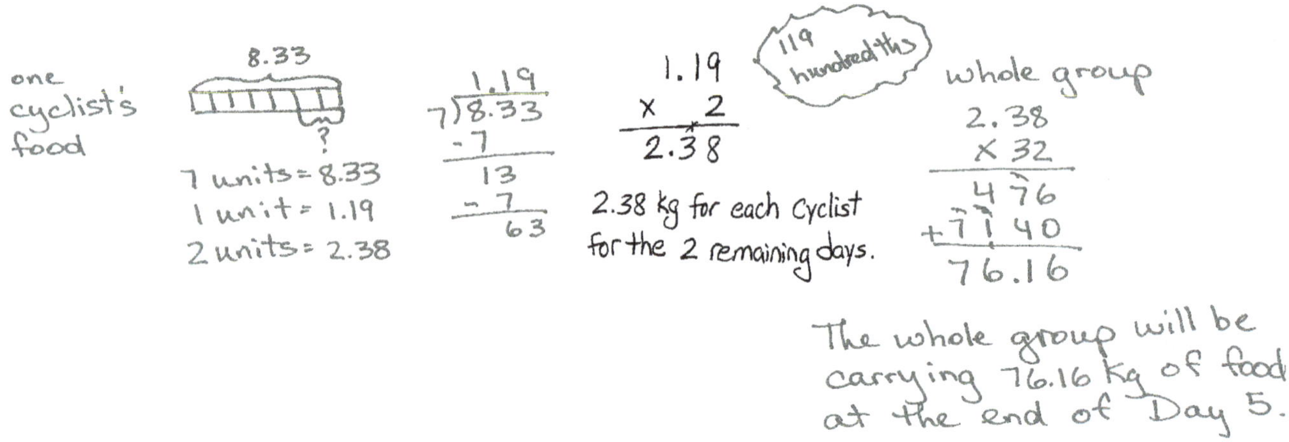


Note: This problem asks students to divide a decimal by a whole number, a skill learned in Module 1. Students also need to multiply a decimal by a two-digit whole number, which is the focus of today's lesson. Accept any valid approach to solving the problem.

## Concept Development (33 minutes)

The time allotted for Lesson 12's Concept Development can be used to consolidate the learning that has occurred in Lessons 10 and 11. Three sets of problems have been provided for students who are ready to extend their decimal multiplication knowledge. The teaching sequence from the aforementioned lessons may be used to guide instruction. Students should be encouraged to imagine the area model while writing the algorithm, as well as verbalize the thinking of multiplying and dividing by 10 and 100.

**NOTES ON MULTIPLE MEANS OF ACTION AND EXPRESSION:**

By this point in the module, students will most certainly differ in their independence with decimal multiplication. Continue to allow students to use area models as a support for finding products. Give students who are comfortable in their knowledge of the algorithm freedom to simply compute the products without drawing the area model.

Note: Problems 7–9 involve decimals less than 1. This is intended to serve as a challenge set for advanced learners.

| Problems 1–3 | Problems 4–6 | Problems 7–9 |
|---|---|---|
| 2.31 × 22 = | 495 × 1.11 = | 2.5 × 51 = |
| 2.31 × 221 = | 0.98 × 495 = | 0.25 × 51 = |
| 2.31 × 201 = | 102.64 × 495 = | 0.56 × 84 = |

### Problem Set (10 minutes)

Students should do their personal best to complete the Problem Set within the allotted 10 minutes. For some classes, it may be appropriate to modify the assignment by specifying which problems they work on first. Some problems do not specify a method for solving. Students should solve these problems using the RDW approach used for Application Problems.

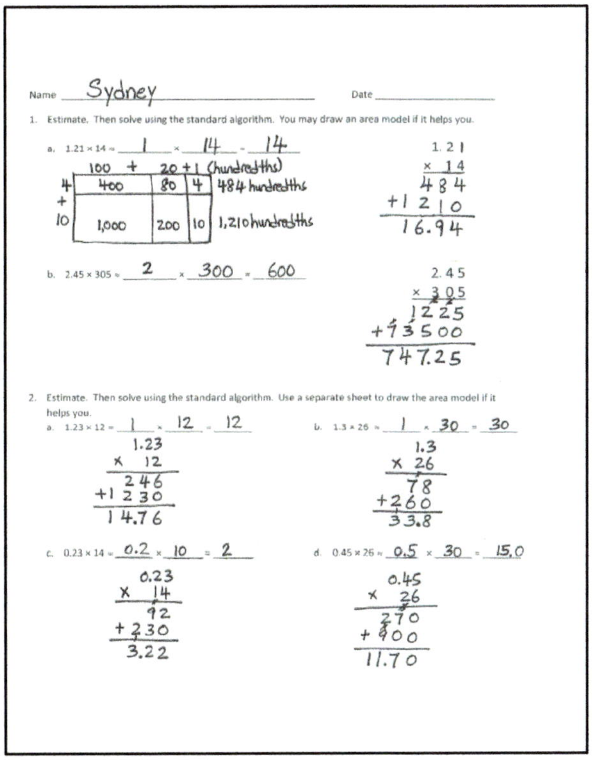
Name Sydney Date

1. Estimate. Then solve using the standard algorithm. You may draw an area model if it helps you.

a. 1.21 × 14 ≈ 1 × 14 = 14

b. 2.45 × 305 ≈ 2 × 300 = 600

2. Estimate. Then solve using the standard algorithm. Use a separate sheet to draw the area model if it helps you.

a. 1.23 × 12 ≈ 1 × 12 = 12

b. 1.3 × 26 ≈ 1 × 30 = 30

c. 0.23 × 14 ≈ 0.2 × 10 = 2

d. 0.45 × 26 ≈ 0.5 × 30 = 15.0

## Student Debrief (10 minutes)

**Lesson Objective:** Reason about the product of a whole number and a decimal with hundredths using place value understanding and estimation.

The Student Debrief is intended to invite reflection and active processing of the total lesson experience.

Invite students to review their solutions for the Problem Set. They should check work by comparing answers with a partner before going over answers as a class. Look for misconceptions or misunderstandings that can be addressed in the Debrief. Guide students in a conversation to debrief the Problem Set and process the lesson.

Any combination of the questions below may be used to lead the discussion.

- Discuss the estimates for Problems 2(b) and 2(d). Have students notice that, in Problem 2(b), 26 is multiplied by a factor a bit more than 1 and in Problem 2(d), by a factor less than 1. What effect does this have on the products?
- Continue to discuss the relationships between the actual problem and the parallel whole number problem they use to obtain the digits of the product. Have them articulate the adjustments that must be made to the products to answer the actual multiplication sentence. (If I think about 1.24 as hundredths, I must multiply by 100, but my product must be adjusted by dividing by 100.)

## Exit Ticket (3 minutes)

After the Student Debrief, instruct students to complete the Exit Ticket. A review of their work will help with assessing students' understanding of the concepts that were presented in today's lesson and planning more effectively for future lessons. The questions may be read aloud to the students.

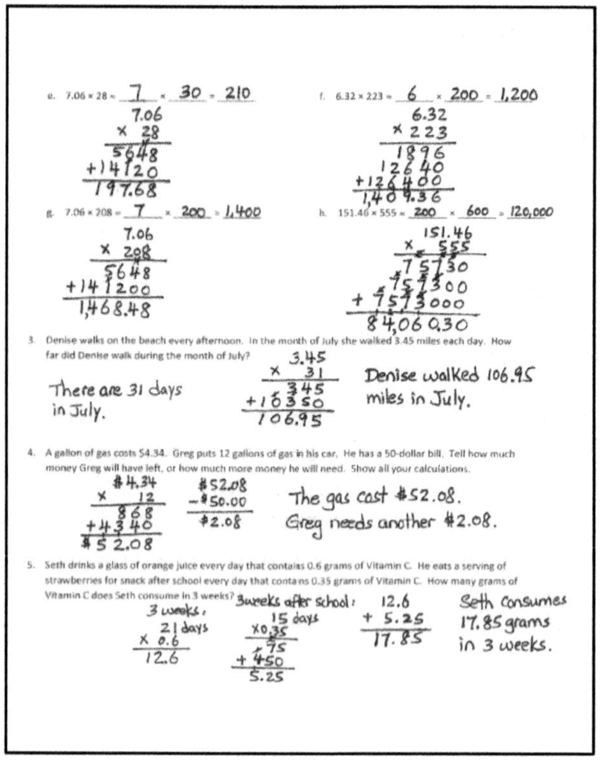

e. 7.06 × 28 ≈ 7 × 30 = 210
7.06 × 28 = 197.68

f. 6.32 × 223 ≈ 6 × 200 = 1,200
6.32 × 223 = 1,409.36

g. 7.06 × 208 ≈ 7 × 200 = 1,400
7.06 × 208 = 1,468.48

h. 151.46 × 555 ≈ 200 × 600 = 120,000
151.46 × 555 = 84,060.30

3. Denise walks on the beach every afternoon. In the month of July she walked 3.45 miles each day. How far did Denise walk during the month of July?

There are 31 days in July. 3.45 × 31 = 106.95

Denise walked 106.95 miles in July.

4. A gallon of gas costs $4.34. Greg puts 12 gallons of gas in his car. He has a 50-dollar bill. Tell how much money Greg will have left, or how much more money he will need. Show all your calculations.

$4.34 × 12 = $52.08 $52.08 − $50.00 = $2.08

The gas cost $52.08. Greg needs another $2.08.

5. Seth drinks a glass of orange juice every day that contains 0.6 grams of Vitamin C. He eats a serving of strawberries for snack after school every day that contains 0.35 grams of Vitamin C. How many grams of Vitamin C does Seth consume in 3 weeks?

3 weeks: 21 days × 0.6 = 12.6

3 weeks after school: 15 days × 0.35 = 5.25

12.6 + 5.25 = 17.85

Seth consumes 17.85 grams in 3 weeks.

Name ______________________________________ Date ____________________

1. Estimate. Then, solve using the standard algorithm. You may draw an area model if it helps you.

   a. 1.21 × 14 ≈ __________ × __________ = __________

$$\begin{array}{r} 1.21 \\ \times\ \underline{\quad 14} \end{array}$$

   b. 2.45 × 305 ≈ __________ × __________ = __________

$$\begin{array}{r} 2.45 \\ \times\ \underline{\ 305} \end{array}$$

2. Estimate. Then, solve using the standard algorithm. Use a separate sheet to draw the area model if it helps you.

a. 1.23 × 12 ≈ _______ × _______ = _______

b. 1.3 × 26 ≈ _______ × _______ = _______

c. 0.23 × 14 ≈ _______ × _______ = _______

d. 0.45 × 26 ≈ _______ × _______ = _______

e. 7.06 × 28 ≈ _______ × _______ = _______

f. 6.32 × 223 ≈ _______ × _______ = _______

g. 7.06 × 208 ≈ _______ × _______ = _______

h. 151.46 × 555 ≈ _______ × _______ = _______

3. Denise walks on the beach every afternoon. In the month of July, she walked 3.45 miles each day. How far did Denise walk during the month of July?

4. A gallon of gas costs $4.34. Greg puts 12 gallons of gas in his car. He has a 50-dollar bill. Tell how much money Greg will have left, or how much more money he will need. Show all your calculations.

5. Seth drinks a glass of orange juice every day that contains 0.6 grams of Vitamin C. He eats a serving of strawberries for snack after school every day that contains 0.35 grams of Vitamin C. How many grams of Vitamin C does Seth consume in 3 weeks?

Name ______________________________ Date ______________

Estimate. Then, solve using the standard algorithm.

a. 3.03 × 402 ≈ _______ × _______ = _______

b. 667 × 1.25 ≈ _______ × _______ = _______

Name ______________________________ Date ______________

1. Estimate. Then, solve using the standard algorithm. You may draw an area model if it helps you.

   a. $24 \times 2.31 \approx$ ________ $\times$ ________ $=$ ________

$$\begin{array}{r} 2.31 \\ \times\ \ 24 \\ \hline \end{array}$$

   b. $5.42 \times 305 \approx$ ________ $\times$ ________ $=$ ________

$$\begin{array}{r} 5.42 \\ \times 305 \\ \hline \end{array}$$

2. Estimate. Then, solve using the standard algorithm. Use a separate sheet to draw the area model if it helps you.

a. 1.23 × 21 ≈ _______ × _______ = _______

b. 3.2 × 41 ≈ _______ × _______ = _______

c. 0.32 × 41 ≈ _______ × _______ = _______

d. 0.54 × 62 ≈ _______ × _______ = _______

e. 6.09 × 28 ≈ _______ × _______ = _______

f. 6.83 × 683 ≈ _______ × _______ = _______

g. 6.09 × 208 ≈ _______ × _______ = _______

h. 171.76 × 555 ≈ _______ × _______ = _______

3. Eric's goal is to walk 2.75 miles to and from the park every day for an entire year. If he meets his goal, how many miles will Eric walk?

4. Art galleries often price paintings by the square inch. If a painting measures 22.5 inches by 34 inches and costs $4.15 per square inch, what is the selling price for the painting?

5. Gerry spends $1.25 each day on lunch at school. On Fridays, she buys an extra snack for $0.55. How much money will she spend in two weeks?

Topic D

# Measurement Word Problems with Whole Number and Decimal Multiplication

**5.NBT.5, 5.NBT.7, 5.MD.1,** 5.NBT.1, 5.NBT.2

| | | |
|---|---|---|
| **Focus Standards:** | 5.NBT.5 | Fluently multiply multi-digit whole numbers using the standard algorithm. |
| | 5.NBT.7 | Add, subtract, multiply, and divide decimals to hundredths, using concrete models or drawings and strategies based on place value, properties of operations, and/or the relationship between addition and subtraction; relate the strategy to a written method and explain the reasoning used. |
| | 5.MD.1 | Convert among different-sized standard measurement units within a given measurement system (e.g., convert 5 cm to 0.05 m), and use these conversions in solving multi-step, real world problems. |
| **Instructional Days:** | 3 | |
| **Coherence -Links from:** | G4–M2 | Unit Conversions and Problem Solving with Metric Measurement |
| **-Links to:** | G5–M4 | Multiplication and Division of Fractions and Decimal Fractions |
| | G6–M1 | Ratios and Unit Rates |

In Topic D, students explore multiplication as a method for expressing equivalent measures. For example, they multiply to convert between meters and centimeters or ounces and cups with measurements in whole number, fraction, and decimal form (**5.MD.1**). These conversions offer opportunities for students to not only apply their newfound knowledge of multi-digit multiplication of both whole and decimal numbers but to also reason deeply about the relationships between unit size and quantity (i.e., how the choice of one affects the other). Students are given the opportunity to review multiplication of a whole number by a fraction, a skill taught in Grade 4.

| A Teaching Sequence Toward Mastery of Measurement Word Problems with Whole Number and Decimal Multiplication | |
|---|---|
| **Objective 1:** | **Use whole number multiplication to express equivalent measurements. (Lesson 13)** |
| **Objective 2:** | **Use fraction and decimal multiplication to express equivalent measurements. (Lesson 14)** |
| **Objective 3:** | **Solve two-step word problems involving measurement conversions. (Lesson 15)** |

# Lesson 13

## Objective: Use whole number multiplication to express equivalent measurements.

### Suggested Lesson Structure

| | |
|---|---|
| Fluency Practice | (12 minutes) |
| Application Problem | (12 minutes) |
| Concept Development | (26 minutes) |
| Student Debrief | (10 minutes) |
| **Total Time** | **(60 minutes)** |

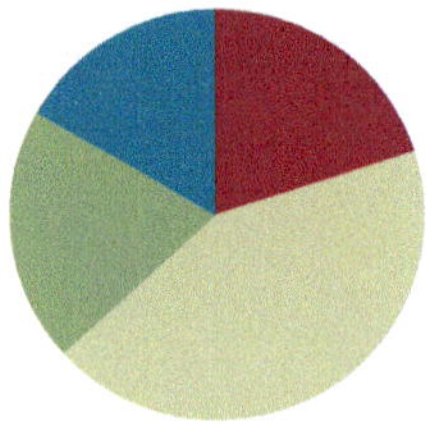

## Fluency Practice (12 minutes)

- Multiply by 0.1, 0.01, and 0.001 **5.NBT.2** (2 minutes)
- Multiply Using the Area Model **5.NBT.2** (7 minutes)
- Unit Conversions **5.MD.1** (3 minutes)

### Multiply by 0.1, 0.01, and 0.001 (2 minutes)

Note: This fluency activity prepares students to multiply by decimal fractions in today's lesson.

T: (Write 30 × 0.1 = ____.) Say the answer.

S: 3.

Repeat the process for the following possible sequence: 300 × 0.01, 3,000 × 0.001, 5,000 × 0.001, 50 × 0.1, 500 × 0.01, 5,000 × 0.01, 3,000 × 0.01, 30,000 × 0.001, 50,000 × 0.001, 40 × 0.1, 400 × 0.1, 4,000 × 0.1, 40,000 × 0.1, 700 × 0.01, 7,000 × 0.01, 70,000 × 0.01, 700,000 × 0.01, and 7,000,000 × 0.001.

### Multiply Using the Area Model (7 minutes)

Follow the same process and procedure as Lesson 6 for the following possible sequence: 5.21 × 34 and 8.35 × 73.

### Unit Conversions (3 minutes)

Materials: (S) Personal white board

Note: Review of this fluency activity builds a foundation for upcoming lessons.

T: (Write 1 ft = ____ in.) 1 foot is the same as how many inches?

S: 12 inches.

Repeat the process for the following possible sequence: 2 ft, 3 ft, 4 ft, 10 ft, 5 ft, 7 ft.

T: (Write 100 cm = ____ m.) 100 centimeters is the same as how many meters?

S: 1 meter.

Repeat the process for the following possible sequence: 200 cm, 300 cm, 600 cm, 800 cm, 900 cm.

## Application Problem (12 minutes)

Materials: (S) Meter strip (Template); one string either 9 cm, 20 cm, 75 cm, or 105 cm

Procedure: Pass out a string to each student so that partners have different-length strings. Have them measure their strings and express the measurement in meters, centimeters, and millimeters, as well as share their measurements with their partners. Record the measurements of the strings in a class chart and revisit the following concepts from Lesson 4:

- Although the number of units has changed, the length of the string is the same.
- 1.05 meters × $10^3$ = 1,050 meters. This equation makes 1,000 copies of 1.05 meters.
- To convert meters to millimeters, we multiply the *number* of meters by $10^3$. 1.05 × $10^3$ = 1,050 to find that 1.05 meters = 1,050 millimeters.

| | m | cm | mm |
|---|---|---|---|
| A | 1.05 m | 105 cm | 1,050 mm |
| B | 0.75 m | 75 cm | 750 mm |
| C | 0.2 m | 20 cm | 200 mm |
| D | 0.09 m | 9 cm | 90 mm |

Note: Today's Application Problem provides a practical, hands-on way for students to experience the conversion reasoning foundational to today's lesson.

## Concept Development (26 minutes)

Materials: (S) Meter strip (Template), personal white board

**Problem 1**

3 weeks = ____ days

3 weeks = 3 × (1 week)

= 3 × (7 days)

= 21 days

T: (Write 3 weeks = 3 × (1 week) on the board.) Explain to your partner why this is true.

S: 3 weeks is the same as 3 units of 1 week. → It's 3 groups of 1 week. → It's like last year we saw that 3 fourths is the same as 3 times 1 fourth

T: What are the two factors?

S: 3 and 1 week.

T: How many days are equal to 1 week?

S: 7 days.

T: So rename 1 week as 7 days.

T: Let's use parentheses to make it clear that this factor, or **conversion factor,** has the same value. (Write 3 × (7 days) directly below 3 × (1 week) so that the equivalence of the two factors is very clear.)

T: 3 times 7 days is how many days?

S: 21 days.

T: So 3 weeks equals how many days?

S: 21 days.

T: On your personal white board, take a moment to convert 3 hours to minutes using the same process. Remember to use the parentheses to clarify the renaming of the conversion factor. Review your conversion with your partner.

T: (Students work.) We converted 3 hours to 180 minutes. Did we convert from a larger unit to a smaller unit or a smaller unit to a larger unit?

3 hours = ____ minutes

3 hours = 3 × (1 hour)

= 3 × (60 minutes)

= 180 minutes

S: Larger to smaller.

T: Yes. An hour is a larger unit than a minute. Since we converted to a smaller unit, a minute, what happened to the amount of time?

S: There are more units but the exact same amount of time. → The number of units increased, but the time stayed the same.

**Problem 2**

1.05 m = ____ cm

1.05 m = 1.05 × (1 m)

= 1.05 × (100 cm)

= 105 cm

T: (Write 1.05 m on the board.) Let's use the same strategy to convert larger units to smaller units, starting with the conversions from our Application Problem.

T: Let's convert 1.05 meters to centimeters. First, let's rename 1.05 meters as a multiplication expression. Consider how we expressed 3 weeks as an expression. Talk to your partner.

S: One factor is 1.05, and the other factor is 1 meter. → 1.05 × 1 meter.

T: (Write 1.05 m = 1.05 × (1 m).) Let's rename the conversion factor in centimeters. 1 meter equals ...?

S: 100 centimeters.

T: (Write 1.05 m = 1.05 × (100 cm).) What is 1.05 times 100 centimeters?

S: 105 centimeters.

T: Is that the correct conversion? Does 1.05 meters equal 105 centimeters? (Hold up the meter strip, and refer to the chart from the Application Problem.)

Have students convert other string measurements: 0.75 meters, 0.2 meters, and 0.09 meters to centimeters and then millimeters using the strategy.

| 0.09 m = ____ cm |
|---|
| 0.09 m = 0.09 × (1 m) |
| = 0.09 × (100 cm) |
| = 9 cm |

| 0.09 m = ____ mm |
|---|
| 0.09 m = 0.09 × (1 m) |
| = 0.09 × (1,000 mm) |
| = 90 mm |

**Problem 3**

A crate of apples weighs 5.7 kilograms. Convert the weight to grams.

A sack holds 56.75 pounds of sand. Convert the weight to ounces.

T: Start the process with students. Write the measurement as an equivalent expression with the unit as a factor.

| 5.7 kg = ____ g |
|---|
| 5.7 kg = 5.7 × (1 kg) |
| = 5.7 × (1,000 g) |
| = 5,700 g |

| 56.75 lb = ____ oz |
|---|
| 56.75 lb = 56.75 × (1 lb) |
| = 56.75 × (16 oz) |
| = 908 oz |

## Problem Set (10 minutes)

Students should do their personal best to complete the Problem Set within the allotted 10 minutes. For some classes, it may be appropriate to modify the assignment by specifying which problems they work on first. Some problems do not specify a method for solving. Students should solve these problems using the RDW approach used for Application Problems.

## Student Debrief (10 minutes)

**Lesson Objective:** Use whole number multiplication to express equivalent measurements.

The Student Debrief is intended to invite reflection and active processing of the total lesson experience.

Invite students to review their solutions for the Problem Set. They should check work by comparing answers with a partner before going over answers as a class. Look for misconceptions or misunderstandings that can be addressed in the Debrief. Guide students in a conversation to debrief the Problem Set and process the lesson.

Any combination of the questions below may be used to lead the discussion.

- In the conversion you completed for Problem 1(d), explain your process as you worked. How did you decide what to multiply by?
- Although we multiplied by 100 to convert 1.05 meters to 105 centimeters, the length remained the same. Why?
- Explain the term **conversion factor**. (The conversion factor is the factor in a multiplication sentence that renames one measurement unit as another equivalent unit.) For example, $14 \times (1 \text{ in}) = 14 \times (\frac{1}{12} \text{ ft})$, 1 in and $\frac{1}{12}$ ft are the conversion factors.
- What would be the conversion factor if we wanted to convert years to days? Years to months? Why isn't there one conversion factor to convert months to days? Why isn't there one conversion factor to convert years to days?
- Can you name some situations in which measurement conversion might be useful and/or necessary?

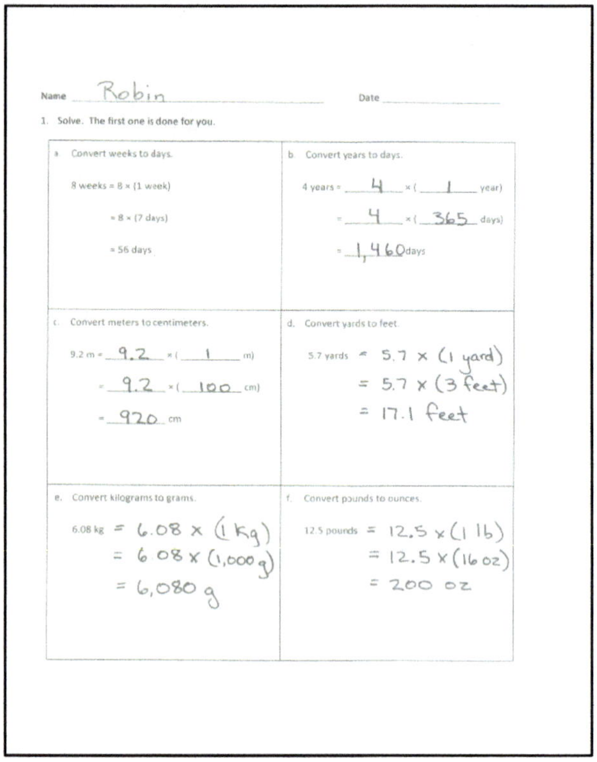
Name Robin Date

1. Solve. The first one is done for you.

| a. Convert weeks to days. 8 weeks = 8 × (1 week) = 8 × (7 days) = 56 days | b. Convert years to days. 4 years = 4 × (1 year) = 4 × (365 days) = 1,460 days |
|---|---|
| c. Convert meters to centimeters. 9.2 m = 9.2 × (1 m) = 9.2 × (100 cm) = 920 cm | d. Convert yards to feet. 5.7 yards = 5.7 × (1 yard) = 5.7 × (3 feet) = 17.1 feet |
| e. Convert kilograms to grams. 6.08 kg = 6.08 × (1 kg) = 6.08 × (1,000 g) = 6,080 g | f. Convert pounds to ounces. 12.5 pounds = 12.5 × (1 lb) = 12.5 × (16 oz) = 200 oz |

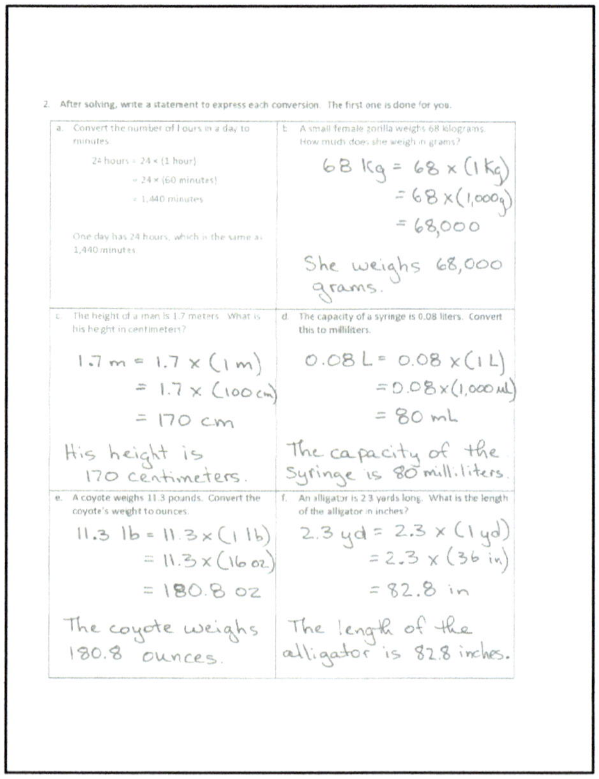
2. After solving, write a statement to express each conversion. The first one is done for you.

| a. Convert the number of hours in a day to minutes. 24 hours = 24 × (1 hour) = 24 × (60 minutes) = 1,440 minutes. One day has 24 hours, which is the same as 1,440 minutes. | b. A small female gorilla weighs 68 kilograms. How much does she weigh in grams? 68 kg = 68 × (1 kg) = 68 × (1,000 g) = 68,000. She weighs 68,000 grams. |
|---|---|
| c. The height of a man is 1.7 meters. What is his height in centimeters? 1.7 m = 1.7 × (1 m) = 1.7 × (100 cm) = 170 cm. His height is 170 centimeters. | d. The capacity of a syringe is 0.08 liters. Convert this to milliliters. 0.08 L = 0.08 × (1 L) = 0.08 × (1,000 mL) = 80 mL. The capacity of the syringe is 80 milliliters. |
| e. A coyote weighs 11.3 pounds. Convert the coyote's weight to ounces. 11.3 lb = 11.3 × (1 lb) = 11.3 × (16 oz) = 180.8 oz. The coyote weighs 180.8 ounces. | f. An alligator is 2.3 yards long. What is the length of the alligator in inches? 2.3 yd = 2.3 × (1 yd) = 2.3 × (36 in) = 82.8 in. The length of the alligator is 82.8 inches. |

## Exit Ticket (3 minutes)

After the Student Debrief, instruct students to complete the Exit Ticket. A review of their work will help with assessing students' understanding of the concepts that were presented in today's lesson and planning more effectively for future lessons. The questions may be read aloud to the students.

Name ______________________________ Date ________________

1. Solve. The first one is done for you.

| a. Convert weeks to days.<br><br>8 weeks = 8 × (1 week)<br>= 8 × (7 days)<br>= 56 days | b. Convert years to days.<br><br>4 years = ________ × ( ________ year)<br>= ________ × ( ________ days)<br>= ________ days |
|---|---|
| c. Convert meters to centimeters.<br><br>9.2 m = ________ × ( ________ m)<br>= ________ × ( ________ cm)<br>= ________ cm | d. Convert yards to feet.<br><br>5.7 yards |
| e. Convert kilograms to grams.<br><br>6.08 kg | f. Convert pounds to ounces.<br><br>12.5 pounds |

2. After solving, write a statement to express each conversion. The first one is done for you.

| | |
|---|---|
| a. Convert the number of hours in a day to minutes.<br>24 hours = 24 × (1 hour)<br>= 24 × (60 minutes)<br>= 1,440 minutes<br><br>One day has 24 hours, which is the same as 1,440 minutes. | b. A small female gorilla weighs 68 kilograms. How much does she weigh in grams? |
| c. The height of a man is 1.7 meters. What is his height in centimeters? | d. The capacity of a syringe is 0.08 liters. Convert this to milliliters. |
| e. A coyote weighs 11.3 pounds. Convert the coyote's weight to ounces. | f. An alligator is 2.3 yards long. What is the length of the alligator in inches? |

Name ___________________________ Date ______________

Solve.

| a. Convert pounds to ounces.<br>(1 pound = 16 ounces)<br><br>14 pounds = ________ × (1 pound)<br><br>= ________ × ( ________ ounces)<br><br>= ________ ounces | b. Convert kilograms to grams.<br><br>18.2 kilograms = ________ × ( ________ )<br><br>= ________ × ( ________ )<br><br>= ________ grams |
|---|---|

Name ____________________________ Date ________________

1. Solve. The first one is done for you.

| a. Convert weeks to days.<br><br>6 weeks = 6 × (1 week)<br>= 6 × (7 days)<br>= 42 days | b. Convert years to days.<br><br>7 years = ________ × ( ________ year)<br>= ________ × ( ________ days)<br>= ________ days |
|---|---|
| c. Convert meters to centimeters.<br><br>4.5 m = ________ × ( ________ m)<br>= ________ × ( ________ cm)<br>= ________ cm | d. Convert pounds to ounces.<br><br>12.6 pounds |
| e. Convert kilograms to grams.<br><br>3.09 kg | f. Convert yards to inches.<br><br>245 yd |

2. After solving, write a statement to express each conversion. The first one is done for you.

| | |
|---|---|
| a. Convert the number of hours in a day to minutes.<br>24 hours = 24 × (1 hour)<br>= 24 × (60 minutes)<br>= 1,440 minutes<br><br>One day has 24 hours, which is the same as 1,440 minutes. | b. A newborn giraffe weighs about 65 kilograms. How much does it weigh in grams? |
| c. The average height of a female giraffe is 4.6 meters. What is her height in centimeters? | d. The capacity of a beaker is 0.1 liter. Convert this to milliliters. |
| e. A pig weighs 9.8 pounds. Convert the pig's weight to ounces. | f. A marker is 0.13 meters long. What is the length in millimeters? |

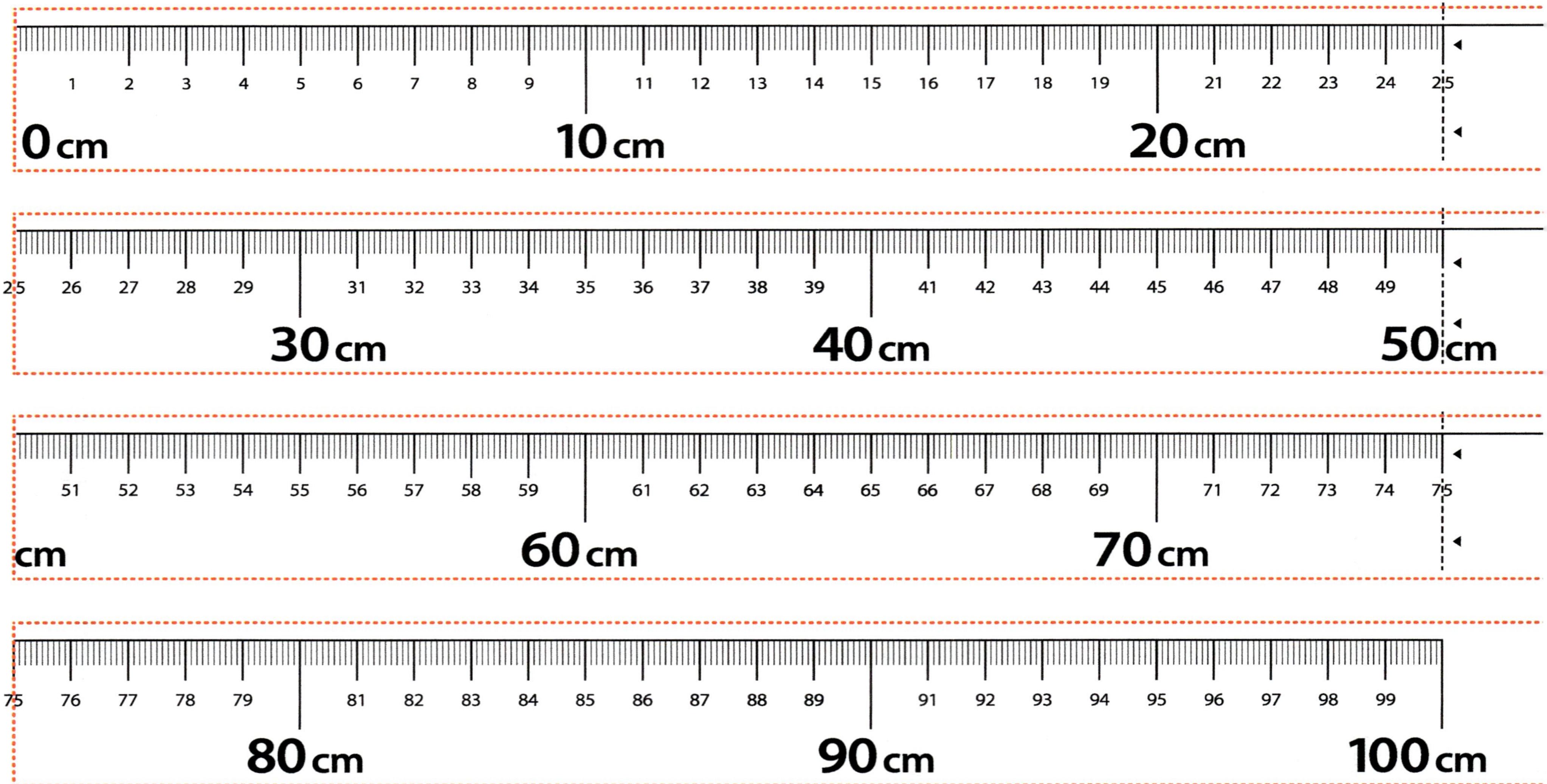


meter strip

**LEGEND** ------ CUT ---- ALIGN EDGE





EUREKA MATH™

# Lesson 14

## Objective: Use fraction and decimal multiplication to express equivalent measurements.

### Suggested Lesson Structure

| | |
|---|---|
| Fluency Practice | (12 minutes) |
| Application Problem | (8 minutes) |
| Concept Development | (30 minutes) |
| Student Debrief | (10 minutes) |
| **Total Time** | **(60 minutes)** |

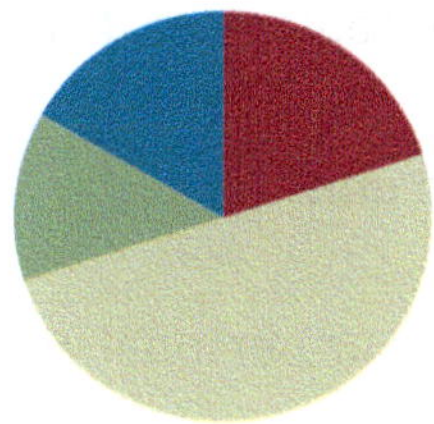

## Fluency Practice (12 minutes)

- Divide by Multiples of 10 **5.NBT.2** (3 minutes)
- Unit Conversions **5.MD.1** (4 minutes)
- Multiply Unit Fractions **4.NF.4** (5 minutes)

### Divide by Multiples of 10 (3 minutes)

Materials: (S) Personal white board

Note: This fluency activity review prepares students to approximate quotients with two-digit divisors in Lesson 17.

T: (Write 420 ÷ 10 = ____.) Say the division sentence.
S: 420 ÷ 10 = 42.
T: (Write 42 ÷ 2 = ____ below 420 ÷ 10 = 42.) Say the division sentence.
S: 42 ÷ 2 = 21.
T: (Write 420 ÷ 20 = ____ below 42 ÷ 2 = 21.) Say 420 ÷ 20 as a three-step division sentence, taking out the ten.
S: 420 ÷ 10 ÷ 2 = 21.
T: (Write 420 ÷ 20 = 21.)

420 ÷ 10 = 42 42 ÷ 2 = 21

420 ÷ 20 = 21

÷ 10 ÷ 2

Direct students to solve using the same method for 960 ÷ 30 and 680 ÷ 20.

## Unit Conversions (4 minutes)

Materials: (S) Personal white board

T: 1 foot is the same as how many inches?
S: 12 inches.
T: (Write 1 ft 1 in = ____ in.) On your personal white board, write the conversion.
S: (Write 1 ft 1 in = 13 in.)

Repeat the process for the following possible sequence: 1 ft 2 in, 1 ft 3 in, 1 ft 10 in, 1 ft 8 in, 2 ft, 2 ft 1 in, 2 ft 10 in, 2 ft 6 in, 3 ft, 3 ft 10 in, 3 ft 4 in.

T: 12 inches is the same as what single unit?
S: 1 foot.
T: (Write 13 in = ____ ft ____ in.) On your personal white board, write the conversion.
S: (Students write 13 in = 1 ft 1 in.)

Repeat the process for the following possible sequence: 14 in, 22 in, 24 in, 34 in, 25 in, 36 in, 46 in, 40 in, 48 in, 47 in, 49 in, 58 in.

## Multiply Unit Fractions (5 minutes)

Materials: (S) Personal white board

Note: This fluency activity reviews the multiplication of unit fractions from Grade 4 to be used in today's Concept Development.

T: (Write 4 × 1 banana.) Say the complete number sentence.
S: 4 × 1 banana = 4 bananas.
T: (Write 4 × 1 seventh.) Say the complete number sentence.
S: 4 × 1 seventh = 4 sevenths.
T: Rewrite the number sentence using fractions.
S: (Write $4 \times \frac{1}{7} = \frac{4}{7}$.)
T: (Write 7 × 1 seventh.) Say the complete number sentence.
S: 7 × 1 seventh = 7 sevenths.
T: Rewrite the number sentence using fractions.
S: (Write $7 \times \frac{1}{7} = \frac{7}{7}$.)
T: Rename 7 sevenths as a whole number.
S: 1.

Continue with 14 × 1 seventh.

T: (Write 8 × 1 fourth.) Say the complete number sentence.
S: 8 × 1 fourth = 8 fourths.
T: Rewrite the number sentence using fractions.
S: (Write $8 \times \frac{1}{4} = \frac{8}{4}$.)

T: Rename 8 fourths as a whole number.

S: 2.

Repeat the process for 12 × 1 fourth, 4 × 1 fourth, 3 × 1 third, 6 × 1 third, and 24 × 1 third.

## Application Problem (8 minutes)

Draw and label a tape diagram to represent each of the following:

1. Express 1 day as a fraction of 1 week.
2. Express 1 foot as a fraction of 1 yard.
3. Express 1 quart as a fraction of 1 gallon.
4. Express 1 centimeter as a fraction of 1 meter. (Decimal form.)
5. Express 1 meter as a fraction of 1 kilometer. (Decimal form.)

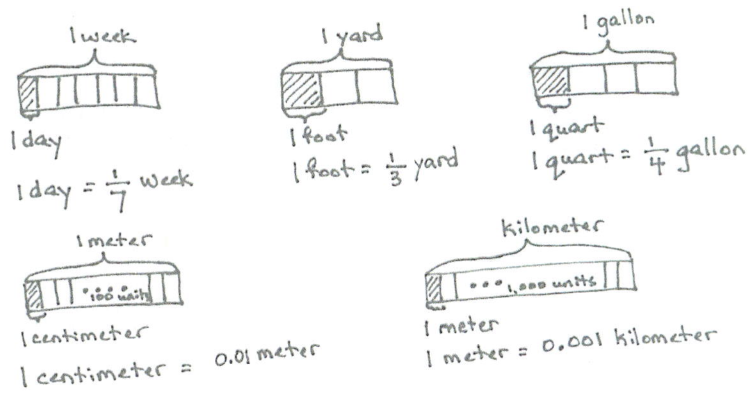


Note: This Application Problem is foundational to the Concept Development wherein students will be multiplying by fractions to convert smaller units to larger units.

## Concept Development (30 minutes)

Materials: (S) Personal white board, meter strip (Lesson 13 Template)

**Problem 1**

14 days = ____ weeks

14 days = 14 × (1 day)

$= 14 \times \left(\frac{1}{7} \text{ week}\right)$

$= \frac{14}{7}$ weeks

= 2 weeks

T: (Write 14 days = 14 × (1 day) on the board.)

T: What are the two factors?

S: 14 and 1 day.

T: What fraction of a week is 1 day?

S: $\frac{1}{7}$.

**NOTES ON MULTIPLE MEANS OF REPRESENTATION:**

Students benefit from seeing fractions in unit form, as in the Fluency Practice, pictorially, as in the Application Problem, and abstractly, as in the Concept Development. Refer back to the unit form and the pictorial to reassure students they can understand and solve fractions.

T: So can I rename 1 day as $\frac{1}{7}$ week?

S: Yes.

T: Let's use parentheses to make it clear that this factor, or conversion factor, has the same value. (Write 14 × ($\frac{1}{7}$ week) directly below 21 × (1 day) so the equivalence of the two factors is very clear.)

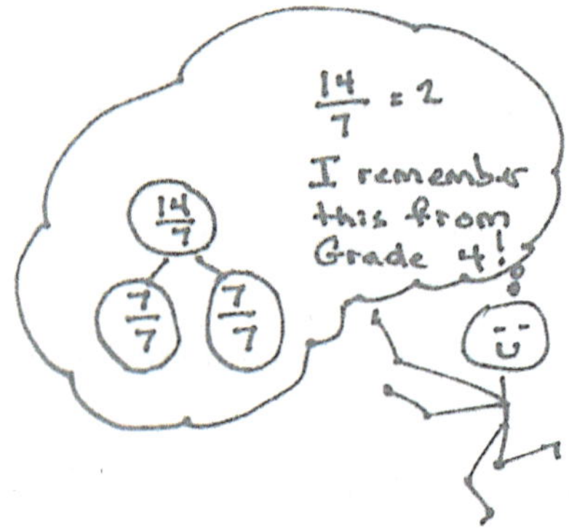


T: What's 14 times $\frac{1}{7}$ week?

S: 14 sevenths week.

T: Let's use a number bond to express 14 sevenths as 2 groups of 7 sevenths. (Draw the number bond pictured to the top right.)

T: How many weeks is $\frac{14}{7}$ weeks?

S: 2 weeks.

T: Did we convert from a larger to smaller unit or smaller to larger unit?

S: Smaller to larger unit.

T: Yes. A day is a smaller unit than a week.

T: On your personal white board, take a moment to convert 24 feet to yards and 24 quarts to gallons using the same process.

24 feet = ____ yards

24 feet = 24 × (1 foot)

$= 24 \times \left(\frac{1}{3}\text{ yard}\right)$

$= \frac{24}{3}$ yards

= 8 yards

24 quarts = ____ gallons

24 quarts = 24 × (1 quart)

$= 24 \times \left(\frac{1}{4}\text{ gallon}\right)$

$= \frac{24}{4}$ gallons

= 6 gallons

**NOTES ON MULTIPLE MEANS OF ACTION AND EXPRESSION:**

Students may benefit from first confirming the conversions using foot rulers (concrete) and then returning to patterns on a T-Chart or list as was used in Grade 4.

12 × (1 inch) = 1 foot.

24 × (1 inch) = 2 feet.

36 × (1 inch) = 3 feet.

Rewrite the chart as fraction multiplication:

$12 \times \frac{1}{12}$ foot = 1 foot.

$24 \times \frac{1}{12}$ foot = 2 feet.

$36 \times \frac{1}{12}$ foot = 3 feet.

This may help solidify the concept of 1 inch being a fraction of a foot and support seeing the simplicity of the fraction multiplication.

**Problem 2**

195 cm = ____ m

195 cm = 195 × (1 cm)

= 195 × (0.01 m)

= 1.95 m

T: (Write 195 cm.) Let's use the same process to convert smaller metric units (point to the centimeters) to larger metric units using decimal numbers. What metric units are larger than centimeters?

S: Meters. → Kilometers.

T: Let's convert 195 centimeters to meters.

T: Just as we have been doing, let's rename 195 centimeters as a multiplication expression with one factor naming the unit. Talk to your partner.

S: Last time, we made the unit a factor, so that means we have 195 groups of 1 centimeter. → One factor is 195, and the other factor is 1 centimeter. → 195 × 1 cm.

T: (Write 195 cm = 195 × (1 cm).) Using the parentheses really helps me see the conversion factor. (Point to 1 cm.)

T: Let's rename the conversion factor as meters. One centimeter is equal to what fraction of a meter?

S: 1 hundredth meter. → 1 one-hundredth meter.

T: Tell me how to write 1 hundredth in decimal notation.

S: Zero point zero 1.

T: (Write 195 cm = 195 × 0.01 m.) What is 195 times 0.01 meter?

S: 1.95 meters.

T: Is that the correct conversion? Does 195 cm equal 1.95 meters? (Hold up a meter stick and model the equivalence at the concrete level to verify.)

Repeat the process with the following possible sequence: convert 4,500 grams to kilograms and convert 578 milliliters to liters.

4,500 grams = ____ kilograms

4,500 grams = 4,500 × (1 gram)

= 4,500 × (0.001 kilogram)

= 4.5 kilograms

578 milliliters = ____ liters

578 milliliters = 578 × (1 milliliter)

= 578 × (0.001 liter)

= 0.578 liter

**Problem 3**

A container holds 16 cups of juice. Convert the capacity to pints. (2 cups = 1 pint.)

A truck weighs 1,675,280 grams. Convert the weight to kilograms.

T: Introduce students to the process, setting up the measurement as an equivalent expression with the unit as a factor.

16 cups = ____ pints

16 cups = 16 × (1 cup)

$= 16 \times \left(\frac{1}{2}\text{ pint}\right) = \frac{16}{2}\text{ pints}$

= 8 pints

1,675,280 grams = ____ kilograms

1,675,280 grams = 1,675,280 × (1 gram)

= 1,675,280 × (0.001 kilogram)

= 1,675.28 kilograms

## Problem Set (10 minutes)

Students should do their personal best to complete the Problem Set within the allotted 10 minutes. For some classes, it may be appropriate to modify the assignment by specifying which problems they work on first. Some problems do not specify a method for solving. Students should solve these problems using the RDW approach used for Application Problems.

## Student Debrief (10 minutes)

**Lesson Objective:** Use fraction and decimal multiplication to express equivalent measurements.

The Student Debrief is intended to invite reflection and active processing of the total lesson experience.

Invite students to review their solutions for the Problem Set. They should check work by comparing answers with a partner before going over answers as a class. Look for misconceptions or misunderstandings that can be addressed in the Debrief. Guide students in a conversation to debrief the Problem Set and process the lesson.

Any combination of the questions below may be used to lead the discussion.

- In each problem, what are the smaller units? What are the conversion factors in each problem?
- 24 feet is 8 yards, while 24 quarts is 6 gallons. Why did we end up with more yards than gallons when both conversions started with 24 units?

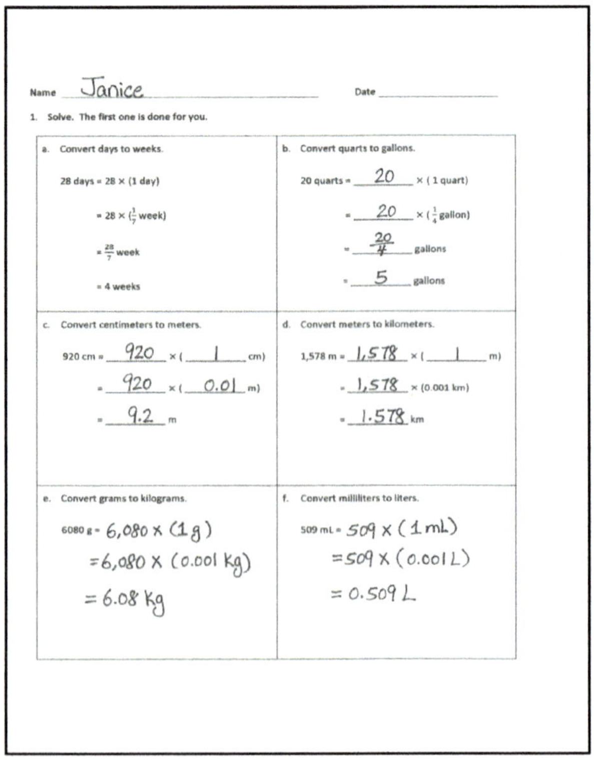
Name Janice Date

1. Solve. The first one is done for you.

a. Convert days to weeks.
28 days = 28 × (1 day)
$= 28 \times (\frac{1}{7}\text{ week})$
$= \frac{28}{7}$ week
= 4 weeks

b. Convert quarts to gallons.
20 quarts = 20 × (1 quart)
$= 20 \times (\frac{1}{4}\text{ gallon})$
$= \frac{20}{4}$ gallons
= 5 gallons

c. Convert centimeters to meters.
920 cm = 920 × (1 cm)
= 920 × (0.01 m)
= 9.2 m

d. Convert meters to kilometers.
1,578 m = 1,578 × (1 m)
= 1,578 × (0.001 km)
= 1.578 km

e. Convert grams to kilograms.
6080 g = 6,080 × (1 g)
= 6,080 × (0.001 kg)
= 6.08 kg

f. Convert milliliters to liters.
509 mL = 509 × (1 mL)
= 509 × (0.001 L)
= 0.509 L

- When our conversion factor is a fraction, we are converting to larger units. When our conversion factor is a whole number, we are converting to smaller units. Explain this using examples from your Problem Set and memory.
- Whether we are converting small units to large units or large units to small units, we are multiplying. Explain why this is true.

## Exit Ticket (3 minutes)

After the Student Debrief, instruct students to complete the Exit Ticket. A review of their work will help with assessing students' understanding of the concepts that were presented in today's lesson and planning more effectively for future lessons. The questions may be read aloud to the students.

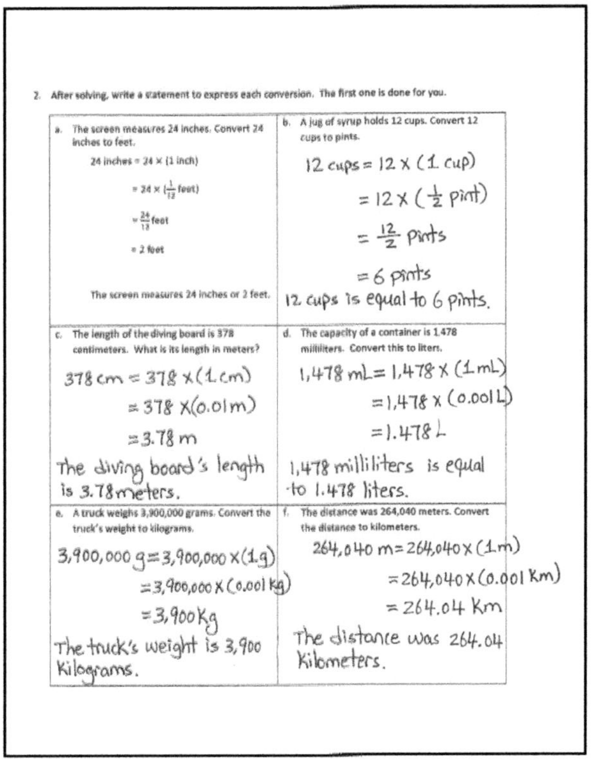

2. After solving, write a statement to express each conversion. The first one is done for you.

| | |
|---|---|
| a. The screen measures 24 inches. Convert 24 inches to feet.<br>24 inches $= 24 \times (1\text{ inch})$<br>$= 24 \times (\frac{1}{12}\text{ feet})$<br>$= \frac{24}{12}\text{ feet}$<br>$= 2\text{ feet}$<br>The screen measures 24 inches or 2 feet. | b. A jug of syrup holds 12 cups. Convert 12 cups to pints.<br>12 cups $= 12 \times (1\text{ cup})$<br>$= 12 \times (\frac{1}{2}\text{ pint})$<br>$= \frac{12}{2}\text{ pints}$<br>$= 6\text{ pints}$<br>12 cups is equal to 6 pints. |
| c. The length of the diving board is 378 centimeters. What is its length in meters?<br>378 cm $= 378 \times (1\text{ cm})$<br>$= 378 \times (0.01\text{ m})$<br>$= 3.78\text{ m}$<br>The diving board's length is 3.78 meters. | d. The capacity of a container is 1,478 milliliters. Convert this to liters.<br>1,478 mL $= 1{,}478 \times (1\text{ mL})$<br>$= 1{,}478 \times (0.001\text{ L})$<br>$= 1.478\text{ L}$<br>1,478 milliliters is equal to 1.478 liters. |
| e. A truck weighs 3,900,000 grams. Convert the truck's weight to kilograms.<br>3,900,000 g $= 3{,}900{,}000 \times (1\text{ g})$<br>$= 3{,}900{,}000 \times (0.001\text{ Kg})$<br>$= 3{,}900\text{ Kg}$<br>The truck's weight is 3,900 Kilograms. | f. The distance was 264,040 meters. Convert the distance to kilometers.<br>264,040 m $= 264{,}040 \times (1\text{ m})$<br>$= 264{,}040 \times (0.001\text{ Km})$<br>$= 264.04\text{ Km}$<br>The distance was 264.04 Kilometers. |

Name ______________________________ Date ________________

1. Solve. The first one is done for you.

<table>
<tr><td>a. Convert days to weeks.<br><br>28 days = 28 × (1 day)<br><br>$= 28 \times \left(\frac{1}{7}\text{ week}\right)$<br><br>$= \frac{28}{7}$ week<br><br>= 4 weeks</td><td>b. Convert quarts to gallons.<br><br>20 quarts = __________ × ( 1 quart)<br><br>= __________ × $\left(\frac{1}{4}\text{ gallon}\right)$<br><br>= __________ gallons<br><br>= __________ gallons</td></tr>
<tr><td>c. Convert centimeters to meters.<br><br>920 cm = __________ × ( __________ cm)<br><br>= __________ × ( __________ m)<br><br>= __________ m</td><td>d. Convert meters to kilometers.<br><br>1,578 m = __________ × ( __________ m)<br><br>= __________ × (0.001 km)<br><br>= __________ km</td></tr>
<tr><td>e. Convert grams to kilograms.<br><br>6,080 g =</td><td>f. Convert milliliters to liters.<br><br>509 mL =</td></tr>
</table>

2. After solving, write a statement to express each conversion. The first one is done for you.

| | |
|---|---|
| a. The screen measures 24 inches. Convert 24 inches to feet.<br><br>24 inches = 24 × (1 inch)<br>$= 24 \times \left(\frac{1}{12}\text{ feet}\right)$<br>$= \frac{24}{12}$ feet<br>= 2 feet<br><br>The screen measures 24 inches or 2 feet. | b. A jug of syrup holds 12 cups. Convert 12 cups to pints. |
| c. The length of the diving board is 378 centimeters. What is its length in meters? | d. The capacity of a container is 1,478 milliliters. Convert this to liters. |
| e. A truck weighs 3,900,000 grams. Convert the truck's weight to kilograms. | f. The distance was 264,040 meters. Convert the distance to kilometers. |

Name ______________________________ Date ______________

1. Convert days to weeks by completing the number sentences.

35 days = __________ × ( __________ day)

= __________ × ( __________ week)

=

=

2. Convert grams to kilograms by completing the number sentences.

4,567 grams = __________ × __________

= __________ × __________

=

=

Name ______________________________ Date ________________

1. Solve. The first one is done for you.

| | |
|---|---|
| a. Convert days to weeks.<br><br>42 days = 42 × (1 day)<br><br>$= 42 \times \left(\frac{1}{7}\text{ week}\right)$<br><br>$= \frac{42}{7}$ week<br><br>= 6 weeks | b. Convert quarts to gallons.<br><br>36 quarts = __________ × ( 1 quart)<br><br>$=$ __________ $\times \left(\frac{1}{4}\text{ gallon}\right)$<br><br>= __________ gallons<br><br>= __________ gallons |
| c. Convert centimeters to meters.<br><br>760 cm = __________ × ( __________ cm)<br><br>= __________ × ( __________ m)<br><br>= __________ m | d. Convert meters to kilometers.<br><br>2,485 m = __________ × ( __________ m)<br><br>= __________ × (0.001 km)<br><br>= __________ km |
| e. Convert grams to kilograms.<br><br>3,090 g = | f. Convert milliliters to liters.<br><br>205 mL = |

2. After solving, write a statement to express each conversion. The first one is done for you.

<table>
<tr><td>a. The screen measures 36 inches. Convert 36 inches to feet.<br><br>36 inches = 36 × (1 inch)<br>$= 36 \times \left(\frac{1}{12}\text{ feet}\right)$<br>$= \frac{36}{12}$ feet<br>= 3 feet<br><br>The screen measures 36 inches or 3 feet.</td><td>b. A jug of juice holds 8 cups. Convert 8 cups to pints.</td></tr>
<tr><td>c. The length of the flower garden is 529 centimeters. What is its length in meters?</td><td>d. The capacity of a container is 2,060 milliliters. Convert this to liters.</td></tr>
<tr><td>e. A hippopotamus weighs 1,560,000 grams. Convert the hippopotamus' weight to kilograms.</td><td>f. The distance was 372,060 meters. Convert the distance to kilometers.</td></tr>
</table>

# Lesson 15

## Objective: Solve two-step word problems involving measurement conversions.

### Suggested Lesson Structure

| | |
|---|---|
| ■ Fluency Practice | (12 minutes) |
| ■ Concept Development | (38 minutes) |
| ■ Student Debrief | (10 minutes) |
| **Total Time** | **(60 minutes)** |

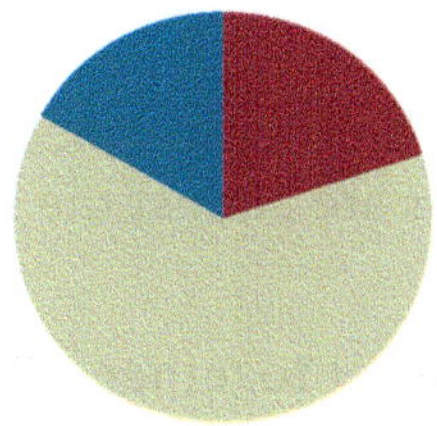

## Fluency Practice (12 minutes)

- Sprint: Convert Inches to Feet and Inches **5.MD.1** (9 minutes)
- Divide by Multiples of 10 and 100 **5.NBT.2** (3 minutes)

### Sprint: Convert Inches to Feet and Inches (9 minutes)

Materials: (S) Convert Inches to Feet and Inches Sprint

### Divide by Multiples of 10 and 100 (3 minutes)

Note: This fluency review prepares students to approximate quotients with two-digit divisors in Lesson 17.

Follow the same process and procedure from Lesson 14 for the following possible sequence: 480 ÷ 20, 6,480 ÷ 20, 690 ÷ 300, and 8,480 ÷ 400.

480 ÷ 10 = 48    48 ÷ 2 = 24

480 ÷ 20 = 24

÷ 10    ÷ 2

**NOTES ON MULTIPLE MEANS OF ACTION AND EXPRESSION:**

For students who struggle with conversion from inches to feet (from smaller to larger units), have them instead begin by making a simple T-chart, while placing 2 rulers end to end. While others work on the Sprint, they can add more data to their charts.

| Inches | Feet |
|---|---|
| 12 in | 1 ft 0 in |
| 13 in | 1 ft 1 in |
| 14 in | 1 ft 2 in |
| 15 in | 1 ft 3 in |

Then, if they are motivated to do so, have them take the Sprint home.

## Concept Development (38 minutes)

Materials: (T/S) Problem Set

Note: Problems for this section are found in this lesson's Problem Set.

**Problems 1 and 2: Convert from larger to smaller units.**

*Problem 1*

Liza's cat had six kittens! When Liza and her brother weighed all of the kittens together, they weighed 4 pounds 2 ounces. Because all of the kittens are almost the same size, about how many ounces does each kitten weigh?

T: Let's do Problem 1 on your Problem Set together. (Project Problem 1 on the board.) Let's read the problem aloud.

S: (Read chorally.)

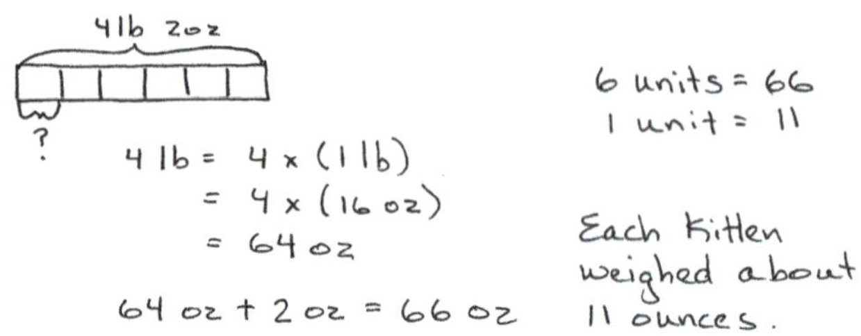


T: Now, let's re-read the problem sentence by sentence and draw as we go.

S: (Read the first sentence.)

T: What do you see? Can you draw something? Share your thinking.

S: I can draw 6 units representing 6 kittens.

T: Read the next sentence. (Students read.) What is the important information, and how can we show it in our drawing?

MP.2

S: The total weight for all 6 kittens is equal to 4 pounds 2 ounces. We can draw 6 equal units with the total of 4 pounds 2 ounces. → We know that 6 units equal 4 pounds 2 ounces.

T: Let's read the question.

T: What are we trying to find? What is missing in our drawing?

S: One kitten's weight, in ounces.

T: I'll put a question mark by one of our 6 units to show what we are trying to find.

T: How do we solve this problem? Turn and talk.

S: We have to divide. → We have use the total weight and divide by 6 to get 1 kitten's weight. → We first have to convert 4 lb 2 oz into ounces, and then we can divide by 6.

T: We were given the total weight of 4 lb 2 oz. Let's convert it into ounces. Work with a partner.

**NOTES ON MULTIPLE MEANS OF REPRESENTATION:**

Problem 2 is an additive comparison problem. The tape diagram's accurate representation of the story can be verified using much smaller numbers.

*A container of oregano is 3 ounces heavier than a container of peppercorns. Their total weight is 11 ounces... etc.*

Immediately follow the use of the smaller numbers with a return to the original problem so that students generalize to the more complex situation.

T: What is the total weight in ounces?

S: 66 oz.

T: Have we answered the question?

S: No. We need to divide the total weight of 66 ounces by 6 to find the weight of 1 kitten.

T: Solve.

T: Say the division sentence with the answer.

S: 66 oz ÷ 6 = 11 oz.

T: Express your answer in a sentence.

S: Each kitten weighed about 11 oz.

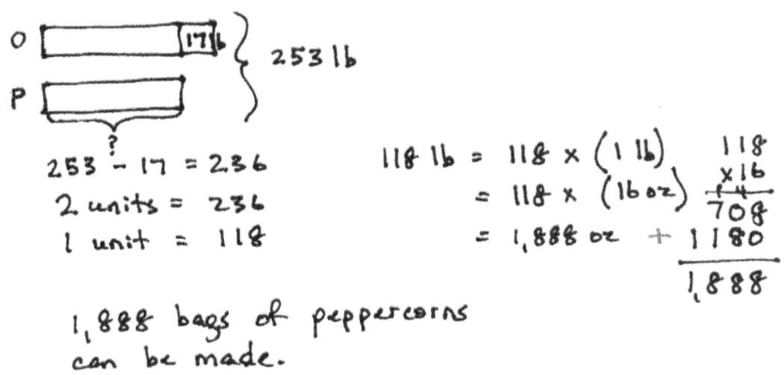


*Problem 2*

A container of oregano is 17 pounds heavier than a container of peppercorns. Their total weight is 253 pounds. The peppercorns will be sold in one-ounce bags. How many bags of peppercorns can be made?

**Problems 3 and 4: Convert from smaller to larger units.**

*Problem 3*

Each costume needs 46 centimeters of red ribbon and 3 times as much yellow ribbon. What is the total length of ribbon needed for 64 costumes? Express your answer in meters.

T: (Have students read the problem chorally, in pairs, or in silence.) Can you draw something? What can you draw?

S: 1 unit to show the red ribbon and 3 units to show the yellow ribbon.

T: Go ahead. Draw and label your tape diagrams.

T: (After drawing.) What conclusions can you make from your diagrams?

S: To find the total length of the ribbon needed for one costume, we have to multiply 46 by 4. → We know the total length of ribbon for 1 costume, but now we have to find the total length of ribbon for 64 costumes. → We have to express our answer in meters.

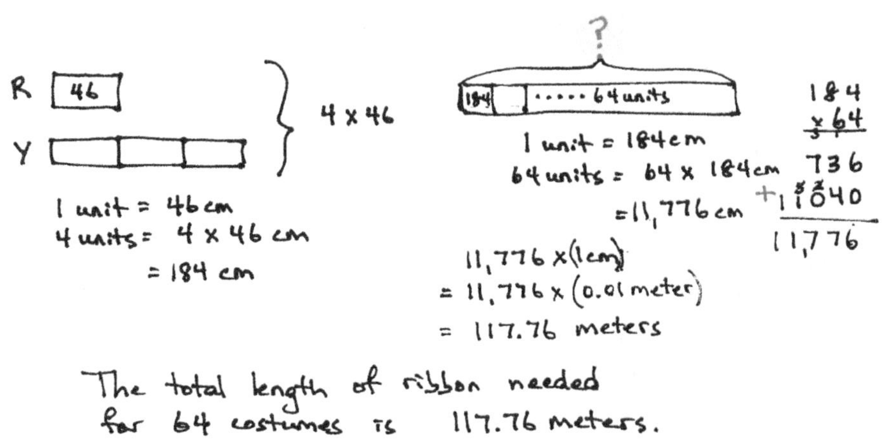

*Problem 4*

When making a batch of orange juice for her basketball team, Jackie used 5 times as much water as concentrate. There were 32 more cups of water than concentrate,

a. How much juice did she make in total?

b. She poured the juice into quart containers. How many containers could she fill?

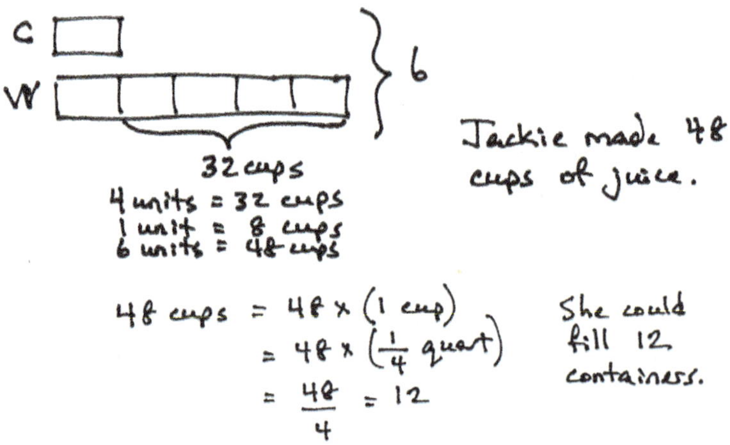


## Problem Set (10 minutes)

Students should do their personal best to complete the Problem Set within the allotted 10 minutes. For some classes, it may be appropriate to modify the assignment by specifying which problems they work on first. Some problems do not specify a method for solving. Students should solve these problems using the RDW approach used for Application Problems.

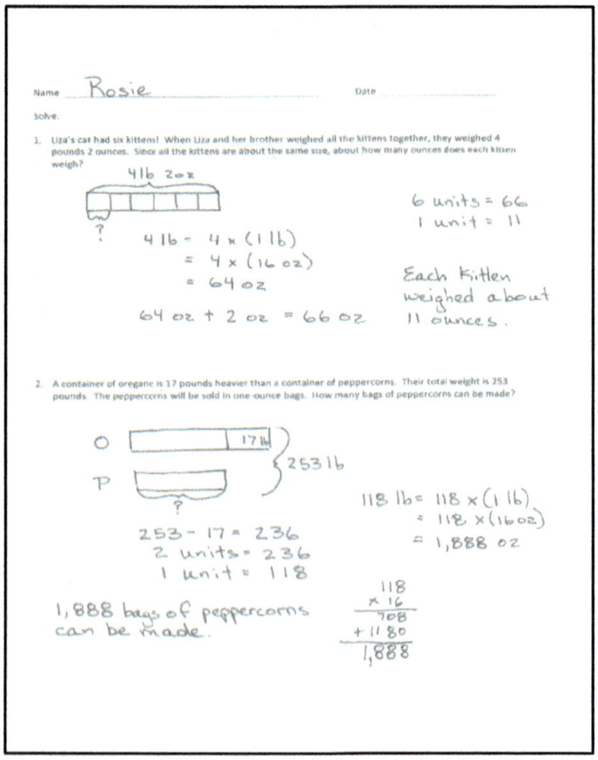

Name Rosie Date

Solve.

1. Liza's cat had six kittens! When Liza and her brother weighed all the kittens together, they weighed 4 pounds 2 ounces. Since all the kittens are about the same size, about how many ounces does each kitten weigh?



2. A container of oregano is 17 pounds heavier than a container of peppercorns. Their total weight is 253 pounds. The peppercorns will be sold in one-ounce bags. How many bags of peppercorns can be made?



## Student Debrief (10 minutes)

**Lesson Objective:** Solve two-step word problems involving measurement conversions.

The Student Debrief is intended to invite reflection and active processing of the total lesson experience.

Invite students to review their solutions for the Problem Set. They should check work by comparing answers with a partner before going over answers as a class. Look for misconceptions or misunderstandings that can be addressed in the Debrief. Guide students in a conversation to debrief the Problem Set and process the lesson.

Any combination of the questions below may be used to lead the discussion.

- Look back at Problem 1. The weight of the kittens was given in both pounds and ounces. What other mixed units do we often see?
- Look back at Problem 4(b). Would it have been possible to answer Part (b) before answering Part (a)? Is there another way to solve? (Students could convert 32 cups to 8 quarts to find that each unit was equal to 2 quarts.)
- Which problems involved converting from larger to smaller units, and which involved converting smaller to larger units? Which conversion is more challenging for you?

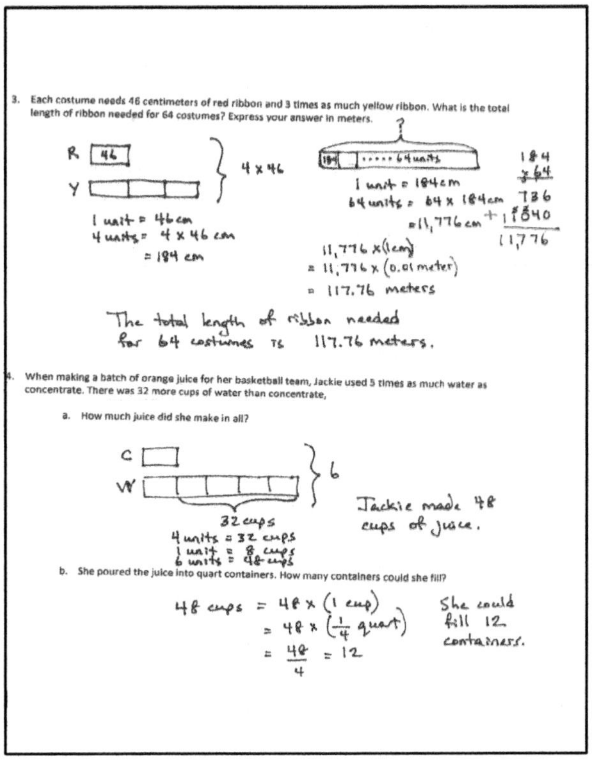

3. Each costume needs 46 centimeters of red ribbon and 3 times as much yellow ribbon. What is the total length of ribbon needed for 64 costumes? Express your answer in meters.

1 unit = 46 cm
4 units = 4 × 46 cm
= 184 cm

1 unit = 184 cm
64 units = 64 × 184 cm
= 11,776 cm

11,776 × (1 cm)
= 11,776 × (0.01 meter)
= 117.76 meters

The total length of ribbon needed for 64 costumes is 117.76 meters.

4. When making a batch of orange juice for her basketball team, Jackie used 5 times as much water as concentrate. There was 32 more cups of water than concentrate,

a. How much juice did she make in all?

4 units = 32 cups
1 unit = 8 cups
6 units = 48 cups

Jackie made 48 cups of juice.

b. She poured the juice into quart containers. How many containers could she fill?

48 cups = 48 × (1 cup)
= 48 × ($\frac{1}{4}$ quart)
= $\frac{48}{4}$ = 12

She could fill 12 containers.

## Exit Ticket (3 minutes)

After the Student Debrief, instruct students to complete the Exit Ticket. A review of their work will help with assessing students' understanding of the concepts that were presented in today's lesson and planning more effectively for future lessons. The questions may be read aloud to the students.

# A

Number Correct: _______

Convert Inches to Feet and Inches

| | | | |
|---|---|---|---|
| 1. | 12 in. = | ft. | in. |
| 2. | 13 in. = | ft. | in. |
| 3. | 14 in. = | ft. | in. |
| 4. | 15 in. = | ft. | in. |
| 5. | 22 in. = | ft. | in. |
| 6. | 20 in. = | ft. | in. |
| 7. | 24 in. = | ft. | in. |
| 8. | 25 in. = | ft. | in. |
| 9. | 26 in. = | ft. | in. |
| 10. | 30 in. = | ft. | in. |
| 11. | 34 in. = | ft. | in. |
| 12. | 35 in. = | ft. | in. |
| 13. | 36 in. = | ft. | in. |
| 14. | 37 in. = | ft. | in. |
| 15. | 46 in. = | ft. | in. |
| 16. | 40 in. = | ft. | in. |
| 17. | 48 in. = | ft. | in. |
| 18. | 58 in. = | ft. | in. |
| 19. | 49 in. = | ft. | in. |
| 20. | 47 in. = | ft. | in. |
| 21. | 50 in. = | ft. | in. |
| 22. | 12 in. = | ft. | in. |

| | | | |
|---|---|---|---|
| 23. | 17 in. = | ft. | in. |
| 24. | 24 in. = | ft. | in. |
| 25. | 28 in. = | ft. | in. |
| 26. | 36 in. = | ft. | in. |
| 27. | 45 in. = | ft. | in. |
| 28. | 48 in. = | ft. | in. |
| 29. | 59 in. = | ft. | in. |
| 30. | 60 in. = | ft. | in. |
| 31. | 64 in. = | ft. | in. |
| 32. | 68 in. = | ft. | in. |
| 33. | 71 in. = | ft. | in. |
| 34. | 73 in. = | ft. | in. |
| 35. | 72 in. = | ft. | in. |
| 36. | 80 in. = | ft. | in. |
| 37. | 84 in. = | ft. | in. |
| 38. | 90 in. = | ft. | in. |
| 39. | 96 in. = | ft. | in. |
| 40. | 100 in. = | ft. | in. |
| 41. | 108 in. = | ft. | in. |
| 42. | 117 in. = | ft. | in. |
| 43. | 104 in. = | ft. | in. |
| 44. | 93 in. = | ft. | in. |

# B

Number Correct: _______

Improvement: _______

Convert Inches to Feet and Inches

| | | | | | | | |
|---|---|---|---|---|---|---|---|
| 1. | 120 in. = | ft. | in. | 23. | 16 in. = | ft. | in. |
| 2. | 12 in. = | ft. | in. | 24. | 24 in. = | ft. | in. |
| 3. | 13 in. = | ft. | in. | 25. | 29 in. = | ft. | in. |
| 4. | 14 in. = | ft. | in. | 26. | 36 in. = | ft. | in. |
| 5. | 20 in. = | ft. | in. | 27. | 42 in. = | ft. | in. |
| 6. | 22 in. = | ft. | in. | 28. | 48 in. = | ft. | in. |
| 7. | 24 in. = | ft. | in. | 29. | 59 in. = | ft. | in. |
| 8. | 25 in. = | ft. | in. | 30. | 60 in. = | ft. | in. |
| 9. | 26 in. = | ft. | in. | 31. | 63 in. = | ft. | in. |
| 10. | 34 in. = | ft. | in. | 32. | 67 in. = | ft. | in. |
| 11. | 30 in. = | ft. | in. | 33. | 70 in. = | ft. | in. |
| 12. | 35 in. = | ft. | in. | 34. | 73 in. = | ft. | in. |
| 13. | 36 in. = | ft. | in. | 35. | 72 in. = | ft. | in. |
| 14. | 46 in. = | ft. | in. | 36. | 77 in. = | ft. | in. |
| 15. | 37 in. = | ft. | in. | 37. | 84 in. = | ft. | in. |
| 16. | 40 in. = | ft. | in. | 38. | 89 in. = | ft. | in. |
| 17. | 48 in. = | ft. | in. | 39. | 96 in. = | ft. | in. |
| 18. | 49 in. = | ft. | in. | 40. | 99 in. = | ft. | in. |
| 19. | 58 in. = | ft. | in. | 41. | 108 in. = | ft. | in. |
| 20. | 47 in. = | ft. | in. | 42. | 115 in. = | ft. | in. |
| 21. | 50 in. = | ft. | in. | 43. | 103 in. = | ft. | in. |
| 22. | 12 in. = | ft. | in. | 44. | 95 in. = | ft. | in. |

Name ______________________________ Date ________________

Solve.

1. Liza's cat had six kittens! When Liza and her brother weighed all the kittens together, they weighed 4 pounds 2 ounces. Since all the kittens are about the same size, about how many ounces does each kitten weigh?

2. A container of oregano is 17 pounds heavier than a container of peppercorns. Their total weight is 253 pounds. The peppercorns will be sold in one-ounce bags. How many bags of peppercorns can be made?

3. Each costume needs 46 centimeters of red ribbon and 3 times as much yellow ribbon. What is the total length of ribbon needed for 64 costumes? Express your answer in meters.

4. When making a batch of orange juice for her basketball team, Jackie used 5 times as much water as concentrate. There were 32 more cups of water than concentrate.

   a. How much juice did she make in all?

   b. She poured the juice into quart containers. How many containers could she fill?

Name ______________________________ Date ________________

Solve.

To practice for an Ironman competition, John swam 0.86 kilometer each day for 3 weeks. How many meters did he swim in those 3 weeks?

Name ______________________________ Date ________________

Solve.

1. Tia cut a 4-meter 8-centimeter wire into 10 equal pieces. Marta cut a 540-centimeter wire into 9 equal pieces. How much longer is one of Marta's wires than one of Tia's?

2. Jay needs 19 quarts more paint for the outside of his barn than for the inside. If he uses 107 quarts in all, how many gallons of paint will be used to paint the inside of the barn?

3. String A is 35 centimeters long. String B is 5 times as long as String A. Both are necessary to create a decorative bottle. Find the total length of string needed for 17 identical decorative bottles. Express your answer in meters.

4. A pineapple is 7 times as heavy as an orange. The pineapple also weighs 870 grams more than the orange.

   a. What is the total weight in grams for the pineapple and orange?

   b. Express the total weight of the pineapple and orange in kilograms.



EUREKA MATH™

Name ______________________________ Date ________________

1. Fill in the chart.

| Words | Expression | The Value of the Expression |
|---|---|---|
| a. 50 times the sum of 64 and 36 | | |
| b. Divide the difference between 1,200 and 700 by 5 | | |
| c. The sum of 3 fifteens and 17 fifteens | | |
| d. 15 times the sum of 14 and 6 | | |
| e. | 10 × (250 + 45) | |
| f. | (560 + 440) × 14 | |

2. Compare the two expressions using < , > , or = . For each, explain how you can determine the answer without calculating.

   a. 100 × 8 ◯ 25 × (4 × 9)

   b. 48 × 12 ◯ 50 twelves – 3 twelves

   c. 24 × 36 ◯ 18 twenty-fours, doubled

3. Solve. Use words, numbers, or pictures to explain how your answers to Parts (a) and (b) are related.

   a. 25 × 30 = ___________  b. 2.5 × 30 = ___________ tenths × 30 = ___________

4. Multiply using the standard algorithm. Show your work below each problem. Write the product in the blank.

   a. 514 × 33 = ___________  b. 546 × 405 = ___________

5. For a field trip, the school bought 47 sandwiches for $4.60 each and 39 bags of chips for $1.25 each. How much did the school spend in all?

6. Jeanne makes hair bows to sell at the craft fair. Each bow requires 1.5 yards of ribbon.

   a. At the fabric store, ribbon is sold by the foot. If Jeanne wants to make 84 bows, how many feet of ribbon must she buy? Show all your work.

   b. If the ribbon costs 10¢ per foot, what is the total cost of the ribbon in dollars? Explain your reasoning, including how you decided where to place the decimal.

   c. A manufacturer is making 1,000 times as many bows as Jeanne to sell in stores nationwide. Write an expression using exponents to show how many yards of ribbon the manufacturer will need. Do not calculate the total.

| Mid-Module Assessment Task Standards Addressed | Topics A–D |
|---|---|

**Write and interpret numerical expressions.**

**5.OA.1** Use parentheses, brackets, or braces in numerical expressions, and evaluate expressions with these symbols.

**5.OA.2** Write simple expressions that record calculations with numbers, and interpret numerical expressions without evaluating them. *For example, express the calculation "add 8 and 7, then multiply by 2" as 2 × (8 + 7). Recognize that 3 × (18932 + 921) is three times as large as 18932 + 921, without having to calculate the indicated sum or product.*

**Understand the place value system.**

**5.NBT.1** Recognize that in a multi-digit number, a digit in one place represents 10 times as much as it represents in the place to its right and 1/10 of what it represents in the place to its left.

**5.NBT.2** Explain patterns in the number of zeros of the product when multiplying a number by powers of 10, and explain patterns in the placement of the decimal point when a decimal is multiplied or divided by a power of 10. Use whole-number exponents to denote powers of 10.

**Perform operations with multi-digit whole numbers and with decimals to hundredths.**

**5.NBT.5** Fluently multiply multi-digit whole numbers using the standard algorithm.

**5.NBT.7** Add, subtract, multiply, and divide decimals to hundredths, using concrete models or drawings and strategies based on place value, properties of operations, and/or the relationship between addition and subtraction; relate the strategy to a written method and explain the reasoning used.

**Convert like measurement units within a given measurement system.**

**5.MD.1** Convert among different-sized standard measurement units within a given measurement system (e.g., convert 5 cm to 0.05 m), and use these conversions in solving multi-step, real world problems.

## Evaluating Student Learning Outcomes

A Progression Toward Mastery is provided to describe steps that illuminate the gradually increasing understandings that students develop on their way to proficiency. In this chart, this progress is presented from left (Step 1) to right (Step 4). The learning goal for students is to achieve Step 4 mastery. These steps are meant to help teachers and students identify and celebrate what the students CAN do now and what they need to work on next.

| A Progression Toward Mastery | | | | |
|---|---|---|---|---|
| **Assessment Task Item** | **STEP 1 Little evidence of reasoning without a correct answer.** <br> **(1 Point)** | **STEP 2 Evidence of some reasoning without a correct answer.** <br> **(2 Points)** | **STEP 3 Evidence of some reasoning with a correct answer or evidence of solid reasoning with an incorrect answer.** <br> **(3 Points)** | **STEP 4 Evidence of solid reasoning with a correct answer.** <br> **(4 Points)** |
| **1** <br> **5.OA.1** <br> **5.OA.2** | Student is able to answer one to three items correctly. | Student is able to answer four to seven items correctly. | Student is able to answer eight to eleven items correctly. | Student is able to answer all 12 items correctly. |
| **2** <br> **5.OA.2** | Student is unable to compare the expressions. | Student is able to correctly compare at least two pairs of expressions but is unable to explain reasoning. | Student is able to correctly compare at least two pairs of expressions and is able to explain reasoning on some parts of the task. | Student correctly compares all pairs of expressions and is able to explain reasoning for all parts of the task. |
| **3** <br> **5.NBT.1** <br> **5.NBT.2** <br> **5.NBT.7** | Student is unable to correctly multiply either Part (a) or (b) and makes no attempt to explain the relationship between products. | Student is able to multiply either Part (a) or (b) correctly but makes no attempt to explain the relationship between the products. | Student is able to correctly multiply both Parts (a) and (b) and provides some explanation of the relationship between the products. | Student correctly multiplies both parts of the task and provides a complete explanation of the relationship between the products using words, numbers, or pictures. <br> a. 750 <br> b. 75 |
| **4** <br> **5.NBT.5** | Student does not use the standard algorithm or any strategy to multiply either Part (a) or (b). | Student does not use the standard algorithm but uses another strategy to multiply Part (a) and/or Part (b). | Student uses the standard algorithm to multiply but makes errors in the partial products or the final product. | Student uses the standard algorithm to correctly multiply both Parts (a) and (b). <br> a. 16,962 <br> b. 221,130 |

| A Progression Toward Mastery | | | | |
|---|---|---|---|---|
| **5**<br><br>**5.NBT.5**<br>**5.NBT.7** | Student uses incorrect reasoning and neither multiplies nor adds. | Student uses partially correct reasoning (multiplies but does not add, or adds but does not multiply) and makes calculation errors. | Student uses correct reasoning but makes calculation errors. | Student uses correct reasoning and also calculates the total correctly as $264.95. |
| **6**<br><br>**5.OA.1**<br>**5.OA.2**<br>**5.NBT.1**<br>**5.NBT.2**<br>**5.NBT.5**<br>**5.NBT.7**<br>**5.MD.1** | Student uses incorrect reasoning in most parts of the task and is unable to correctly convert, calculate, and/or write an accurate expression. | Student uses some correct reasoning and is able to answer one part of the task. | Student uses correct reasoning but makes calculation errors on part of the task or writes an incorrect expression. | Student uses correct reasoning, correctly calculates all parts of the task, and writes a correct expression.<br>a. 378 ft<br>b. $37.80<br>c. $84 \times 1.5 \times 10^3$<br>or<br>$84 \times 10^3 \times 1.5$ |

Name Charlie Date ____________

1. Fill in the chart.

| Words | Expression | The Value of the Expression |
|---|---|---|
| a. 50 times the sum of 64 and 36 | 50×(64+36) | 5,000 |
| b. Divide the difference between 1,200 and 700 by 5 | (1,200−700)÷5 | 100 |
| c. The sum of 3 fifteens and 17 fifteens | (3×15)+(17×15) | 300 |
| d. 15 times the sum of 14 and 6 | 15×(14+6) | 300 |
| e. 10 times the sum of 250 and 45 | 10 × (250 + 45) | 2,950 |
| f. 14 times the sum of 560 and 440 | (560 + 440) × 14 | 14,000 |

2. Compare the two expressions using < , > , or =. For each, explain how you can determine the answer without calculating.

a. 100 × 8 (<) 25 × (4 × 9)

The product here is 800.

The product of this part is 100, so 100×9 is equal to 900.

b. 48 × 12 (>) 50 twelves – 3 twelves

This is 48 twelves.

This is 47 twelves
The other side is 1 more group of twelve.

c. 24 × 36 (=) 18 twenty-fours, doubled

Double 18 is 36, so it's 36 twenty-fours on both sides.

3. Solve. Use words, numbers or pictures to explain how your answers to parts (a) and (b) are related.

a. 25 × 30 = 750

b. 2.5 × 30 = 25 tenths × 30 = 750 tenths = 75.0

The digits are exactly the same. But the units in (b) are smaller so the answer is smaller. Ones are 10 times as large as tenths so the answer to (a) is ten times larger than (b)

4. Multiply using the standard algorithm. Show your work below each problem. Write the product in the blank.

a. 514 × 33 = 16,962

```
   514
 ×  33
 -----
  1542
+15420
 -----
16,962
```

b. 546 × 405 = 221,130

```
    546
  × 405
 ------
   2730
+218400
 ------
221,130
```

5. For a field trip, the school bought 47 sandwiches for $4.60 each and 39 bags of chips for $1.25 each. How much did the school spend in all?

```
   460 cents          125 cents
  ×  47              ×  39
  -----              -----
   3220               1125
 +18400              +3750
  -----              -----
21,620 cents         4875 cents
$216.20              $48.75

  216.20
+  48.75
  ------
  264.95
```

The school spent $264.95 in all.

6. Jeanne makes hair bows to sell at the craft fair. Each bow requires 1.5 yards of ribbon.

a. At the fabric store, ribbon is sold by the foot. If Jeanne wants to make 84 bows, how many feet of ribbon must she buy? Show all your work.

$1.5 \text{ yd} = 1.5 \times (1 \text{ yd})$
$= 1.5 \times (3 \text{ ft})$
$= 4.5 \text{ ft}$

45 tenths

$$\begin{array}{r} \times\ 84 \\ \hline 180 \\ +\ 3600 \\ \hline 378.0 \end{array}$$

Jeanne has to buy 378 feet of ribbon.

b. If the ribbon costs 10¢ per foot, what is the total cost of the ribbon in dollars? Explain your reasoning, including how you decided where to place the decimal.

378 × 10¢ = 3780¢ = $37.80

When I multiplied by 10, all the digits got 10 times larger and moved one place to the left. That was 3,780 cents. To find dollars, I divided by 100 which moved my digits back 2 places to the left, so my decimal point went between the 7 and 8.

c. A manufacturer is making 1,000 times as many bows as Jeanne to sell in stores nationwide. Write an expression using exponents to show how many yards of ribbon the manufacturer will need. Do not calculate the total.

$84 \times 10^3 \times 1.5$

Topic E

# Mental Strategies for Multi-Digit Whole Number Division

**5.NBT.1, 5.NBT.2, 5.NBT.6**

| | | |
|---|---|---|
| **Focus Standards:** | 5.NBT.1 | Recognize that in a multi-digit number, a digit in one place represents 10 times as much as it represents in the place to its right and 1/10 of what it represents in the place to its left. |
| | 5.NBT.2 | Explain patterns in the number of zeros of the product when multiplying a number by powers of 10, and explain patterns in the placement of the decimal point when a decimal is multiplied or divided by a power of 10. Use whole-number exponents to denote power of 10. |
| | 5.NBT.6 | Find whole-number quotients of whole numbers with up to four-digit dividends and two-digit divisors, using strategies based on place value, the properties of operations, and/or the relationship between multiplication and division. Illustrate and explain the calculation by using equations, rectangular arrays, and/or area models. |
| **Instructional Days:** | 3 | |
| **Coherence -Links from:** | G4–M3 | Multi-Digit Multiplication and Division |
| | G5–M1 | Place Value and Decimal Fractions |
| **-Links to:** | G5–M4 | Multiplication and Division of Fractions and Decimal Fractions |
| | G6–M2 | Arithmetic Operations Including Division of Fractions |

Topics E through H provide a parallel sequence for division to that offered in Topics A to D for multiplication. Topic E begins concretely with place value disks as an introduction to division with multi-digit whole numbers (5.NBT.6). In Lesson 17, $420 \div 60$ is interpreted as $420 \div 10 \div 6$. Next, students round dividends and 2-digit divisors to nearby multiples of ten in order to estimate single digit quotients (e.g., $431 \div 58 \approx 420 \div 60 = 7$) and then multi-digit quotients. This work is done horizontally, outside the context of the written vertical method.

| A Teaching Sequence Toward Mastery of Mental Strategies for Multi-Digit Whole Number Division | |
|---|---|
| **Objective 1:** | **Use *divide by 10* patterns for multi-digit whole number division.**<br>**(Lesson 16)** |
| **Objective 2:** | **Use basic facts to approximate quotients with two-digit divisors.**<br>**(Lessons 17–18)** |

# Lesson 16

## Objective: Use *divide by 10* patterns for multi-digit whole number division.

### Suggested Lesson Structure

| | |
|---|---|
| Fluency Practice | (12 minutes) |
| Application Problem | (5 minutes) |
| Concept Development | (33 minutes) |
| Student Debrief | (10 minutes) |
| **Total Time** | **(60 minutes)** |

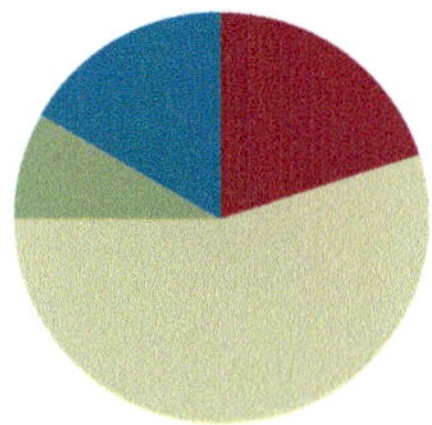

## Fluency Practice (12 minutes)

- Sprint: Divide by Multiples of 10 and 100 **5.NBT.2** (7 minutes)
- Round to the Nearest Ten **5.NBT.4** (2 minutes)
- Group Count by Multiples of 10 **5.NBT.2** (3 minutes)

### Sprint: Divide by Multiples of 10 and 100 (7 minutes)

Materials: (S) Divide by Multiples of 10 and 100 Sprint

Note: This Sprint prepares students for the Concept Development.

### Round to the Nearest Ten (2 minutes)

Note: Rounding to the nearest ten prepares students to estimate quotients.

T: (Write 32 ≈ ____.) What's 32 rounded to the nearest ten?

S: 30.

Repeat the process for 47, 18, 52, 74, 85, and 15.

### Group Count by Multiples of 10 (3 minutes)

Note: Counting by multiples of 10 prepares students for Lesson 17's Concept Development.

T: Count by threes.

S: 3, 6, 9, 12, 15, 18, 21, 24, 27, 30.

T: Count by 3 tens. When I raise my hand, stop counting.

S: 3 tens, 6 tens, 9 tens.

T: (Raise hand.) Say 9 tens in standard form.

S: 90.

Continue the process, stopping at 15 tens, 24 tens, and 30 tens.

Repeat the process with 6 tens, stopping periodically.

## Application Problem (5 minutes)

The area of a rectangular vegetable garden is 200 $ft^2$. The width is 10 ft. What is the length of the vegetable garden?

Note: This problem provides a nice opportunity to quickly address area concepts and division by a power of ten, allowing for a smooth transition into the day's Concept Development. While solving, students should be encouraged to draw a picture of a rectangle to support their work.

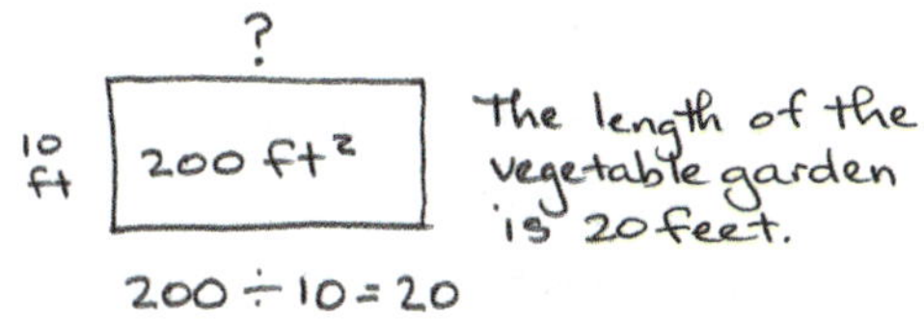


## Concept Development (33 minutes)

Materials: (S) Personal white board

**NOTES ON MULTIPLE MEANS OF ENGAGEMENT:**

The disk representations used here are a shorthand version of the work done in Grade 5 Module 1 with place value charts. Some students may need to see the division on the chart using arrows before moving directly to drawn disks.

**Problem 1: 420 ÷ 10**

T: (Write 420 ÷ 10 horizontally on board.) Let's use place value disks to solve this problem. Work with a partner to show 420 using the disks.

T/S: (Draw 4 hundred disks and 2 ten disks, as shown to the lower right.)

T: Say 420 in unit form.

S: 4 hundreds 2 tens.

T: Let's divide. What is 1 hundred divided by 10?

S: 10.

T: If 1 hundred divided by 10 is 1 ten, what is 4 hundreds divided by 10?

S: 4 tens.

T: I'll show that division with my place value disks. You do the same. (Draw an arrow showing ÷ *10* and 4 tens disks.)

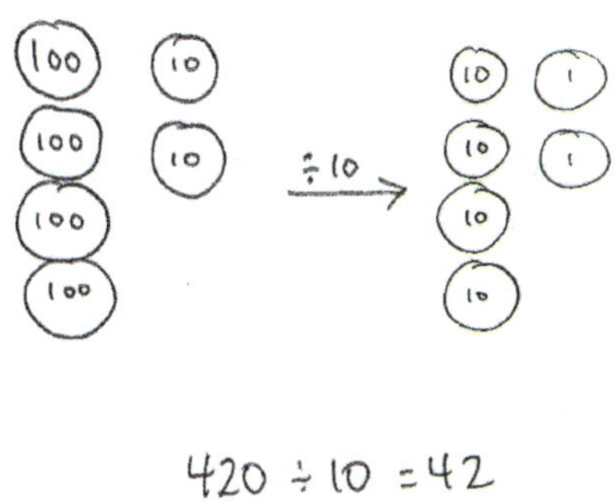


S: (Draw.)

T: What is 1 ten divided by 10?

S: 1.

T: If 1 ten divided by 10 is 1 one, what is 2 tens divided by 10?

S: 2 ones.

T: Show that division with place value disks.

T: (Point to the original problem.) Read the division sentence with the solution.

S: 420 ÷ 10 = 42.

T: Let's solve this problem again using our place value charts. Show 420 in numerical form on your chart.

S: (Write 420 on the place value chart.)

T: When we divide this whole number by 10, will the quotient be greater than or less than 420?

S: Less than 420.

T: Therefore, in which direction will the digits shift when we divide by 10?

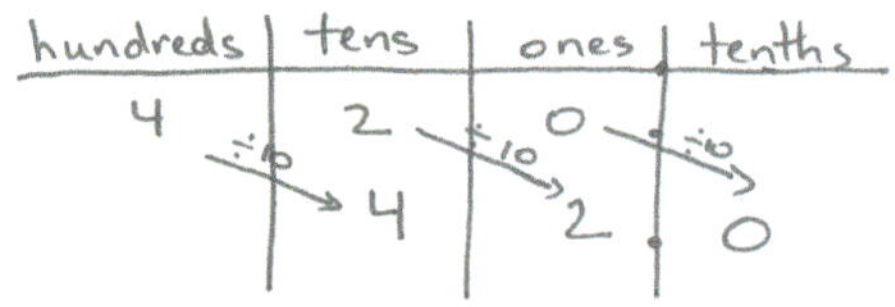


S: To the right.

T: How many places to the right?

S: One place to the right.

T: Use arrows to show the shifting of digits. Show your neighbor when you're finished, and then discuss whether this happens every time we divide a number by 10.

S: (Work and share.)

T: Say the division sentence, or the division equation, you just completed on your place value chart.

S: 420 ÷ 10 = 42.

**Problem 2: 1,600 ÷ 100**

T: (Write 1,600 ÷ 100 horizontally on the board.) Work with a partner to solve. Partner A will use place value disks to solve, and Partner B will use the place value chart to solve.

S: (Draw and solve.)

T: (Point to the board.) Say the division sentence with the solution.

1,600 ÷ 100
= 16 hundreds ÷ 1 hundred
= 16

S: 1,600 divided by 100 equals 16.

T: Let's try to solve this problem now using our knowledge of place value. Say 1,600 in unit form. How many hundreds in 1,600?

S: 16 hundreds.

T: (Write 16 hundreds beneath 1,600. Then, point to the original problem.) So we have 16 hundreds divided by what?

S: 1 hundred.

MP.2

T: (Write 1 hundred beneath 100.) Visualize what will happen to the digits in 1,600 when we divide by 100. Tell your neighbor what will happen.

S: The digits will all move two places to the right.

T: What math term could I say other than *division sentence?*

S: You could say *division equation*.

T: Read the complete division equation in unit form.

S: 16 hundreds divided by 1 hundred equals 16.

MP.7

T: Why did our unit change from hundreds to ones?

S: 1 hundred divided by 1 hundred is just 1. So, 16 hundreds divided by 1 hundred is 16 ones. → If you make as many groups of 100 as you can out of 1,600, you will be able to make 16 groups. → You could also think about putting 1,600 into 100 equal groups. If you do that, then there would be 16 in each group. → I know that it takes 16 copies of 1 hundred to make 16 hundreds, or 16 × 100 = 1,600.

**Problem 3: 24,000 ÷ 600**

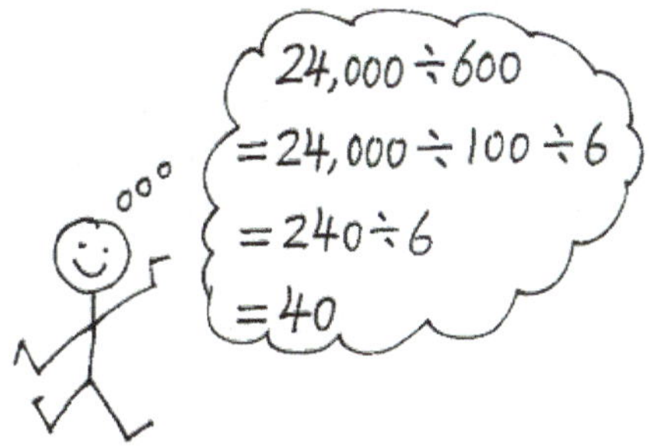


T: (Write 24,000 ÷ 600 horizontally on the board.) How is this problem different than the others we've solved? Turn and talk.

S: I know 24 divided by 6 equals 4. → We're still dividing with many zeros, but there are 6 hundreds rather than 1 hundred. → It looks different, but we can still just think of dividing by 600 as dividing by 6 hundreds.

T: Our divisor this time is 600. Can you decompose 600 with 100 as a factor?

S: Yes, 100 × 6 = 600.

T: So, let's rewrite this problem. (Write 24,000 ÷ 600 = 24,000 ÷ 100 ÷ 6.) Turn and tell your neighbor what 24,000 divided by 100 is. If necessary, you may use your place value chart or visualize what happens when dividing by 100.

T: What is 24,000 divided by 100?

S: 240.

T: Are we finished?

S: No, we still need to divide by 6.

T: Say the division sentence that we now have to solve.

S: 240 divided by 6.

T: Solve it on your personal white board.

T: Say the original division equation with the quotient.

S: 24,000 divided by 600 equals 40.

**NOTES ON MULTIPLE MEANS OF REPRESENTATION:**

There are two distinct interpretations for division. Although the quotients are the same, the approaches are different.

- Partitive Division: 15 apples were placed equally into 3 bags. How many apples were in each bag?
- Measurement Division: 15 apples were put in bags with 3 apples in each bag. How many bags were needed?

**Problem 4: 180,000 ÷ 9,000**

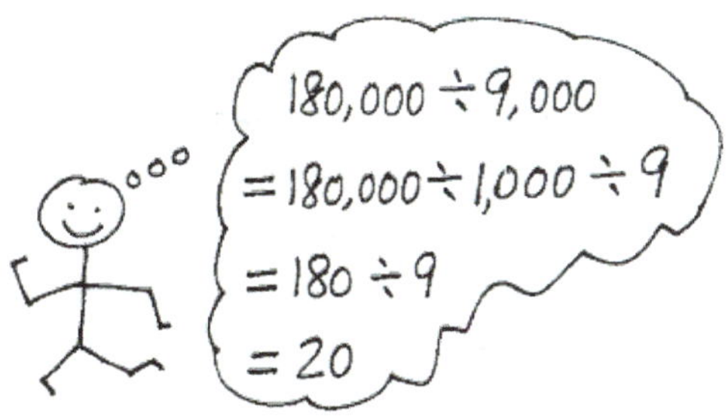


T: (Write 180,000 ÷ 9,000 horizontally on the board.) How can we rewrite this division problem so the 9,000 is decomposed with 1,000 as a factor? Turn and share.

T: Say the division problem you discussed.

S: 180,000 ÷ 1,000 ÷ 9.

T: Work with a partner to solve. If you want, you may use a place value chart to help.

T: Say the original division equation with the quotient.

S: 180,000 ÷ 9,000 = 20.

## Problem Set (10 minutes)

Students should do their personal best to complete the Problem Set within the allotted 10 minutes. For some classes, it may be appropriate to modify the assignment by specifying which problems they work on first. Some problems do not specify a method for solving. Students should solve these problems using the RDW approach used for Application Problems.

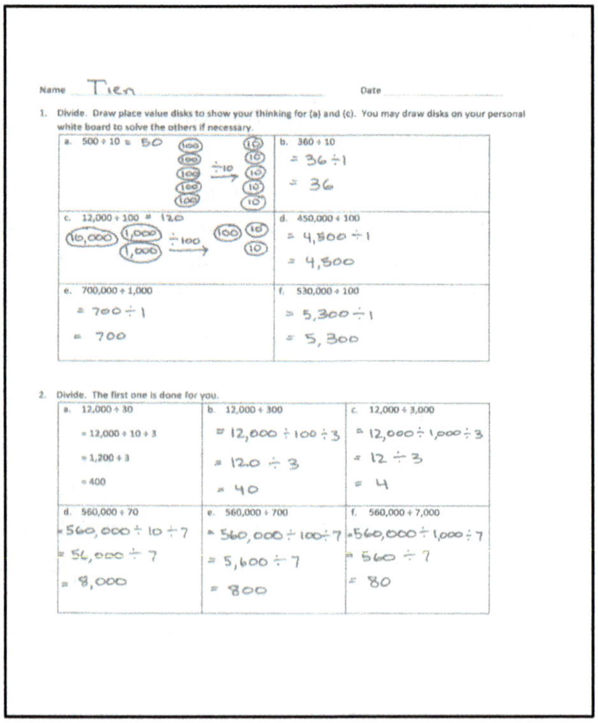

Name Tien Date

1. Divide. Draw place value disks to show your thinking for (a) and (c). You may draw disks on your personal white board to solve the others if necessary.

| a. $500 \div 10 = 50$ | b. $360 \div 10$<br>$= 36 \div 1$<br>$= 36$ |
|---|---|
| c. $12{,}000 \div 100 = 120$ | d. $450{,}000 \div 100$<br>$= 4{,}500 \div 1$<br>$= 4{,}500$ |
| e. $700{,}000 \div 1{,}000$<br>$= 700 \div 1$<br>$= 700$ | f. $530{,}000 \div 100$<br>$= 5{,}300 \div 1$<br>$= 5{,}300$ |

2. Divide. The first one is done for you.

| a. $12{,}000 \div 30$<br>$= 12{,}000 \div 10 \div 3$<br>$= 1{,}200 \div 3$<br>$= 400$ | b. $12{,}000 \div 300$<br>$= 12{,}000 \div 100 \div 3$<br>$= 120 \div 3$<br>$= 40$ | c. $12{,}000 \div 3{,}000$<br>$= 12{,}000 \div 1{,}000 \div 3$<br>$= 12 \div 3$<br>$= 4$ |
|---|---|---|
| d. $560{,}000 \div 70$<br>$= 560{,}000 \div 10 \div 7$<br>$= 56{,}000 \div 7$<br>$= 8{,}000$ | e. $560{,}000 \div 700$<br>$= 560{,}000 \div 100 \div 7$<br>$= 5{,}600 \div 7$<br>$= 800$ | f. $560{,}000 \div 7{,}000$<br>$= 560{,}000 \div 1{,}000 \div 7$<br>$= 560 \div 7$<br>$= 80$ |

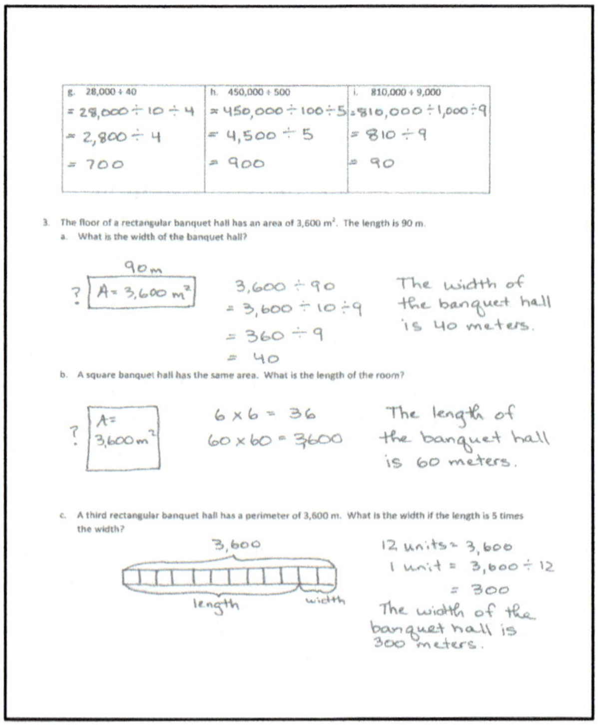

| g. $28{,}000 \div 40$<br>$= 28{,}000 \div 10 \div 4$<br>$= 2{,}800 \div 4$<br>$= 700$ | h. $450{,}000 \div 500$<br>$= 450{,}000 \div 100 \div 5$<br>$= 4{,}500 \div 5$<br>$= 900$ | i. $810{,}000 \div 9{,}000$<br>$= 810{,}000 \div 1{,}000 \div 9$<br>$= 810 \div 9$<br>$= 90$ |
|---|---|---|

3. The floor of a rectangular banquet hall has an area of 3,600 m². The length is 90 m.

a. What is the width of the banquet hall?

$3{,}600 \div 90 = 3{,}600 \div 10 \div 9 = 360 \div 9 = 40$

The width of the banquet hall is 40 meters.

b. A square banquet hall has the same area. What is the length of the room?

$6 \times 6 = 36$
$60 \times 60 = 3600$

The length of the banquet hall is 60 meters.

c. A third rectangular banquet hall has a perimeter of 3,600 m. What is the width if the length is 5 times the width?

12 units = 3,600
1 unit = 3,600 ÷ 12
= 300

The width of the banquet hall is 300 meters.

## Student Debrief (10 minutes)

**Lesson Objective:** Use *divide by 10* patterns for multi-digit whole number division.

The Student Debrief is intended to invite reflection and active processing of the total lesson experience.

Invite students to review their solutions for the Problem Set. They should check work by comparing answers with a partner before going over answers as a class. Look for misconceptions or misunderstandings that can be addressed in the Debrief. Guide students in a conversation to debrief the Problem Set and process the lesson.

Any combination of the questions below may be used to lead the discussion.

- Were place value disks helpful when solving the questions in Problem 1? Why or why not?
- Look back at your solutions to Problem 2(a–f). What pattern did you find? Can you explain the relationship between the quotients?
- How did your knowledge of basic facts help you as you solved the questions in Problem 2?
- Talk with your neighbor about your thought process as you solved Problem 3(b).
- Look back at Problem 4. What did you notice about the correct answer in Kim and Carter's problem and the quotient in Part (b)? Can you create a similar division problem that would yield the same quotient? What about a problem with a quotient that is 10 times greater? 100 times greater? 1 tenth as large?
- Use Problem 4 to generate a word problem where the quotient (500) represents the number of groups of 400 that can be made from 8,000. Then, generate a situation where the quotient (500) represents the size of each of 400 groups.

## Exit Ticket (3 minutes)

After the Student Debrief, instruct students to complete the Exit Ticket. A review of their work will help with assessing students' understanding of the concepts that were presented in today's lesson and planning more effectively for future lessons. The questions may be read aloud to the students.

4. Two fifth graders solved 400,000 divided by 800. Carter said the answer is 500, while Kim said the answer is 5000.
   a. Who has the correct answer? Explain your thinking.

$400{,}000 \div 800$
$= 400{,}000 \div 100 \div 8$
$= 4{,}000 \div 8$
$= 500$

Carter was Correct because as it showed on the last step, 4 thousands divided by 8 is equal to 5 hundreds, not 5 thousands.

   b. What if the problem is 4,000,000 divided by 8,000. What is the quotient?

$4{,}000{,}000 \div 8000$
$= 4{,}000{,}000 \div 1{,}000 \div 8$
$= 4{,}000 \div 8$
$= 500$

The quotient was 500.

# A

Number Correct: _______

Divide by Multiples of 10 and 100

| | | | | | |
|---|---|---|---|---|---|
| 1. | 30 ÷ 10 = | | 23. | 480 ÷ 4 = | |
| 2. | 430 ÷ 10 = | | 24. | 480 ÷ 40 = | |
| 3. | 4,300 ÷ 10 = | | 25. | 6,300 ÷ 3 = | |
| 4. | 4,300 ÷ 100 = | | 26. | 6,300 ÷ 30 = | |
| 5. | 43,000 ÷ 100 = | | 27. | 6,300 ÷ 300 = | |
| 6. | 50 ÷ 10 = | | 28. | 8,400 ÷ 2 = | |
| 7. | 850 ÷ 10 = | | 29. | 8,400 ÷ 20 = | |
| 8. | 8,500 ÷ 10 = | | 30. | 8,400 ÷ 200 = | |
| 9. | 8,500 ÷ 100 = | | 31. | 96,000 ÷ 3 = | |
| 10. | 85,000 ÷ 100 = | | 32. | 96,000 ÷ 300 = | |
| 11. | 600 ÷ 10 = | | 33. | 96,000 ÷ 30 = | |
| 12. | 60 ÷ 3 = | | 34. | 900 ÷ 30 = | |
| 13. | 600 ÷ 30 = | | 35. | 1,200 ÷ 30 = | |
| 14. | 4,000 ÷ 100 = | | 36. | 1,290 ÷ 30 = | |
| 15. | 40 ÷ 2 = | | 37. | 1,800 ÷ 300 = | |
| 16. | 4,000 ÷ 200 = | | 38. | 8,000 ÷ 200 = | |
| 17. | 240 ÷ 10 = | | 39. | 12,000 ÷ 200 = | |
| 18. | 24 ÷ 2 = | | 40. | 12,800 ÷ 200 = | |
| 19. | 240 ÷ 20 = | | 41. | 2,240 ÷ 70 = | |
| 20. | 3,600 ÷ 100 = | | 42. | 18,400 ÷ 800 = | |
| 21. | 36 ÷ 3 = | | 43. | 21,600 ÷ 90 = | |
| 22. | 3,600 ÷ 300 = | | 44. | 25,200 ÷ 600 = | |

# B

Number Correct: _______

Improvement: _______

Divide by Multiples of 10 and 100

| | | |
|---|---|---|
| 1. | 20 ÷ 10 = | |
| 2. | 420 ÷ 10 = | |
| 3. | 4,200 ÷ 10 = | |
| 4. | 4,200 ÷ 100 = | |
| 5. | 42,000 ÷ 100 = | |
| 6. | 40 ÷ 10 = | |
| 7. | 840 ÷ 10 = | |
| 8. | 8,400 ÷ 10 = | |
| 9. | 8,400 ÷ 100 = | |
| 10. | 84,000 ÷ 100 = | |
| 11. | 900 ÷ 10 = | |
| 12. | 90 ÷ 3 = | |
| 13. | 900 ÷ 30 = | |
| 14. | 6,000 ÷ 100 = | |
| 15. | 60 ÷ 2 = | |
| 16. | 6,000 ÷ 200 = | |
| 17. | 240 ÷ 10 = | |
| 18. | 24 ÷ 2 = | |
| 19. | 240 ÷ 20 = | |
| 20. | 6,300 ÷ 100 = | |
| 21. | 63 ÷ 3 = | |
| 22. | 6,300 ÷ 300 = | |

| | | |
|---|---|---|
| 23. | 840 ÷ 4 = | |
| 24. | 840 ÷ 40 = | |
| 25. | 3,600 ÷ 3 = | |
| 26. | 3,600 ÷ 30 = | |
| 27. | 3,600 ÷ 300 = | |
| 28. | 4,800 ÷ 2 = | |
| 29. | 4,800 ÷ 20 = | |
| 30. | 4,800 ÷ 200 = | |
| 31. | 69,000 ÷ 3 = | |
| 32. | 69,000 ÷ 300 = | |
| 33. | 69,000 ÷ 30 = | |
| 34. | 800 ÷ 40 = | |
| 35. | 1,200 ÷ 40 = | |
| 36. | 1,280 ÷ 40 = | |
| 37. | 1,600 ÷ 400 = | |
| 38. | 8,000 ÷ 200 = | |
| 39. | 14,000 ÷ 200 = | |
| 40. | 14,600 ÷ 200 = | |
| 41. | 2,560 ÷ 80 = | |
| 42. | 16,100 ÷ 700 = | |
| 43. | 14,400 ÷ 60 = | |
| 44. | 37,800 ÷ 900 = | |

Name ______________________________ Date ______________

1. Divide. Draw place value disks to show your thinking for (a) and (c). You may draw disks on your personal white board to solve the others if necessary.

| | |
|---|---|
| a. 500 ÷ 10 | b. 360 ÷ 10 |
| c. 12,000 ÷ 100 | d. 450,000 ÷ 100 |
| e. 700,000 ÷ 1,000 | f. 530,000 ÷ 100 |

2. Divide. The first one is done for you.

| | | |
|---|---|---|
| a. 12,000 ÷ 30<br>= 12,000 ÷ 10 ÷ 3<br>= 1,200 ÷ 3<br>= 400 | b. 12,000 ÷ 300 | c. 12,000 ÷ 3,000 |
| d. 560,000 ÷ 70 | e. 560,000 ÷ 700 | f. 560,000 ÷ 7,000 |
| g. 28,000 ÷ 40 | h. 450,000 ÷ 500 | i. 810,000 ÷ 9,000 |

3. The floor of a rectangular banquet hall has an area of 3,600 $m^2$. The length is 90 m.

   a. What is the width of the banquet hall?

   b. A square banquet hall has the same area. What is the length of the room?

   c. A third rectangular banquet hall has a perimeter of 3,600 m. What is the width if the length is 5 times the width?

4. Two fifth graders solved 400,000 divided by 800. Carter said the answer is 500, while Kim said the answer is 5,000.

   a. Who has the correct answer? Explain your thinking.

   b. What if the problem is 4,000,000 divided by 8,000? What is the quotient?

Name ______________________________ Date ______________

Divide. Show your thinking.

| a. 17,000 ÷ 100 | b. 59,000 ÷ 1,000 |
|---|---|
| c. 12,000 ÷ 40 | d. 480,000 ÷ 600 |

Name ______________________________ Date ______________

1. Divide. Draw place value disks to show your thinking for (a) and (c). You may draw disks on your personal white board to solve the others if necessary.

| | |
|---|---|
| a. 300 ÷ 10 | b. 450 ÷ 10 |
| c. 18,000 ÷ 100 | d. 730,000 ÷ 100 |
| e. 900,000 ÷ 1,000 | f. 680,000 ÷ 1,000 |

2. Divide. The first one is done for you.

| | | |
|---|---|---|
| a. 18,000 ÷ 20<br>= 18,000 ÷ 10 ÷ 2<br>= 1,800 ÷ 2<br>= 900 | b. 18,000 ÷ 200 | c. 18,000 ÷ 2,000 |
| d. 420,000 ÷ 60 | e. 420,000 ÷ 600 | f. 420,000 ÷ 6,000 |
| g. 24,000 ÷ 30 | h. 560,000 ÷ 700 | i. 450,000 ÷ 9,000 |

3. A stadium holds 50,000 people. The stadium is divided into 250 different seating sections. How many seats are in each section?

4. Over the course of a year, a tractor trailer commutes 160,000 miles across America.

   a. Assuming a trucker changes his tires every 40,000 miles, and that he starts with a brand new set of tires, how many sets of tires will he use in a year?

   b. If the trucker changes the oil every 10,000 miles, and he starts the year with a fresh oil change, how many times will he change the oil in a year?

# Lesson 17

## Objective: Use basic facts to approximate quotients with two-digit divisors.

### Suggested Lesson Structure

| | |
|---|---|
| Fluency Practice | (12 minutes) |
| Application Problem | (5 minutes) |
| Concept Development | (33 minutes) |
| Student Debrief | (10 minutes) |
| **Total Time** | **(60 minutes)** |

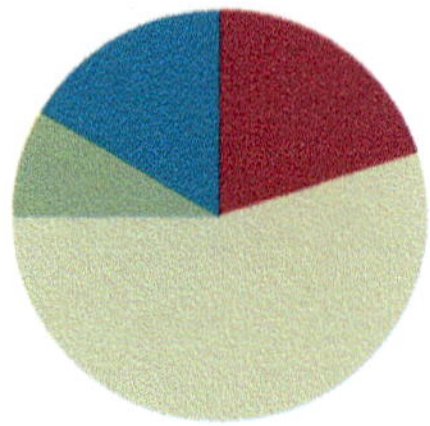

## Fluency Practice (12 minutes)

- Group Count by Multiples of 10 **5.NBT.2** (5 minutes)
- Round to the Nearest Ten **5.NBT.4** (2 minutes)
- Divide by Multiples of 10, 100, and 1,000 **5.NBT.6** (5 minutes)

### Group Count by Multiples of 10 (5 minutes)

Note: Counting by multiples of 10 helps students estimate quotients with two-digit divisors, which students practice during this lesson's Concept Development.

Repeat the process in Lesson 16 for 4 tens, 5 tens, and 7 tens.

### Round to the Nearest Ten (2 minutes)

Note: Rounding to whole numbers and one decimal place prepares students to estimate quotients.

Repeat the process in Lesson 16 for the following possible sequence: 21, 37, 16, 54, 73, 65, 25.

### Divide by Multiples of 10, 100, and 1,000 (5 minutes)

Materials: (S) Personal white board

Note: This fluency activity reviews content from Lesson 16.

T: (Write 700 ÷ 10.) Say the division sentence.

S: 700 ÷ 10 = 70.

T: (Write 800 ÷ 20.) Write 800 ÷ 20 as a two-step division sentence taking out the ten.

S: (Write = 800 ÷ 10 ÷ 2.)

T: Below the two-step division sentence, rewrite it in one step after solving the first division problem.

S: (Write = 80 ÷ 2.)

T: Write the answer below 80 ÷ 2.

S: (Write = 40.)

Repeat the process for the following possible sequence: 15,000 ÷ 30, 15,000 ÷ 300, 15,000 ÷ 3,000, 450,000 ÷ 50, and 21,000 ÷ 300.

## Application Problem (5 minutes)

852 pounds of grapes were packed equally into 3 boxes for shipping. How many pounds of grapes were there in 2 boxes?

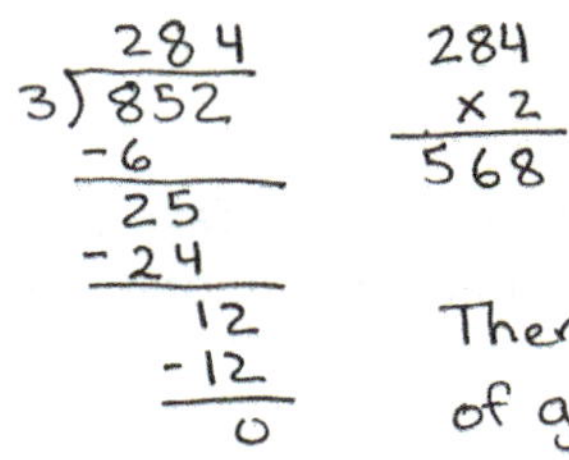


There were 568 pounds of grapes in two boxes.

Note: The focus of this Application Problem is division with a one-digit divisor. This review encompasses both the meaning of and skill with division, which helps students as they learn to work with two-digit divisors.

## Concept Development (33 minutes)

Materials: (S) Personal white board

Note: The word *whole* is used throughout the module to indicate the dividend. The choice of this term is two-fold. First, *whole* provides a natural scaffold for the fraction work that is to come in Modules 3 and 4. Second, the words *dividend* and *divisor* are easily confused. While the word *dividend* can certainly be included as well, students may find *whole* to be a more meaningful term.

**NOTES ON MULTIPLE MEANS OF ENGAGEMENT:**

Allow students to continue to use place value disks or charts to represent the division by multiples of 10 if this scaffold is necessary.

Additionally, students may need to continue to record the division sentences in two steps similar to the fluency activity. For example, Problem 1's estimate could be written as 400 ÷ 20 = 400 ÷ 10 ÷ 2 = 40 ÷ 2 = 20.

**Problem 1: 402 ÷ 19**

T: (Write 402 ÷ 19 on the board.) What's the whole in this problem?

S: 402.

T: What's the divisor?

S: 19.

T: Let's round the divisor first. What is 19 rounded to the nearest ten?

S: 20.

T: Let's record our estimation. (Under the original problem, write ≈ ____ ÷ 20 on the board.) We need to round our whole, 402, to a number that can easily be divided by 20. Turn and share your ideas with your partner. (Allow time for students to share.)

T: How can we round 402?

S: I can round 402 to 400. → I can use mental math to divide 400 by 20.

T: (Fill in the blank to get ≈ 400 ÷ 20.) What is 400 divided by 20? Turn and share.

S: 400 divided by 10 is 40. 40 divided by 2 is 20. → 400 divided by 2 equals 200. 200 divided by 10 equals 20.

T: Yes. We know that 400 divided by 20 is equal to 40 divided by 2. (Write 40 ÷ 2 below 400 ÷ 20.) What is our estimated quotient?

S: 20.

**Problem 2: 149 ÷ 71**

T: (Write 149 ÷ 71 on the board.) Take out your personal white board. To estimate the quotient, what do we do first?

S: Round the divisor.

T: Do it now.

S: (Round to 70.)

T: (Write ≈ ____ ÷ 70 on the board.) Now, round the whole, 149. Usually when we round, we round to place value units. But, since we're dividing, it'd be really nice if our whole was a multiple of the divisor. So, let's round it to the nearest multiple of 70. Count with me by seventies. (Write the multiples as students count out loud.)

S: 70, 140, 210, 280.

T: Our whole is between which of these?

S: 140 and 210.

T: We are rounding to the *nearest* multiple of 70. 149 is closest to which multiple of 70?

S: 140.

**NOTES ON MULTIPLE MEANS OF ACTION AND EXPRESSION:**

Having students write the multiples in a vertical or horizontal list as they count can provide helpful support.

T: (Fill in ≈ 140 ÷ 70.) 140 divided by 70 is the same as 14 divided by what? Say the division sentence.

S: 14 ÷ 7 = 2.

T: How do you know?

S: 140 ÷ 70 = 140 ÷ 10 ÷ 7. → Dividing by 70 is the same as dividing by 10 and then dividing by 7. → If I put parentheses like this (140 ÷ 10) ÷ 7, it's easy to see the two expressions are equal.

T: So, what's 140 ÷ 70?

S: 2.

T: (Record = 14 ÷ 7 = 2.) Good. Our estimated quotient is 2.

**Problem 3: 427 ÷ 58**

T: (Write 427 ÷ 58 on the board.) Work with a partner to round the divisor and the whole. Then, estimate the quotient.

T: What's the estimated divisor?

S: 60.

T: So, 60 will be our unit. Let's count by sixties and stop when we find a multiple near 427.

S: 60, 120, 180, 240, 300, 360, 420, ... stop!

T: To what number should we round the whole?

S: 420.

T: What's the next multiple of 60?

S: 480.

T: Is 427 closer to 420 or 480?

S: 420.

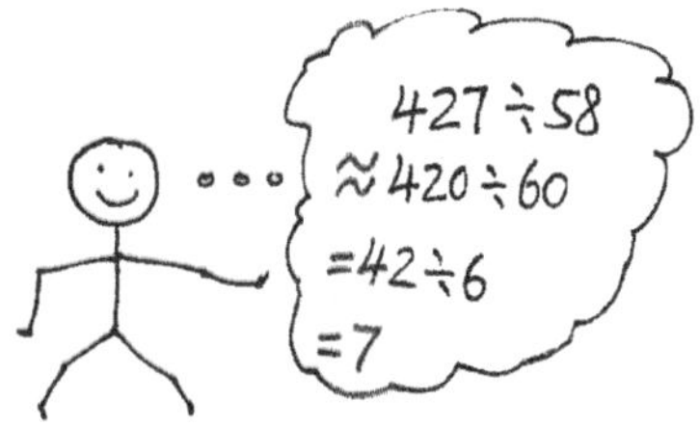


T: Then, let's use 420 as our estimated whole. 420 divided by 60 is the same as what division equation?

S: 42 ÷ 6 = 7.

T: Share with your partner how you know.

S: 420 ÷ 60 = 420 ÷ 10 ÷ 6. → It's 42 tens divided by 6 tens. It's like 10 divided by 10 is 1, so you are just left with 42 divided by 6. → If I write 42 × 10 ÷ 10 ÷ 6, it's easy to see that multiplying by 10 and dividing by 10 equals 1, so we are left with 42 divided by 6. → 420 ÷ 60 = 42 ÷ 6.

T: What is our estimated quotient? What's 420 divided by 60?

S: 7.

T: Yes. The estimated quotient is 7.

**Problem 4: 293 ÷ 42**

T: (Write 293 ÷ 42 on the board.) Round the divisor.

S: 40.

T: (Write ≈ _____ ÷ 40 on the board.) So, 40 is our unit. Round the whole to a multiple of 40. Whisper the multiples of 40 to your neighbor. Stop when you hear a multiple that is near our whole.

S: 40, 80, 120, 160, 200, 240, 280, 320.

T: I see you went past 293. Our total is between which two multiples of 40?

S: 280 and 320.

T: Visualize a number line. Which multiple is closer to 293?

S: 280.

T: Finish the division problem on your board. Compare your work with a neighbor.

T: Tell me how to estimate the quotient.

S: 280 ÷ 40 = 28 ÷ 4 = 7.

**Problem 5: 751 ÷ 93**

T: (Write 751 ÷ 93 on the board.) Work independently to round the divisor and the whole. Then, estimate the quotient.

T: Share your work with a neighbor.

T: What was your estimated divisor?

S: 90.

T: And the estimated whole?

S: 720.

T: Tell me how to estimate the quotient.

S: 720 ÷ 90 = 72 ÷ 9 = 8.

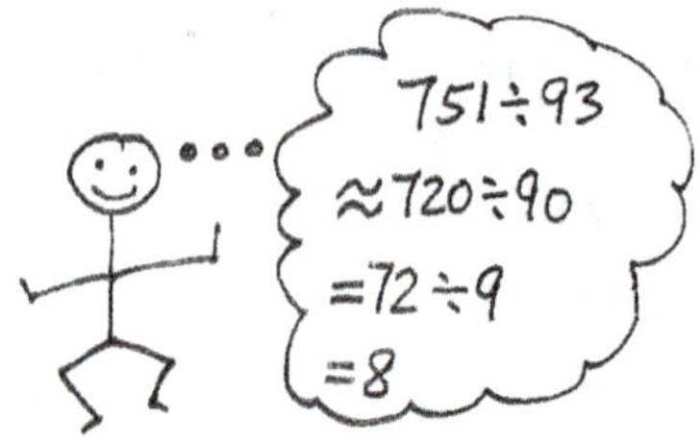


## Problem Set (10 minutes)

Students should do their personal best to complete the Problem Set within the allotted 10 minutes. For some classes, it may be appropriate to modify the assignment by specifying which problems they work on first. Some problems do not specify a method for solving. Students should solve these problems using the RDW approach used for Application Problems.

## Student Debrief (10 minutes)

**Lesson Objective:** Use basic facts to approximate quotients with two-digit divisors.

The Student Debrief is intended to invite reflection and active processing of the total lesson experience.

Invite students to review their solutions for the Problem Set. They should check work by comparing answers with a partner before going over answers as a class. Look for misconceptions or misunderstandings that can be addressed in the Debrief. Guide students in a conversation to debrief the Problem Set and process the lesson.

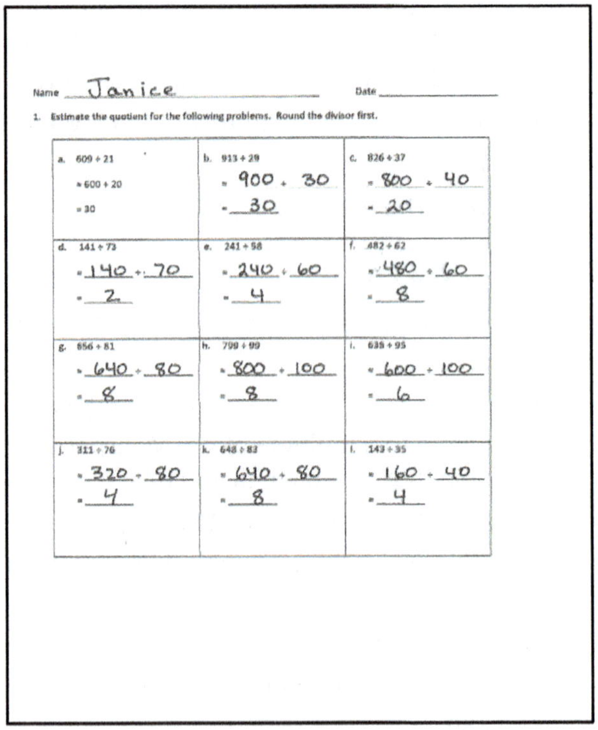

Name Janice Date

1. Estimate the quotient for the following problems. Round the divisor first.

| | | |
|---|---|---|
| a. 609 ÷ 21<br>≈ 600 ÷ 20<br>= 30 | b. 913 ÷ 29<br>≈ 900 ÷ 30<br>= 30 | c. 826 ÷ 37<br>≈ 800 ÷ 40<br>= 20 |
| d. 141 ÷ 73<br>≈ 140 ÷ 70<br>= 2 | e. 241 ÷ 58<br>≈ 240 ÷ 60<br>= 4 | f. 482 ÷ 62<br>≈ 480 ÷ 60<br>= 8 |
| g. 656 ÷ 81<br>≈ 640 ÷ 80<br>= 8 | h. 799 ÷ 99<br>≈ 800 ÷ 100<br>= 8 | i. 635 ÷ 95<br>≈ 600 ÷ 100<br>= 6 |
| j. 311 ÷ 76<br>≈ 320 ÷ 80<br>= 4 | k. 648 ÷ 83<br>≈ 640 ÷ 80<br>= 8 | l. 143 ÷ 35<br>≈ 160 ÷ 40<br>= 4 |

Any combination of the questions below may be used to lead the discussion.

- Look back at the divisors in Problem 1 (l), (m), and (n). What did you notice about them? How did the 5 in the ones place affect the way you rounded?
- In Problem 1(o), did anyone leave the divisor 11 unrounded? Is it always necessary to round? (There are several correct estimations including 660 ÷ 10, 660 ÷ 11, or 600 ÷ 10, as shown in student work.)

- Do we follow our typical rounding rules when estimating with division? Why not? (We do not always follow our typical rules of rounding to certain place value units because we are looking for easy multiples of our divisor. Sometimes that means we choose a number that is farther away from our actual whole than rounding by place value would produce.)
- How can you use place value disks or two-step division sentences to justify your answer in Problem 3? (This problem provides an opportunity for students to discuss division by multiples of 10.)

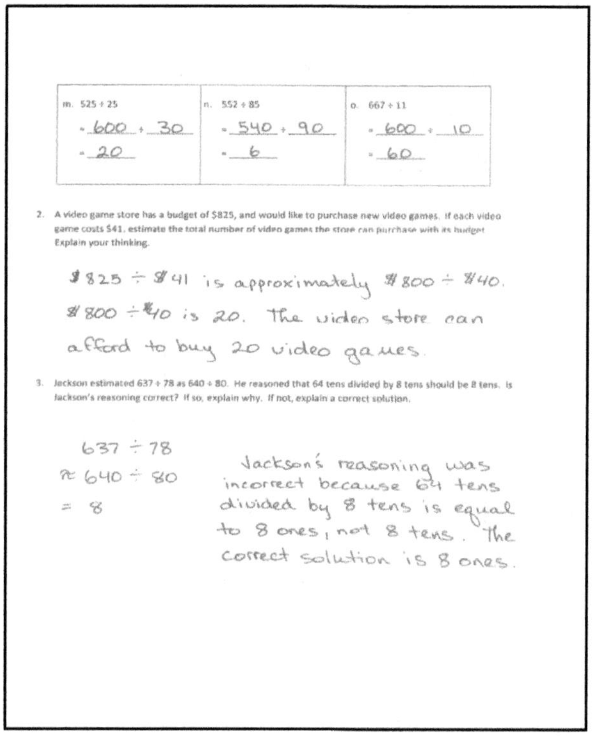

| m. 525 ÷ 25 | n. 552 ÷ 85 | o. 667 ÷ 11 |
|---|---|---|
| ≈ 600 ÷ 30 = 20 | ≈ 540 ÷ 90 = 6 | ≈ 600 ÷ 10 = 60 |

2. A video game store has a budget of $825, and would like to purchase new video games. If each video game costs $41, estimate the total number of video games the store can purchase with its budget. Explain your thinking.

$825 ÷ $41 is approximately $800 ÷ $40. $800 ÷ $40 is 20. The video store can afford to buy 20 video games.

3. Jackson estimated 637 ÷ 78 as 640 ÷ 80. He reasoned that 64 tens divided by 8 tens should be 8 tens. Is Jackson's reasoning correct? If so, explain why. If not, explain a correct solution.

637 ÷ 78
≈ 640 ÷ 80
= 8

Jackson's reasoning was incorrect because 64 tens divided by 8 tens is equal to 8 ones, not 8 tens. The correct solution is 8 ones.

### Exit Ticket (3 minutes)

After the Student Debrief, instruct students to complete the Exit Ticket. A review of their work will help with assessing students' understanding of the concepts that were presented in today's lesson and planning more effectively for future lessons. The questions may be read aloud to the students.

Name ______________________________ Date ______________

1. Estimate the quotient for the following problems. Round the divisor first.

| | | |
|---|---|---|
| a. 609 ÷ 21<br>≈ 600 ÷ 20<br>= 30 | b. 913 ÷ 29<br>≈ ______ ÷ ______<br>= ______ | c. 826 ÷ 37<br>≈ ______ ÷ ______<br>= ______ |
| d. 141 ÷ 73<br>≈ ______ ÷ ______<br>= ______ | e. 241 ÷ 58<br>≈ ______ ÷ ______<br>= ______ | f. 482 ÷ 62<br>≈ ______ ÷ ______<br>= ______ |
| g. 656 ÷ 81<br>≈ ______ ÷ ______<br>= ______ | h. 799 ÷ 99<br>≈ ______ ÷ ______<br>= ______ | i. 635 ÷ 95<br>≈ ______ ÷ ______<br>= ______ |
| j. 311 ÷ 76<br>≈ ______ ÷ ______<br>= ______ | k. 648 ÷ 83<br>≈ ______ ÷ ______<br>= ______ | l. 143 ÷ 35<br>≈ ______ ÷ ______<br>= ______ |
| m. 525 ÷ 25<br>≈ ______ ÷ ______<br>= ______ | n. 552 ÷ 85<br>≈ ______ ÷ ______<br>= ______ | o. 667 ÷ 11<br>≈ ______ ÷ ______<br>= ______ |

2. A video game store has a budget of $825, and would like to purchase new video games. If each video game costs $41, estimate the total number of video games the store can purchase with its budget. Explain your thinking.

3. Jackson estimated 637 ÷ 78 as 640 ÷ 80. He reasoned that 64 tens divided by 8 tens should be 8 tens. Is Jackson's reasoning correct? If so, explain why. If not, explain a correct solution.

Name ______________________________ Date ______________

Estimate the quotient for the following problems.

| | |
|---|---|
| a. $608 \div 23$<br><br>$\approx$ ________ $\div$ ________<br><br>$=$ ________ | b. $913 \div 31$<br><br>$\approx$ ________ $\div$ ________<br><br>$=$ ________ |
| c. $151 \div 39$<br><br>$\approx$ ________ $\div$ ________<br><br>$=$ ________ | d. $481 \div 68$<br><br>$\approx$ ________ $\div$ ________<br><br>$=$ ________ |

Name ______________________________ Date ______________

1. Estimate the quotient for the following problems. The first one is done for you.

| | | |
|---|---|---|
| a. 821 ÷ 41<br>≈ 800 ÷ 40<br>= 20 | b. 617 ÷ 23<br>≈ ______ ÷ ______<br>= ______ | c. 821 ÷ 39<br>≈ ______ ÷ ______<br>= ______ |
| d. 482 ÷ 52<br>≈ ______ ÷ ______<br>= ______ | e. 531 ÷ 48<br>≈ ______ ÷ ______<br>= ______ | f. 141 ÷ 73<br>≈ ______ ÷ ______<br>= ______ |
| g. 476 ÷ 81<br>≈ ______ ÷ ______<br>= ______ | h. 645 ÷ 69<br>≈ ______ ÷ ______<br>= ______ | i. 599 ÷ 99<br>≈ ______ ÷ ______<br>= ______ |
| j. 301 ÷ 26<br>≈ ______ ÷ ______<br>= ______ | k. 729 ÷ 81<br>≈ ______ ÷ ______<br>= ______ | l. 636 ÷ 25<br>≈ ______ ÷ ______<br>= ______ |
| m. 835 ÷ 89<br>≈ ______ ÷ ______<br>= ______ | n. 345 ÷ 72<br>≈ ______ ÷ ______<br>= ______ | o. 559 ÷ 11<br>≈ ______ ÷ ______<br>= ______ |

2. Mrs. Johnson spent $611 buying lunch for 78 students. If all the lunches cost the same, about how much did she spend on each lunch?

3. An oil well produces 172 gallons of oil every day. A standard oil barrel holds 42 gallons of oil. About how many barrels of oil will the well produce in one day? Explain your thinking.

# Lesson 18

## Objective: Use basic facts to approximate quotients with two-digit divisors.

### Suggested Lesson Structure

| | |
|---|---|
| Fluency Practice | (12 minutes) |
| Application Problem | (6 minutes) |
| Concept Development | (32 minutes) |
| Student Debrief | (10 minutes) |
| **Total Time** | **(60 minutes)** |

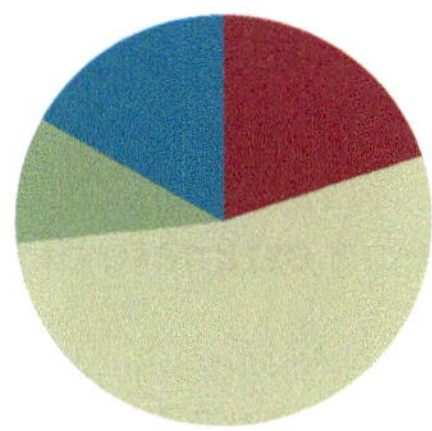

## Fluency Practice (12 minutes)

- Group Count by Multiples of 10 **5.NBT.2** (4 minutes)
- Divide by Multiples of 10, 100, and 1,000 **5.NBT.6** (4 minutes)
- Estimate and Divide **5.NBT.2** (4 minutes)

### Group Count by Multiples of 10 (4 minutes)

Note: Counting by multiples of 10 helps students estimate quotients with two-digit divisors, which students do during this lesson's Concept Development.

Repeat the process in Lesson 16 for 6 tens, 8 tens, and 9 tens.

### Divide by Multiples of 10, 100, and 1,000 (4 minutes)

Materials: (S) Personal white board

Note: This fluency activity reviews content from Lesson 16 and helps estimate quotients.

Repeat the process in Lesson 17 for the following possible sequence: 12,000 ÷ 40, 12,000 ÷ 400, 12,000 ÷ 4,000, 360,000 ÷ 6,000, 490,000 ÷ 700, and 640,000 ÷ 80.

### Estimate and Divide (4 minutes)

Materials: (S) Personal white board

Note: This fluency activity reviews content from Lesson 17.

T: (Write 812 ÷ 39.) Say the divisor rounded to the nearest ten.

S: 40.

T: Name the multiple of 40 that's closest to 812.

S: 800.

T: On your personal white board, write a division problem that will estimate the value.

S: (Write ≈ 800 ÷ 40.)

T: Below 800 ÷ 40, write the answer.

S: (Write = 20.)

Repeat the process for the following possible sequence: 183 ÷ 31, 437 ÷ 72, 823 ÷ 97, and 8,191 ÷ 92.

## Application Problem (6 minutes)

Sandra bought 38 DVD movies for $874. Give an estimate of the cost of each DVD movie.

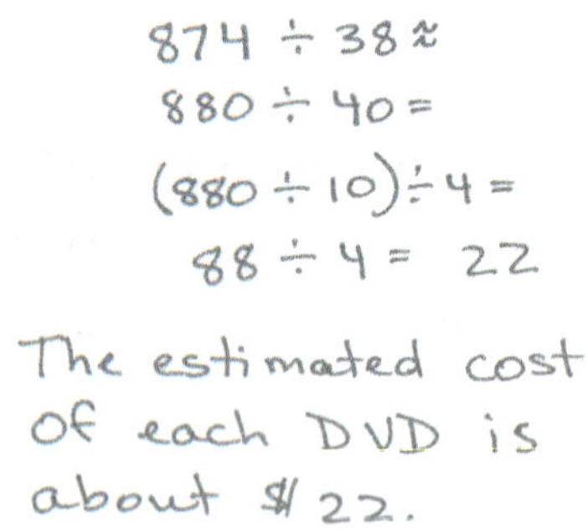


Note: This Application Problem is a review of Lesson 17, which focused on estimation of a three-digit whole by a two-digit divisor. In this lesson, students estimate the division of a four-digit number by a two-digit divisor.

## Concept Development (32 minutes)

Materials: (S) Personal white board

**Problem 1: 8,095 ÷ 23**

T: (Write 8,095 ÷ 23 horizontally on the board.) Which number should we round first, the whole or divisor? Why?

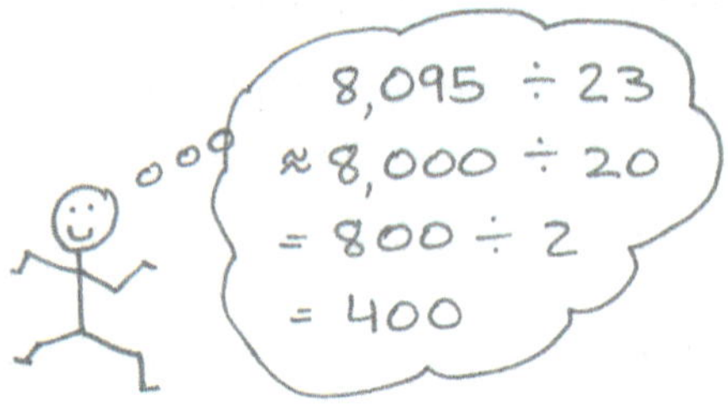


S: We round the divisor first so we know what our unit is. → Round the divisor first. This helps us know what multiples to look for when we are rounding the whole.

T: Good. What's 23 rounded to the nearest ten?

S: 20.

T: Let's round 8,095 so it is easy to divide by 20. How would we do that? Turn and share with your partner.

S: I see that 8 can easily be divided by 2, and 8,000 is easy to divide by 2. So, I can round 8,095 to 8,000. → 8 divided by 2 is an easy fact, so I can round 8,095 to 8,000. I can solve 8,000 divided by 2 easily.

T: Yes. I see an easy fact of 8 divided by 2, so let's round 8,095 to 8,000. Tell me the expression to estimate the quotient.

S: 8,000 divided by 20.

T: (Write ≈ 8,000 ÷ 20.) How can we solve 8,000 divided by 20? Turn and discuss with your partner.

S: I know that 8,000 divided by 10 equals 800, and 800 divided by 2 equals 400. → 8,000 divided by 2 equals 4,000. 4,000 divided by 10 equals 400. → 8,000 divided by 20 is the same as 800 divided by 2, and the answer is 400.

T: 8,000 divided by 20 is the same as 800 divided by what? Say the expression.

S: 800 ÷ 2.

T: (Write = 800 ÷ 2.) Excellent. So what's the answer?

S: 400.

T: Say the two division sentences. (Point at the two sentences on the board as students read aloud.)

S: 8,000 ÷ 20 = 800 ÷ 2 = 400.

**Problem 2: 2,691 ÷ 48**

T: (Write 2,691 ÷ 48 horizontally on the board.) Take out your personal white board. Let's first round the divisor 48 to the nearest ten.

T: (Write ≈ ____ ÷ 50 on the board.) Let's round 2,691 so it is easy to divide by 50. Remember! You can think of multiples of 5 to help you! Turn and talk.

S: I see an easy fact of 5 times 5 is equal to 25. → 26 hundreds is close to 30 hundreds, and that's easy to divide by 5. I can round 2,691 to 3,000. → 3,000 is an easy multiple of 50. → 26 hundreds is close to 25 hundreds. That's easy to divide by 5 or 50. → I can round 2,691 to 2,500 because 2,500 divided by 50 is 50.

T: I heard both 2,500 and 3,000.

T: Tell your partner how to estimate the quotient using 3,000 as the rounded whole.

S: 3,000 ÷ 50 = 300 ÷ 5 = 60. → It's ten times more than 30 ÷ 5. → I divided 3,000 by 10 and then divided by 5.

T: What is 3,000 divided by 50?

S: 60.

T: I heard another rounded whole, 2,500. How would we estimate the quotient using the rounded whole of 2,500?

S: 2,500 ÷ 50 = 250 ÷ 5 = 50.

T: We have two estimated quotients, 60 and 50.

T: Our estimates are different. What does that mean?

S: Well, they're just estimates. Our actual answer would be somewhere around 50 or 60.

MP.2

T: Right, one is probably closer to the actual answer than the other. Since they're both pretty close, we could use either one if we only want to estimate our answer.

**NOTES ON MULTIPLE MEANS OF REPRESENTATION:**

Unit form is a powerful means of representing these 4-digit dividends so that students can more easily *see* the multiples of the divisor. A model exchange might be:

T: (Rewrite 2,691 as 26 hundreds + 91 ones.) Is 26 hundreds a multiple of 5? Can you think of a multiple of 5 close to 26 hundreds?

S: 25 hundreds and 30 hundreds.

T: Are these also multiples of 50?

S: Yes!

T: Let's write them in standard form.

S: 2,500 and 3,000.

**Problem 3: 5,484 ÷ 71**

T: (Write 5,484 ÷ 71 horizontally on the board.) Work independently to estimate the quotient.

T: What's the estimated divisor?

S: 70.

T: What's the estimated whole that's also an easy multiple of 70?

S: 5,600.

T: Say the equation to find the rounded quotient.

S: 5,600 ÷ 70 = 560 ÷ 7 = 80.

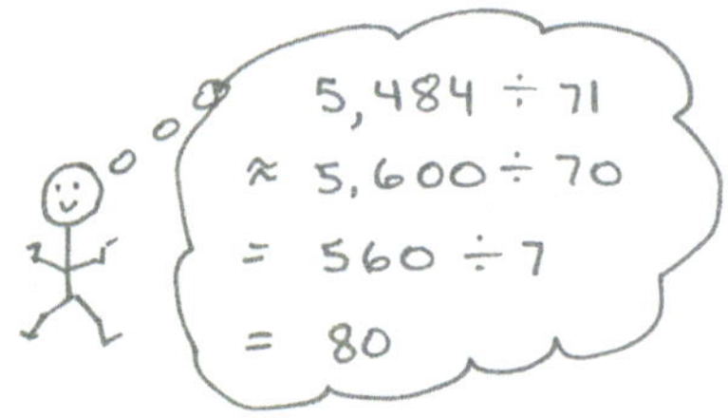


**Problem 4: 9,215 ÷ 95**

T: (Write 9,215 ÷ 95 horizontally on the board.) Let's estimate. The divisor is 95. Should I round up or round down?

S: If we're rounding to the nearest ten or hundred, we'd round up to 100. But, we're dividing, and we don't always round according to place value when we do that. 95 is halfway between 90 and 100, so maybe either 90 or 100 would work.

T: Very good. I want you to work with a partner. Partner A will round 95 down to 90, and solve. Partner B will round 95 up to 100 and solve. When you're finished, compare your answer with your partner's answer.

T: Partner A, how would you find the rounded quotient?

S: 9,000 ÷ 90 = 900 ÷ 9 = 100.

T: Partner B, how would you find the rounded quotient?

S: 9,000 ÷ 100 = 90 ÷ 1 = 90.

T: Let's consider another possibility. Let's not round our divisor at all. Can you find a multiple of 95 that's close to our whole? Turn and talk.

S: 9,500 would work! → We could just think about 9,215 as 9,500, and then divide.

T: So, what is 9,500 ÷ 95?

S: 100.

T: Will our actual quotient be greater than or less than 100? Explain your thinking.

S: The actual quotient will be less than 100. For the quotient to be 100, our whole would need to be at least 9,500, and the whole is less than that.

**NOTES ON MULTIPLE MEANS OF ENGAGEMENT:**

Students should reason about how the estimation of the divisors and dividends affect the quotients. For example, if both the dividend and the divisor are rounded down, the estimated quotient will be less than the actual quotient. Whether the actual quotient is greater than or less than the estimated quotient can be harder to predict when the divisor is rounded up and the dividend is rounded down, or vice versa. How much each number (dividend or divisor) was rounded also affects whether the estimated quotient is greater than or less than the actual quotient. After a problem is completed, ask students to compare the estimated quotients to the actual quotients and reason about the differences.

## Problem Set (10 minutes)

Students should do their personal best to complete the Problem Set within the allotted 10 minutes. For some classes, it may be appropriate to modify the assignment by specifying which problems they work on first. Some problems do not specify a method for solving. Students should solve these problems using the RDW approach used for Application Problems.

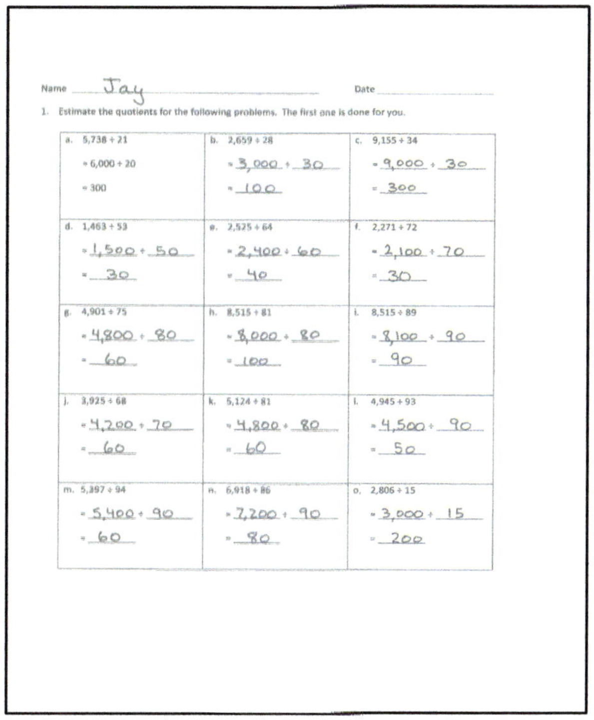

Name Jay Date

1. Estimate the quotients for the following problems. The first one is done for you.

| | | |
|---|---|---|
| a. 5,738 ÷ 21<br>≈ 6,000 ÷ 20<br>= 300 | b. 2,659 ÷ 28<br>≈ 3,000 ÷ 30<br>= 100 | c. 9,155 ÷ 34<br>≈ 9,000 ÷ 30<br>= 300 |
| d. 1,463 ÷ 53<br>≈ 1,500 ÷ 50<br>= 30 | e. 2,525 ÷ 64<br>≈ 2,400 ÷ 60<br>= 40 | f. 2,271 ÷ 72<br>≈ 2,100 ÷ 70<br>= 30 |
| g. 4,901 ÷ 75<br>≈ 4,800 ÷ 80<br>= 60 | h. 8,515 ÷ 81<br>≈ 8,000 ÷ 80<br>= 100 | i. 8,515 ÷ 89<br>≈ 8,100 ÷ 90<br>= 90 |
| j. 3,925 ÷ 68<br>≈ 4,200 ÷ 70<br>= 60 | k. 5,124 ÷ 81<br>≈ 4,800 ÷ 80<br>= 60 | l. 4,945 ÷ 93<br>≈ 4,500 ÷ 90<br>= 50 |
| m. 5,397 ÷ 94<br>≈ 5,400 ÷ 90<br>= 60 | n. 6,918 ÷ 86<br>≈ 7,200 ÷ 90<br>= 80 | o. 2,806 ÷ 15<br>≈ 3,000 ÷ 15<br>= 200 |

## Student Debrief (10 minutes)

**Lesson Objective:** Use basic facts to approximate quotients with two-digit divisors.

The Student Debrief is intended to invite reflection and active processing of the total lesson experience.

Invite students to review their solutions for the Problem Set. They should check work by comparing answers with a partner before going over answers as a class. Look for misconceptions or misunderstandings that can be addressed in the Debrief. Guide students in a conversation to debrief the Problem Set and process the lesson.

Any combination of the questions below may be used to lead the discussion.

- When estimating Problem 1(g), what did you choose for your unit, 70 or 80? Why? How did it affect the way you estimated the whole?
- Look back at Problems 2 and 4. How was your approach similar in solving? How was it different?
- When solving Problem 3(b), could a mental math approach be used to solve for the exact product? Explain. (14 apps × \$2 – 14¢)
- Talk in groups about how you rounded the divisor in Problem 5. Why did you choose the unit you did? Could a quotient sometimes be estimated without rounding the divisor? How?

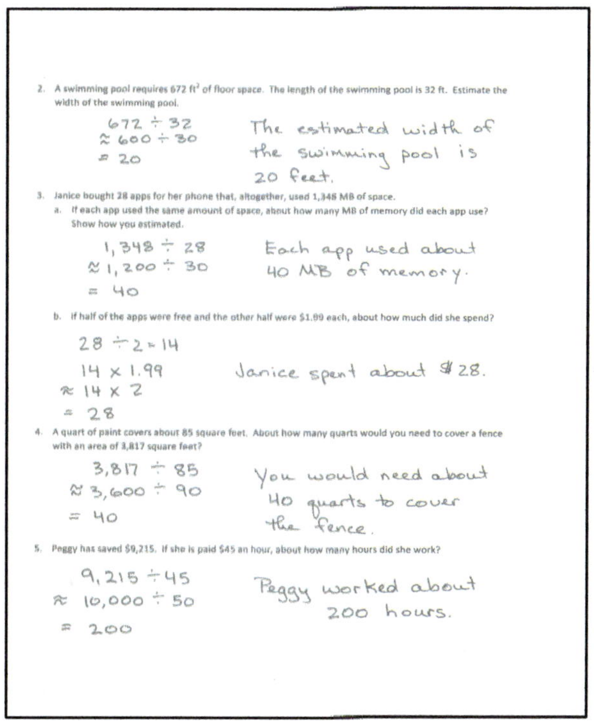

2. A swimming pool requires 672 $ft^2$ of floor space. The length of the swimming pool is 32 ft. Estimate the width of the swimming pool.

672 ÷ 32
≈ 600 ÷ 30
= 20

The estimated width of the swimming pool is 20 feet.

3. Janice bought 28 apps for her phone that, altogether, used 1,348 MB of space.
   a. If each app used the same amount of space, about how many MB of memory did each app use? Show how you estimated.

1,348 ÷ 28
≈ 1,200 ÷ 30
= 40

Each app used about 40 MB of memory.

   b. If half of the apps were free and the other half were \$1.99 each, about how much did she spend?

28 ÷ 2 = 14
14 × 1.99
≈ 14 × 2
= 28

Janice spent about \$28.

4. A quart of paint covers about 85 square feet. About how many quarts would you need to cover a fence with an area of 3,817 square feet?

3,817 ÷ 85
≈ 3,600 ÷ 90
= 40

You would need about 40 quarts to cover the fence.

5. Peggy has saved \$9,215. If she is paid \$45 an hour, about how many hours did she work?

9,215 ÷ 45
≈ 10,000 ÷ 50
= 200

Peggy worked about 200 hours.

## Exit Ticket (3 minutes)

After the Student Debrief, instruct students to complete the Exit Ticket. A review of their work will help with assessing students' understanding of the concepts that were presented in today's lesson and planning more effectively for future lessons. The questions may be read aloud to the students.

Name ______________________________ Date ______________

1. Estimate the quotients for the following problems. The first one is done for you.

| | | |
|---|---|---|
| a. 5,738 ÷ 21<br>≈ 6,000 ÷ 20<br>= 300 | b. 2,659 ÷ 28<br>≈ ________ ÷ ________<br>= ________ | c. 9,155 ÷ 34<br>≈ ________ ÷ ________<br>= ________ |
| d. 1,463 ÷ 53<br>≈ ________ ÷ ________<br>= ________ | e. 2,525 ÷ 64<br>≈ ________ ÷ ________<br>= ________ | f. 2,271 ÷ 72<br>≈ ________ ÷ ________<br>= ________ |
| g. 4,901 ÷ 75<br>≈ ________ ÷ ________<br>= ________ | h. 8,515 ÷ 81<br>≈ ________ ÷ ________<br>= ________ | i. 8,515 ÷ 89<br>≈ ________ ÷ ________<br>= ________ |
| j. 3,925 ÷ 68<br>≈ ________ ÷ ________<br>= ________ | k. 5,124 ÷ 81<br>≈ ________ ÷ ________<br>= ________ | l. 4,945 ÷ 93<br>≈ ________ ÷ ________<br>= ________ |
| m. 5,397 ÷ 94<br>≈ ________ ÷ ________<br>= ________ | n. 6,918 ÷ 86<br>≈ ________ ÷ ________<br>= ________ | o. 2,806 ÷ 15<br>≈ ________ ÷ ________<br>= ________ |

2. A swimming pool requires 672 $ft^2$ of floor space. The length of the swimming pool is 32 ft. Estimate the width of the swimming pool.

3. Janice bought 28 apps for her phone that, altogether, used 1,348 MB of space.

   a. If each app used the same amount of space, about how many MB of memory did each app use? Show how you estimated.

   b. If half of the apps were free and the other half were $1.99 each, about how much did she spend?

4. A quart of paint covers about 85 square feet. About how many quarts would you need to cover a fence with an area of 3,817 square feet?

5. Peggy has saved $9,215. If she is paid $45 an hour, about how many hours did she work?

Name ______________________________ Date ________________

Estimate the quotients for the following problems.

| | |
|---|---|
| a. 6,523 ÷ 21<br><br>≈ ________ ÷ ________<br><br>= ________ | b. 8,491 ÷ 37<br><br>≈ ________ ÷ ________<br><br>= ________ |
| c. 3,704 ÷ 53<br><br>≈ ________ ÷ ________<br><br>= ________ | d. 4,819 ÷ 68<br><br>≈ ________ ÷ ________<br><br>= ________ |

Name ________________________________ Date ________________

1. Estimate the quotients for the following problems. The first one is done for you.

| | | |
|---|---|---|
| a. 8,328 ÷ 41<br>≈ 8,000 ÷ 40<br>= 200 | b. 2,109 ÷ 23<br>≈ ________ ÷ ________<br>= ________ | c. 8,215 ÷ 38<br>≈ ________ ÷ ________<br>= ________ |
| d. 3,861 ÷ 59<br>≈ ________ ÷ ________<br>= ________ | e. 2,899 ÷ 66<br>≈ ________ ÷ ________<br>= ________ | f. 5,576 ÷ 92<br>≈ ________ ÷ ________<br>= ________ |
| g. 5,086 ÷ 73<br>≈ ________ ÷ ________<br>= ________ | h. 8,432 ÷ 81<br>≈ ________ ÷ ________<br>= ________ | i. 9,032 ÷ 89<br>≈ ________ ÷ ________<br>= ________ |
| j. 2,759 ÷ 48<br>≈ ________ ÷ ________<br>= ________ | k. 8,194 ÷ 91<br>≈ ________ ÷ ________<br>= ________ | l. 4,368 ÷ 63<br>≈ ________ ÷ ________<br>= ________ |
| m. 6,537 ÷ 74<br>≈ ________ ÷ ________<br>= ________ | n. 4,998 ÷ 48<br>≈ ________ ÷ ________<br>= ________ | o. 6,106 ÷ 25<br>≈ ________ ÷ ________<br>= ________ |

2. 91 boxes of apples hold a total of 2,605 apples. Assuming each box has about the same number of apples, estimate the number of apples in each box.

3. A wild tiger can eat up to 55 pounds of meat in a day. About how many days would it take for a tiger to eat the following prey?

| Prey | Weight of Prey | Number of Days |
|---|---|---|
| Eland Antelope | 1,754 pounds | |
| Boar | 661 pounds | |
| Chital Deer | 183 pounds | |
| Water Buffalo | 2,322 pounds | |

Topic F

# Partial Quotients and Multi-Digit Whole Number Division

**5.NBT.6**

| | | |
|---|---|---|
| **Focus Standard:** | 5.NBT.6 | Find whole-number quotients of whole numbers with up to four-digit dividends and two-digit divisors, using strategies based on place value, the properties of operations, and/or the relationship between multiplication and division. Illustrate and explain the calculation by using equations, rectangular arrays, and/or area models. |
| **Instructional Days:** | 5 | |
| **Coherence -Links from:** | G4–M3 | Multi-Digit Multiplication and Division |
| **-Links to:** | G6–M2 | Arithmetic Operations Including Division of Fractions |

The series of lessons in Topic F lead students to divide multi-digit dividends by two-digit divisors using the written vertical method. Each lesson moves to a new level of difficulty with a sequence beginning with divisors that are multiples of 10 to non-multiples of 10. Two instructional days are devoted to single-digit quotients with and without remainders before progressing to two- and three-digit quotients (**5.NBT.6**).

**A Teaching Sequence Toward Mastery of Partial Quotients and Multi-Digit Whole Number Division**

**Objective 1: Divide two- and three-digit dividends by multiples of 10 with single-digit quotients, and make connections to a written method.**
**(Lesson 19)**

**Objective 2: Divide two- and three-digit dividends by two-digit divisors with single-digit quotients, and make connections to a written method.**
**(Lessons 20–21)**

**Objective 3: Divide three- and four-digit dividends by two-digit divisors resulting in two- and three-digit quotients, reasoning about the decomposition of successive remainders in each place value.**
**(Lessons 22–23)**

# Lesson 19

## Objective: Divide two- and three-digit dividends by multiples of 10 with single-digit quotients, and make connections to a written method.

### Suggested Lesson Structure

| | |
|---|---|
| Fluency Practice | (12 minutes) |
| Application Problem | (7 minutes) |
| Concept Development | (31 minutes) |
| Student Debrief | (10 minutes) |
| **Total Time** | **(60 minutes)** |

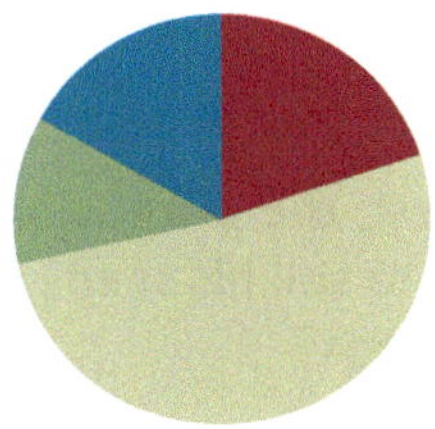

## Fluency Practice (12 minutes)

- Estimate and Divide **5.NBT.6** (5 minutes)
- Group Count by Multiples of 10 **5.NBT.2** (3 minutes)
- Group Count by Multi-Digit Numbers **5.NBT.6** (4 minutes)

### Estimate and Divide (5 minutes)

Materials: (S) Personal white board

Note: This fluency activity reviews content from Lessons 17 and 18.

Repeat the process from Lesson 18 with the following possible sequence: 908 ÷ 28; 152 ÷ 33; 398 ÷ 98; and 7,272 ÷ 81.

### Group Count by Multiples of 10 (3 minutes)

Note: Counting by multiples of 10 helps students estimate quotients with two-digit divisors, which students do during this lesson's Concept Development.

Repeat the process from Lesson 16 for various multiples of 10.

### Group Count by Multi-Digit Numbers (4 minutes)

Materials: (S) Personal white board

Note: This fluency activity prepares students for Lesson 20's Concept Development.

T: I'm going to call out a number. I want you to write down the multiples of that number. You have one minute. Ready? 21.

S: (Write down multiples of 21.)

T: Stop. Let's correct your work. (On the board, write down multiples from 21 to 210 as students check their multiples.) Let's skip count again by twenty-ones. This time we'll count out loud. Try not to look at the board as I guide you.

Stand away from the board. Direct the students to count by 5–10 multiples of 21 forward and backward, occasionally changing directions.

Repeat the process for 43.

## Application Problem (7 minutes)

At the Highland Falls pumpkin-growing contest, the prize winning pumpkin contains 360 seeds. The proud farmer plans to sell his seeds in packs of 12. How many packs can he make using all the seeds?

Note: Although students have not yet divided three-digit dividends by two-digit divisors, this problem has the basic fact 12 × 3 = 36 embedded in it, and it is similar to problems encountered in Lesson 18's Concept Development.

$360 \div 12 =$
$(36 \times 10) \div 12 =$
$(36 \div 12) \times 10 =$
$3 \times 10 = 30$

The farmer can make 30 packs of seeds.

## Concept Development (31 minutes)

Materials: (S) Personal white board

**NOTES ON MULTIPLE MEANS OF ENGAGEMENT:**

This is the first lesson where students are solving division problems using the standard algorithm. It is imperative at this point that students know basic facts and understand what it means to divide. If students are not ready for double-digit divisors, then consider reviewing Grade 3 Module 1, Grade 3 Module 3, and Grade 4 Module 3.

**Problem 1: 70 ÷ 30**

T: (Write 70 ÷ 30 on the board.) The divisor is…?

S: 30.

T: We need a multiple of 30 to make the division easy. How should we estimate the quotient? Turn and share with your partner.

S: I see an easy fact of 6 divided by 3 is equal to 2. → Yeah, 6 tens divided by 3 tens is 2. → I can estimate 70 to 60, because I can easily divide 30 into 60.

T: On your personal white board, show me how to estimate the quotient.

S: (Show 60 ÷ 30 = 6 ÷ 3 = 2.)

T: (Write and set up the standard algorithm below 70 ÷ 30 on the board.) Our estimated quotient is 2, which means that I should be able to distribute 2 × 30. (Record 2 in the quotient.) What's 2 × 30?

S: 60.

T: (Record 60 below the 70.) I distributed 60. The difference between 60 and 70 is…?

S: 10.

70 ÷ 30

| estimate | solve | check |
|---|---|---|
| ≈ 60 ÷ 30<br>= 6 ÷ 3<br>= 2 | 30 ⟌ 70 = 2<br>− 60<br>10 | 30 × 2 = 60<br>60 + 10 = 70 |

T: What does this 10 mean?

S: 10 is the remainder. → 10 is the leftover from the original total of 70. → We started with 70, made 2 groups of 30, used up 60, and were left with 10. → We have 10 left over, but we need 20 more in order to make 1 more group of 30.

T: Can we make another group of 30 with our remainder?

S: No, 10 is not enough to make a group of 30.

T: How might we know that our quotient is correct?

S: We can check our answer to see if our quotient is correct.

T: Yes! Let's multiply: 30 times 2. (Write 30 × 2 = on the board.) What's the answer?

S: 60.

T: We started with 70, and 60 ≠ 70. Does this mean we made an error? What else must we do? Turn and discuss.

S: Oh, no! We made a mistake because 60 doesn't equal 70. → We have to add the remainder of 10. Then, the total will be 70. → Our thinking is correct. We could make 2 groups of 30, but there were 10 left over. They are part of the original whole. We need to add the 10 to the 60 that were put into groups.

**NOTES ON MULTIPLE MEANS OF REPRESENTATION:**

It may be beneficial for some learners to see a tape diagram as they are working through their checks. In the visual model below, students are able to see when dividing that nothing is being added or subtracted; the dividends are simply being grouped in a new way. In this case, they started out with 70, and they still ended with 70.

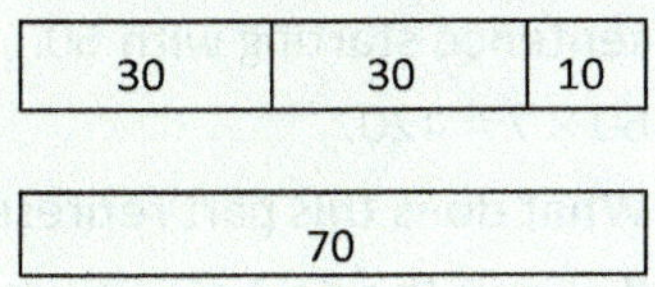


T: Yes. (Draw the number bond.) One part is made of groups of 30. The other part is the remainder.

T: What's 60 plus 10? (Write 60 + 10 = on the board below 30 × 2 = 60.)

S: 70.

T: Yes. We did it. We solved the division problem correctly. Today, we got a precise answer with a quotient and remainder. In the previous lessons, we merely estimated the quotient.

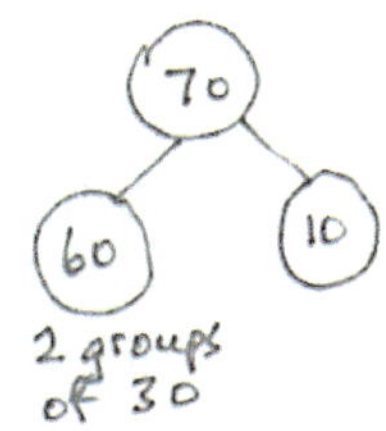


**Problem 2: 430 ÷ 60**

T: (Write 430 ÷ 60 on the board.) What's our whole?

S: 430.

T: Again, we need a multiple of 60 to make the division easy. Show me how to estimate the quotient.

S: (Show 420 ÷ 60 = 42 ÷ 6 = 7.)

T: Let's record this division sentence in the vertical algorithm. You do the same on your board. (Write and set up the standard algorithm below 430 ÷ 60 on the board.) Our estimate was 7, which means that there should be 7 groups of 60 in 430. Let's divide and see if that's true.

T: Let's record the 7 in our quotient. (Record 7.) Why is the 7 recorded above the zero in the vertical algorithm?

S: 7 represents 7 ones, so it must be recorded in the ones place directly above the ones place in the whole. → 420 divided by 60 is just 42 tens divided by 6 tens. The answer is just 7, not 7 tens.

T: What's 7 times 60?

S: 420.

T: (Record 420 below 430.) Was it possible to make 7 groups of 60 from 430? How do you know?

S: Yes, we distributed 420 and still have some left.

T: How many are remaining after making the groups?

S: 10.

T: What does this remainder of 10 mean?

S: 10 is what is left over after making groups from the whole. We don't have enough to make another group of 60. → We need 60 to make 1 group, so we'll need 50 more in order to make another group of 60.

T: There are 7 units of 60 in 430 and 10 remaining. Now, work with a partner and check the answer.

T: Look at your checking equation. Say the multiplication sentence starting with 60.

S: 60 × 7 = 420.

**NOTES ON MULTIPLE MEANS OF REPRESENTATION:**

Make sure when doing the two-step check that students are writing the equations correctly. Students have a couple of options. They may record their work in two separate equations. They must take the product and start a new equation for the addition. It is *not acceptable* to write

$40 \times 2 = 80 + 12 = 92$ because

$40 \times 2 \neq 92$.

Alternatively, they may record their work with a single, *valid* equation:

$40 \times 2 + 12 = 80 + 12 = 92$.

T: What does this part represent?

S: It shows the part of our whole that was put into groups of 60. (Draw a number bond similar to the one pictured to the right.)

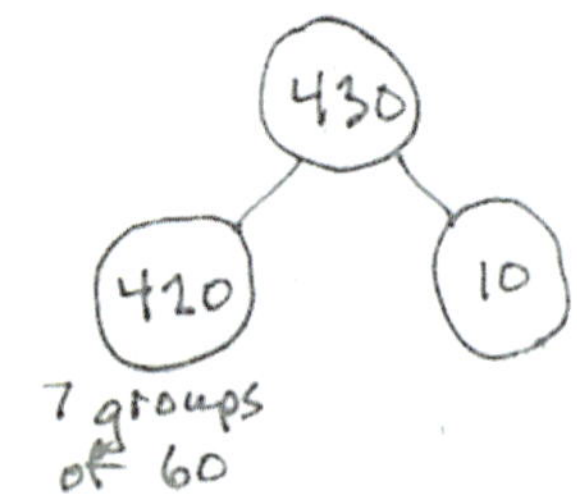


T: (Write 60 × 7 = 420 on the board.) Say the equation to complete the original whole.

S: 420 + 10 = 430.

T: (Write 420 + 10 = 430 on the board below 60 × 7 = 420.) What does this part of our check represent?

S: This shows the part of the total that we could put into groups added to the part that we couldn't put into groups. Together it is all that we had to distribute.

**Problem 3: 572 ÷ 90**

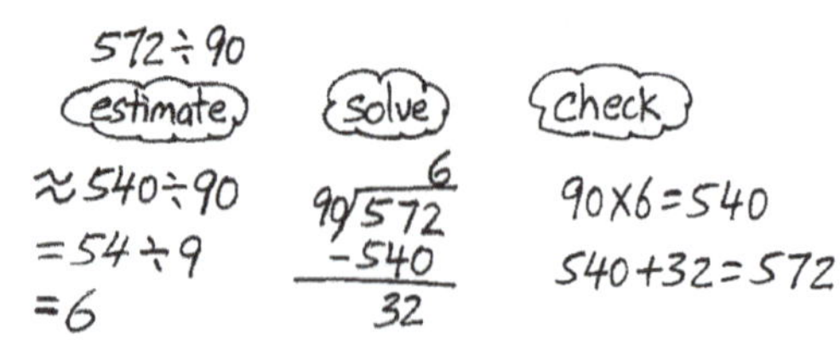


T: (Write 572 ÷ 90 horizontally on the board.) We're trying to make groups of 90. What multiple of 90 is closest to 572 and would make this division easy? Show me how to estimate the quotient.

S: (Show 540 ÷ 90 = 54 ÷ 9 = 6.)

T: Our estimated quotient is 6. With a partner, find the actual quotient using the standard algorithm, and check the answer. When you're finished, check your answer with another group.

T: How many nineties are there in 572? (Record the algorithm.)

S: 6.

T: Where is this recorded in the algorithm?

S: In the ones place above the ones place in the whole.

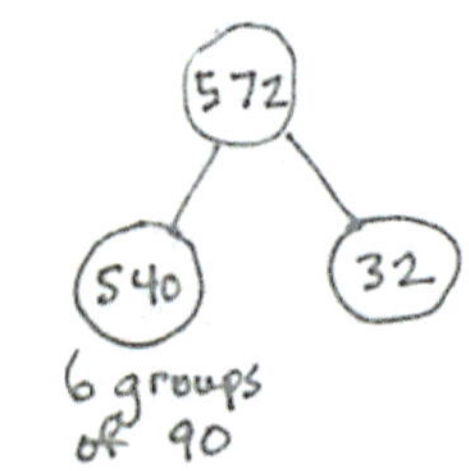






EUREKA MATH™

T: How many are remaining?

S: 32.

T: Is this enough to make another ninety?

S: No.

T: What are the equations for checking the problem?

S: 90 × 6 = 540 and 540 + 32 = 572.

## Problem Set (10 minutes)

Students should do their personal best to complete the Problem Set within the allotted 10 minutes. For some classes, it may be appropriate to modify the assignment by specifying which problems they work on first. Some problems do not specify a method for solving. Students should solve these problems using the RDW approach used for Application Problems.

## Student Debrief (10 minutes)

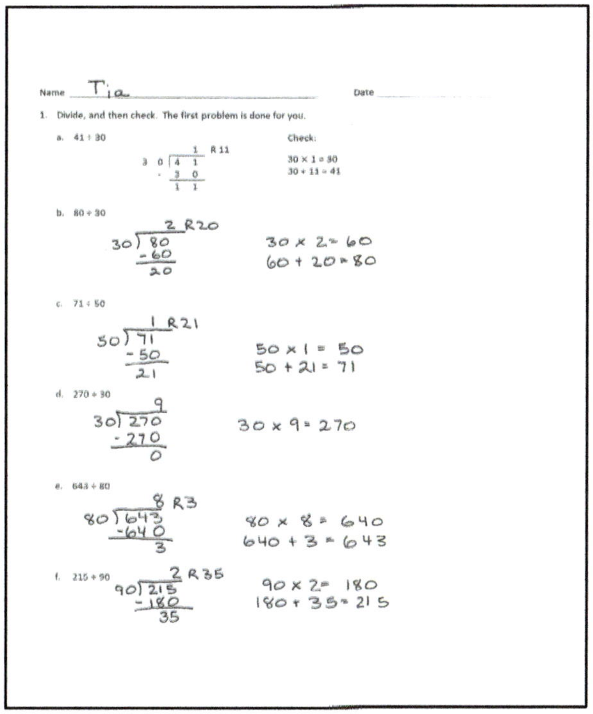
Name Tia Date

1. Divide, and then check. The first problem is done for you.

a. 41 ÷ 30 — 1 R 11; Check: 30 × 1 = 30, 30 + 11 = 41

b. 80 ÷ 30 — 2 R20; 30 × 2 = 60, 60 + 20 = 80

c. 71 ÷ 50 — 1 R21; 50 × 1 = 50, 50 + 21 = 71

d. 270 ÷ 30 — 9; 30 × 9 = 270

e. 643 ÷ 80 — 8 R3; 80 × 8 = 640, 640 + 3 = 643

f. 215 ÷ 90 — 2 R35; 90 × 2 = 180, 180 + 35 = 215

**Lesson Objective:** Divide two- and three-digit dividends by multiples of 10 with single-digit quotients, and make connections to a written method.

The Student Debrief is intended to invite reflection and active processing of the total lesson experience.

Invite students to review their solutions for the Problem Set. They should check work by comparing answers with a partner before going over answers as a class. Look for misconceptions or misunderstandings that can be addressed in the Debrief. Guide students in a conversation to debrief the Problem Set and process the lesson.

Any combination of the questions below may be used to lead the discussion.

- In Problem 1(d), did anyone notice something different? Does it always make sense to use the standard algorithm?
- In Problem 2, what was Terry's mistake? If you had to estimate the quotient, what would you have done? What could he do to correct his quotient without erasing his work so far? (Make sure students recognize that Terry's thinking was accurate, but he stopped making groups too soon. His error can be corrected by simply making another group of 40 and subtracting it from the remaining whole.)
- What if Terry had estimated too large a quotient? What should he do?
- How was solving Problem 3 different from solving all the others? Why?
- Explain your thought process as you solved Problem 4.

- What did all our divisors have in common today? Did this make estimation easier?
- Does a divisor have to be a multiple of 10? Why do you think I chose multiples of 10 for divisors today?

## Exit Ticket (3 minutes)

After the Student Debrief, instruct students to complete the Exit Ticket. A review of their work will help you assess the students' understanding of the concepts that were presented in the lesson today and plan more effectively for future lessons. You may read the questions aloud to the students.

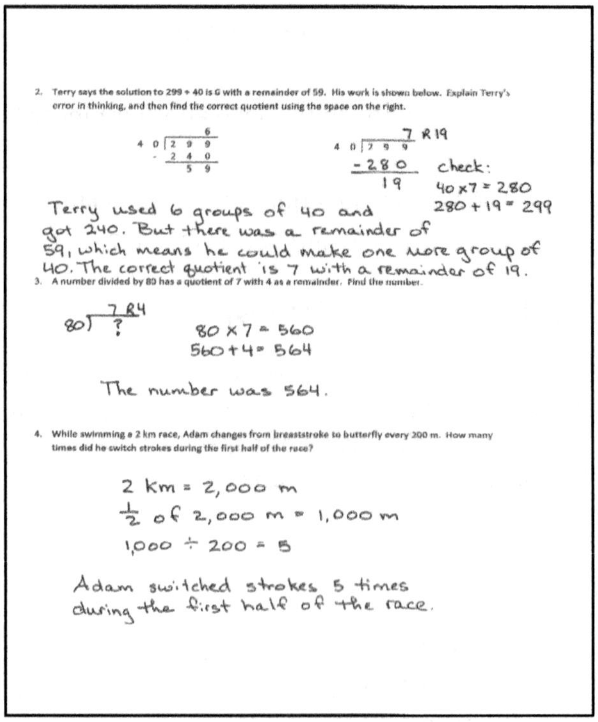

2. Terry says the solution to 299 ÷ 40 is 6 with a remainder of 59. His work is shown below. Explain Terry's error in thinking, and then find the correct quotient using the space on the right.

40 ) 299 → 6; − 240; 59

40 ) 299 → 7 R19; − 280; 19

check: $40 \times 7 = 280$; $280 + 19 = 299$

Terry used 6 groups of 40 and got 240. But there was a remainder of 59, which means he could make one more group of 40. The correct quotient is 7 with a remainder of 19.

3. A number divided by 80 has a quotient of 7 with 4 as a remainder. Find the number.

80 ) ? → 7 R4

$80 \times 7 = 560$

$560 + 4 = 564$

The number was 564.

4. While swimming a 2 km race, Adam changes from breaststroke to butterfly every 200 m. How many times did he switch strokes during the first half of the race?

2 km = 2,000 m

$\frac{1}{2}$ of 2,000 m = 1,000 m

$1{,}000 \div 200 = 5$

Adam switched strokes 5 times during the first half of the race.



EUREKA MATH

Name ______________________________ Date ______________

1. Divide, and then check. The first problem is done for you.

   a. 41 ÷ 30

```
        1  R 11
3 0 | 4 1
  -   3 0
      1 1
```

   *Check:*

   30 × 1 = 30
   30 + 11 = 41

   b. 80 ÷ 30

   c. 71 ÷ 50

   d. 270 ÷ 30

   e. 643 ÷ 80

   f. 215 ÷ 90

2. Terry says the solution to 299 ÷ 40 is 6 with a remainder of 59. His work is shown below. Explain Terry's error in thinking, and then find the correct quotient using the space on the right.

```
          6
  4 0 | 2 9 9            4 0 | 2 9 9
    -   2 4 0
          5 9
```

3. A number divided by 80 has a quotient of 7 with 4 as a remainder. Find the number.

4. While swimming a 2 km race, Adam changes from breaststroke to butterfly every 200 m. How many times does he switch strokes during the first half of the race?

Name ______________________________ Date ____________________

Divide, and then check using multiplication.

a. $73 \div 20$

b. $291 \div 30$

Name ______________________________ Date ________________

1. Divide, and then check using multiplication. The first one is done for you.

   a. 71 ÷ 20

$$\begin{array}{r} 3 \text{ R } 11 \\ 20\,\overline{)\,71} \\ -\,60 \\ \hline 11 \end{array}$$

*Check:*

20 × 3 = 60

60 + 11 = 71

   b. 90 ÷ 40

   c. 95 ÷ 60

   d. 280 ÷ 30

   e. 437 ÷ 60

   f. 346 ÷ 80





EUREKA MATH™

2. A number divided by 40 has a quotient of 6 with a remainder of 16. Find the number.

3. A shipment of 288 reams of paper was delivered. Each of the 30 classrooms received an equal share of the paper. Any extra reams of paper were stored. After the paper was distributed to the classrooms, how many reams of paper were stored?

4. How many groups of sixty are in two hundred forty-four?

# Lesson 20

## Objective: Divide two- and three-digit dividends by two-digit divisors with single-digit quotients, and make connections to a written method.

### Suggested Lesson Structure

| | |
|---|---|
| Fluency Practice | (12 minutes) |
| Application Problem | (5 minutes) |
| Concept Development | (33 minutes) |
| Student Debrief | (10 minutes) |
| **Total Time** | **(60 minutes)** |

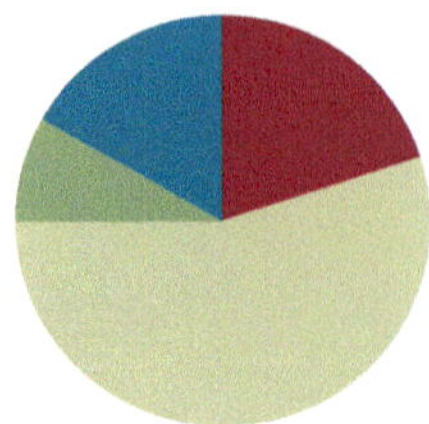

## Fluency Practice (12 minutes)

- Group Count by Multi-Digit Numbers **5.NBT.6** (3 minutes)
- Estimate and Divide **5.NBT.6** (4 minutes)
- Divide by Multiples of Ten with Remainders **5.NBT.6** (5 minutes)

### Group Count by Multi-Digit Numbers (3 minutes)

Materials: (S) Personal white board

Note: This fluency activity prepares students for this lesson's Concept Development.

Direct the students to count by 5–10 multiples of 21 forward and backward, occasionally changing directions and attempting to avoid student frustration.

Repeat process for 43.

### Estimate and Divide (4 minutes)

Materials: (S) Personal white board

Note: This fluency exercise reviews Lesson 17 content.

Repeat the process from Lesson 18 for the following possible sequence: 607 ÷ 19, 123 ÷ 24, 891 ÷ 96, and 5,482 ÷ 62.

### Divide by Multiples of Ten with Remainders (5 minutes)

Materials: (S) Personal white board

Note: This exercise reviews Lesson 19 content.

T: (Write 73 ÷ 50.) On your personal white boards, solve the division problem using the standard algorithm. Check your work using multiplication and addition.

Repeat process for 70 ÷ 30, 157 ÷ 30, and 432 ÷ 70.

## Application Problem (5 minutes)

Billy has 2.4 m of ribbon for crafts. He wants to share it evenly with 12 friends. How many centimeters of ribbon would 7 friends get?

2.4 m = ___ cm
2.4 × 1 m =
2.4 × 100 cm =
240 cm
240 ÷ 12 =
(24 × 10) ÷ 12 =
(24 ÷ 12) × 10 =
2 × 10 = 20

20 × 7 =
(2 × 10) × 7 =
(2 × 7) × 10 =
14 × 10 =
140

Seven friends will get 140 cm of ribbon.

Note: This Application Problem reaches back to concepts taught in Grade 5 Module 1.

**NOTES ON MULTIPLE MEANS OF ACTION AND EXPRESSION:**

Students have a choice of strategies they can use to solve this Application Problem. They can think of 2.4 as 24 tenths. 24 tenths divided by 12 is 2 tenths. They can also compensate. Students can multiply the whole, 2.4, by 10. After they divide 24 by 12, students need to divide the quotient by 10. Both of these methods were explored in Module 1.

## Concept Development (33 minutes)

Materials: (S) Personal white board

**Problem 1: 72 ÷ 21**

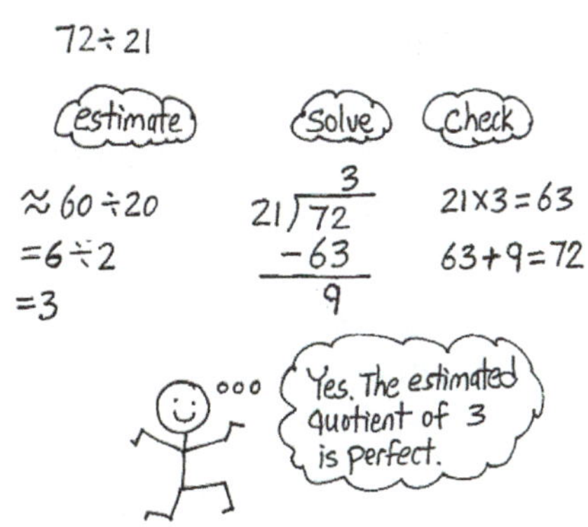


T: (Write 72 ÷ 21 horizontally on the board.) What is our whole?

S: 72.

T: Find a multiple of 20 close to 72 that makes this division easy. Show me how to estimate the quotient on your personal white board.

S: (Show 60 ÷ 20 = 6 ÷ 2 = 3.)

T: I see you chose 60. Why not choose 80 and estimate the quotient as 4?

S: Because 4 × 20 is 80, and that's already too big.

T: Right, so our estimate means that there are about 3 twenty-ones in 72. Let's record that estimate. Where should it be recorded? (Write and set up the standard algorithm below 72 ÷ 21 on the board.)

S: In the ones place.

T: What is 3 × 21?

S: 63.

T: (Record 63 below 72.) So, we've distributed 3 units of 21. How many of the 72 remain? Give me the full subtraction sentence.

S: 72 – 63 = 9.

T: Is 9 enough to make another group of 21?

S: No.

T: How did our estimate help us solve the problem? Turn and share with your partner.

S: We divided 60 by 20 to get our estimate, which was 3 ones. So, that's what we tried first in the quotient. → Our estimated quotient was 3, and it turned out that our actual quotient was 3 with a leftover of 9.

T: Great. Let's check our answer. Whisper the number sentences to your partner.

T: If I have 3 groups of 21 and add 9, what should my total be?

S: 72.

T: If I have 21 groups with 3 in each and 9 more, what should my total be?

S: 72. → It's the same thing: 21 groups of 3 and 3 groups of 21 are both just 3 × 21.

T: Then, that means that when using the algorithm, we can view the divisor as either the number of groups or the size of each group.

**Problem 2: 94 ÷ 43**

T: (Write 94 ÷ 43 horizontally on the board.) Use your personal white boards. Work with a partner.

1. Round the divisor.
2. Find a multiple of the divisor that makes the division easy.
3. Estimate the quotient.
4. Solve using the standard algorithm.

T: Partner A will divide using the standard algorithm, and Partner B will check the answer. (Allow time for students to work.)

T: Partner A, say the quotient and the remainder for 94 ÷ 43.

S: The quotient is 2, and the remainder is 8.

T: What does the quotient, 2, represent?

S: 2 groups of 43. → 43 groups with 2 in each one.

T: What does the remainder of 8 represent?

S: After 43 groups were made, 8 were left over. → We have 8 for the next group. → 8 that couldn't be distributed fairly into 43 groups.

T: Partner B, say your number sentences for checking the problem.

S: 43 × 2 = 86, and 86 + 8 = 94.

T: Again, let's look at our estimated quotient and our actual quotient. Did our estimated quotient turn out to be the actual quotient?

S: Yes.

**Problem 3: 84 ÷ 23**

T: (Write 84 ÷ 23 horizontally on the board.) We need a multiple of 20 that will make this division easy. Show me how to estimate the quotient.

S: 80 ÷ 20 = 8 ÷ 2 = 4.

T: What are other ways of estimating this problem?

S: 90 ÷ 30 = 9 ÷ 3 = 3. → 100 ÷ 25 = 4.

T: These are all good ideas. Let's use our first possibility. (Write 80 ÷ 20 = 4 on the board.) Let's now solve this problem using the standard algorithm. (Write and set up the standard algorithm below 84 ÷ 23 on the board.) Our estimated quotient was 4, so I'll put 4 as the quotient. (Record 4 as the quotient in the ones place in standard algorithm.)

T: What are 4 units of 23?

S: 92.

T: Wait a minute! Let's stop and think. We have 84 in our total. Do we have enough to make 4 units of 23?

S: No.

T: What's happening here? Why didn't our estimated quotient work this time? Turn and discuss with your partner.

S: Our estimation sentence was correct. 84 ÷ 23 becomes 80 ÷ 20 = 4. → We rounded our divisor down from 23 to 20. When we multiply 23 times 4, the product is 92. The product of 20 times 4 is 80. The extra part came from 4 × 3. → I know. We made the divisor smaller. The real divisor was bigger, so that means we are going to make fewer units. → Yeah! If the divisor was just two more, 25, we would have rounded to 30, and then 90 divided by 30 is obviously 3.

MP.2

T: So, if 4 ones is too big to be the quotient, what should we do?

S: Let's try 3.

T: How much is 3 × 23?

S: 69.

T: Take away those that we've distributed.

T: How many ones are remaining?

S: 15.

**NOTES ON MULTIPLE MEANS OF ENGAGEMENT:**

For Problem 3, some students may easily see that 4 is an overestimate. Encourage these students to solve the problem by simply skip-counting by 23. They could also round the whole and keep the divisor unchanged while they estimate. This cultivates their number sense and challenges them appropriately. Students could also be asked, "What is the largest the whole could be and still have a quotient of 3?"

T: What does the remainder of 15 tell us?

S: We don't have enough for a fourth group. Those 15 ones are left over. → We'll need 8 more to make another group of 23.

T: Give me the quotient and remainder for 84 ÷ 23.

S: The quotient is 3, and the remainder is 15.

T: Whisper to your partner what these numbers represent and how we should check this problem.

S: The 3 is 3 groups of 23, and the 15 are the ones that weren't enough to make another group. → We should multiply the quotient and the divisor, and then add the remainder.

T: Say the multiplication sentence starting with 23.

S: 23 × 3 = 69.

T: (Record 23 × 3 = 69 horizontally on the board.) Say the addition sentence starting with 69.

S: 69 + 15 = 84.

T: (Record 69 + 15 = 84 below 23 × 3 = 69 on the board.) Is 84 our original whole?

S: Yes, we solved it correctly.

T: What did we just learn about estimated quotients? Turn and discuss.

S: We should always estimate before we solve, but we may need to adjust it. → If we change the divisor or the whole a lot, it could make our estimate too big or too small.

**Problem 4: 57 ÷ 29**

T: (Write 57 ÷ 29 horizontally on the board.) Use your personal white board. Work on this problem independently. Remember to estimate, divide, and check. Compare your work with a partner when you're finished.

T: Tell me how you estimated.

S: 60 ÷ 30 = 6 ÷ 3 = 2.

T: Can I use the quotient of 2? Discuss with your neighbor.

S: No.

T: Why not? How much is 2 units of 29?

S: 58. 58 is greater than our whole of 57.

T: So, what's the actual quotient?

S: 1.

T: Give me the quotient and remainder for 57 ÷ 29.

S: The quotient is 1 with a remainder of 28.

T: What are the sentences for checking the problem?

S: 29 × 1 = 29 and 29 + 28 = 57.

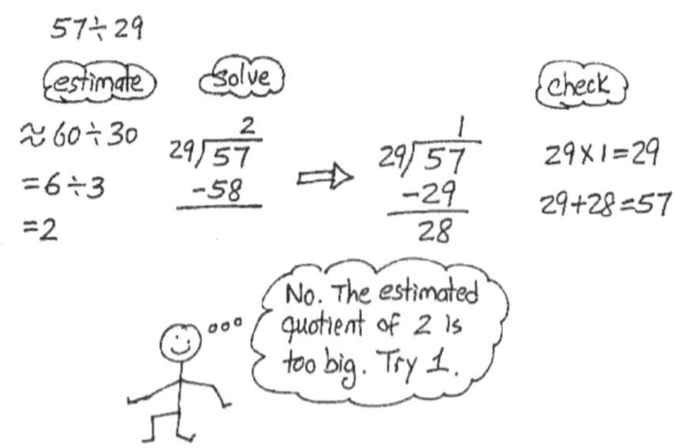


T: Talk to your partner about how we could create another division problem whose quotient is also 1 and whose remainder is 28.

S: Just put any number in place of 29 in the check sentences, and get a new whole. We could use 34 × 1 + 28 = 62. So, 62 ÷ 34 is also 1 R28. → We need 1 group of a number, and then we would add 28 to that. That will give us a new whole.

## Problem Set (10 minutes)

Students should do their personal best to complete the Problem Set within the allotted 10 minutes.
For some classes, it may be appropriate to modify the assignment by specifying which problems they work on first. Some problems do not specify a method for solving. Students should solve these problems using the RDW approach used for Application Problems.

## Student Debrief (10 minutes)

**Lesson Objective:** Divide two- and three-digit dividends by two-digit divisors with single digit quotients, and make connections to a written method.

The Student Debrief is intended to invite reflection and active processing of the total lesson experience.

Invite students to review their solutions for the Problem Set. They should check work by comparing answers with a partner before going over answers as a class. Look for misconceptions or misunderstandings that can be addressed in the Debrief. Guide students in a conversation to debrief the Problem Set and process the lesson.

Any combination of the questions below may be used to lead the discussion.

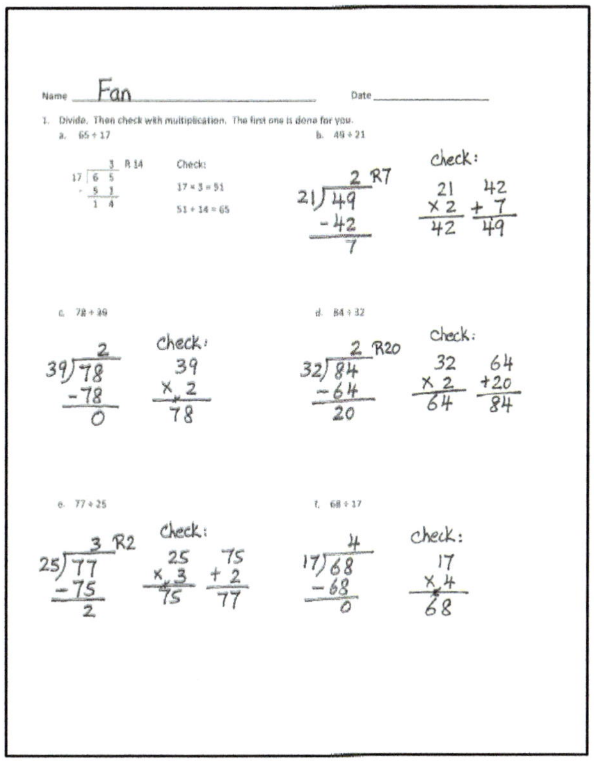


- What pattern did you notice between Problems 1(c) and 1(f)?
- Did your initial estimates work for every example in Problem (1)? Why or why not? What happened in 1(d)?
- In Problem 2, what would you tell Linda in order to help her solve the problem? What lesson does Linda need to learn? What is another way that Linda could have estimated that would have eliminated the issue she encountered in the standard algorithm?
- Explain your thought process as you set up and began to solve Problems 3 and 4. What was challenging or unique about them? (Generating a division problem with the same quotient and remainder appears on the End-of-Module Assessment. Make time to debrief the students' thinking about Problem 4 thoroughly.)
- Talk about the importance of estimation when dividing with two-digit divisors.

## Exit Ticket (3 minutes)

After the Student Debrief, instruct students to complete the Exit Ticket. A review of their work will help with assessing students' understanding of the concepts that were presented in today's lesson and planning more effectively for future lessons. The questions may be read aloud to the students.

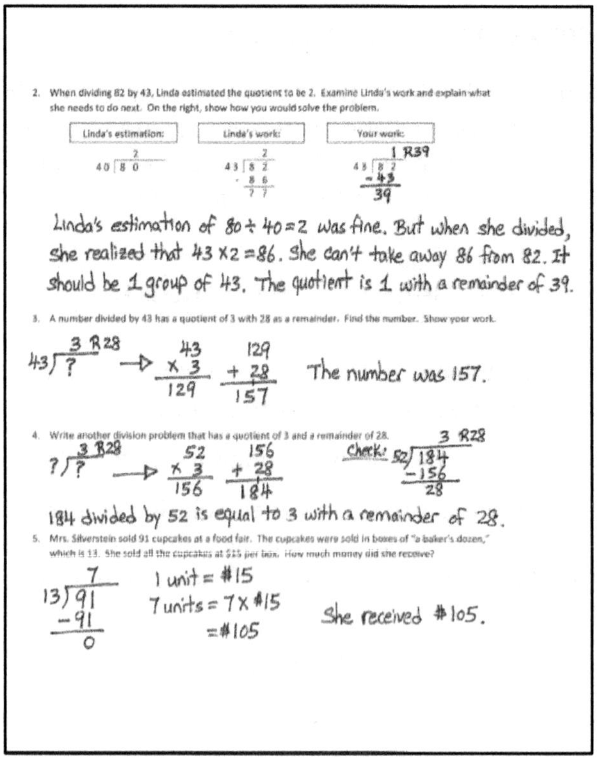

2. When dividing 82 by 43, Linda estimated the quotient to be 2. Examine Linda's work and explain what she needs to do next. On the right, show how you would solve the problem.

| Linda's estimation: | Linda's work: | Your work: |
|---|---|---|
| $40\overline{)80}$ = 2 | $43\overline{)82}$ = 2, − 86, 7 7 | $43\overline{)82}$ = 1 R39, − 43, 39 |

Linda's estimation of 80 ÷ 40 = 2 was fine. But when she divided, she realized that 43 × 2 = 86. She can't take away 86 from 82. It should be 1 group of 43. The quotient is 1 with a remainder of 39.

3. A number divided by 43 has a quotient of 3 with 28 as a remainder. Find the number. Show your work.

$43\overline{)?}$ = 3 R28 → 43 × 3 = 129; 129 + 28 = 157. The number was 157.

4. Write another division problem that has a quotient of 3 and a remainder of 28.

$?\overline{)?}$ = 3 R28 → 52 × 3 = 156; 156 + 28 = 184. Check: $52\overline{)184}$ = 3 R28; 184 − 156 = 28

184 divided by 52 is equal to 3 with a remainder of 28.

5. Mrs. Silverstein sold 91 cupcakes at a food fair. The cupcakes were sold in boxes of "a baker's dozen," which is 13. She sold all the cupcakes at $15 per box. How much money did she receive?

$13\overline{)91}$ = 7; 91 − 91 = 0

1 unit = $15
7 units = 7 × $15
= $105

She received $105.

Name ______________________________ Date ______________

1. Divide. Then, check with multiplication. The first one is done for you.

a. $65 \div 17$

$$\begin{array}{r} 3 \text{ R } 14 \\ 17 \overline{)65} \\ -\ 51 \\ \hline 14 \end{array}$$

*Check:*

$17 \times 3 = 51$

$51 + 14 = 65$

b. $49 \div 21$

c. $78 \div 39$

d. $84 \div 32$

e. $77 \div 25$

f. $68 \div 17$

2. When dividing 82 by 43, Linda estimated the quotient to be 2. Examine Linda's work, and explain what she needs to do next. On the right, show how you would solve the problem.

| Linda's Estimation: | Linda's Work: | Your Work: |
|---|---|---|

$$40 \overline{)\,80\,}\quad 2$$

$$\begin{array}{r} 2 \\ 43\,\overline{)\,82} \\ -\ 86 \\ \hline ?\ ? \end{array}$$

$$43 \overline{)\,82\,}$$

3. A number divided by 43 has a quotient of 3 with 28 as a remainder. Find the number. Show your work.

4. Write another division problem that has a quotient of 3 and a remainder of 28.

5. Mrs. Silverstein sold 91 cupcakes at a food fair. The cupcakes were sold in boxes of "a baker's dozen," which is 13. She sold all the cupcakes at $15 per box. How much money did she receive?

Name ______________________________ Date ______________

Divide. Then, check with multiplication.

a. $78 \div 21$

b. $89 \div 37$

Name ______________________________ Date ______________

1. Divide. Then, check with multiplication. The first one is done for you.

a. 72 ÷ 31

```
      2 R 10
31 | 7 2
   - 6 2
     1 0
```

*Check:*

$31 \times 2 = 62$

$62 + 10 = 72$

b. 89 ÷ 21

c. 94 ÷ 33

d. 67 ÷ 19

e. 79 ÷ 25

f. 83 ÷ 21

2. A 91 square foot bathroom has a length of 13 feet. What is the width of the bathroom?

3. While preparing for a morning conference, Principal Corsetti is laying out 8 dozen bagels on square plates. Each plate can hold 14 bagels.

   a. How many plates of bagels will Mr. Corsetti have?

   b. How many more bagels would be needed to fill the final plate with bagels?

# Lesson 21

## Objective: Divide two- and three-digit dividends by two-digit divisors with single-digit quotients, and make connections to a written method.

### Suggested Lesson Structure

| | | |
|---|---|---|
| ■ | Fluency Practice | (12 minutes) |
| ■ | Application Problem | (5 minutes) |
| ■ | Concept Development | (33 minutes) |
| ■ | Student Debrief | (10 minutes) |
| | **Total Time** | **(60 minutes)** |

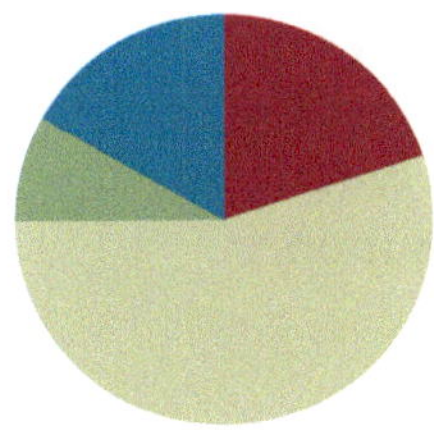

## Fluency Practice (12 minutes)

- Group Count by Multi-Digit Numbers **5.NBT.6** (5 minutes)
- Divide by Two-Digit Numbers **5.NBT.6** (7 minutes)

### Group Count by Multi-Digit Numbers (5 minutes)

Materials: (S) Personal white board

Note: This fluency activity prepares students for this lesson's Concept Development.

Repeat the process from Lesson 19 for 31 and 16.

### Divide by Two-Digit Numbers (7 minutes)

Materials: (S) Personal white board

Notes: This fluency activity reviews Lesson 20 content.

T: (Write 61 ÷ 17.) On your personal white boards, show me how to estimate the quotient.

S: (Write 60 ÷ 20 = 3.)

T: Solve the equation.

S: (Solve and check, as exemplified to the right.)

$60 \div 30 = 2$

$$\begin{array}{r} 3 \\ 17\overline{)61} \\ 51 \\ \hline 10 \end{array}$$

$3 \times 17 + 10 = 61$

Repeat the process using the following possible sequence: 48 ÷ 21, 99 ÷ 32, and 74 ÷ 37.

## Application Problem (5 minutes)

105 students were divided equally into 15 teams.

a. How many players were on each team?

b. If each team had 3 girls, how many boys were there altogether?

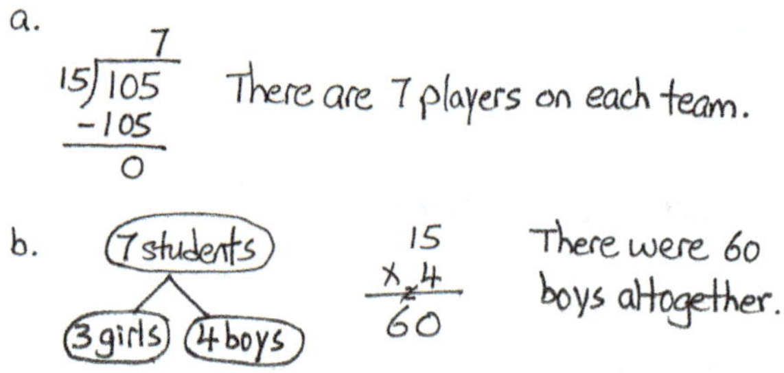


Note: Although yesterday's lesson focused only on two-digit totals, the friendly divisor of 15 makes this problem manageable for students. Students who have difficulty answering Part (a) may need extra support during the Concept Development.

## Concept Development (33 minutes)

Materials: (S) Personal white board

**Problem 1: 256 ÷ 47**

T: (Write 256 ÷ 47 horizontally on the board.) How can we estimate the quotient? Discuss with a partner.

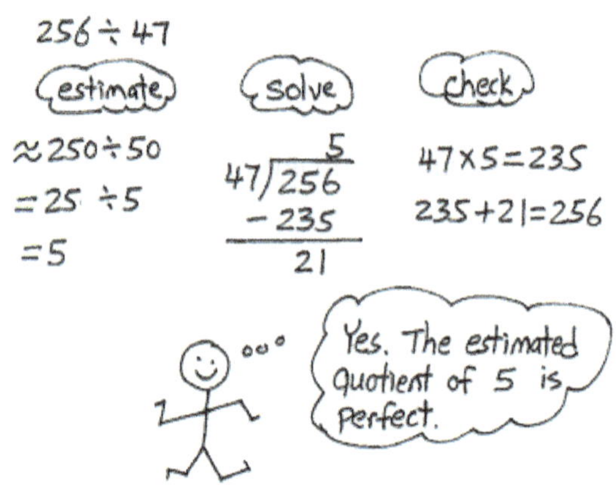


**NOTES ON TRUE EQUATIONS:**

Be careful to record work correctly. When estimating a quotient:

*Correct:*

- 250 ÷ 50 = 25 ÷ 5 = 5
- 256 ÷ 47 ≈ 250 ÷ 50

*Incorrect:*

- 256 ÷ 47 = 250 ÷ 50 = 25 ÷ 5 = 5
- 256 ÷ 47 ≈ 250 ÷ 50 = 25 ÷ 5 = 5

When checking a quotient and remainder:

*Correct:*

- 47 × 5 + 21 = 256

*Incorrect:*

- 47 × 5 = 235 + 21 = 256

S: (Discuss.) We need a multiple of 50 that is close to 256. 250 ÷ 50 = 25 ÷ 5 = 5.

T: Let's use the estimate to help us solve in the standard algorithm. (On the board, write and set up the standard algorithm below 256 ÷ 47.) Our estimated quotient is 5. I'll record that. (Record the quotient 5 in the ones place above 256.) What is 5 × 47? You may solve it on your personal white board if you like.

S: 235.

T: (Record 235 below 256.) How many are remaining?

S: 21.

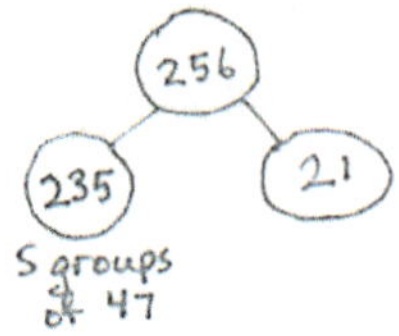




EUREKA MATH

T: (Record 21 in the algorithm.) Do we have enough for another group of 47?

S: No.

T: So, what does the 21 represent? Whisper to your neighbor.

S: This is what is left of our whole after we made all the groups of 47 we could.

T: How did our estimate help us solve?

S: It gave us a starting point for our quotient. → We estimated the quotient to be 5, and our actual quotient is 5 with a remainder of 21. The estimate was just right.

T: This time our estimate did not need to be adjusted. Why do you think that is the case?

S: We estimated 47 to be 50, and the whole was almost a multiple of 50. → Our divisor was smaller than 50, so we didn't go over. → Maybe if it was 54, it wouldn't have worked so well, even though it rounds to 50, too. → Yes, 54 would go over. 54 times 5 is 270, which is over 256.

T: Work with a partner to check the quotient.

T: One part is 5 complete groups of 47. The other part is the 21. What's the whole?

S: 256.

**Problem 2: 236 ÷ 39**

T: (Write 236 ÷ 39 horizontally on the board.) Think on your own. How will you estimate? (Give students time to think.) Tell me how you'll estimate.

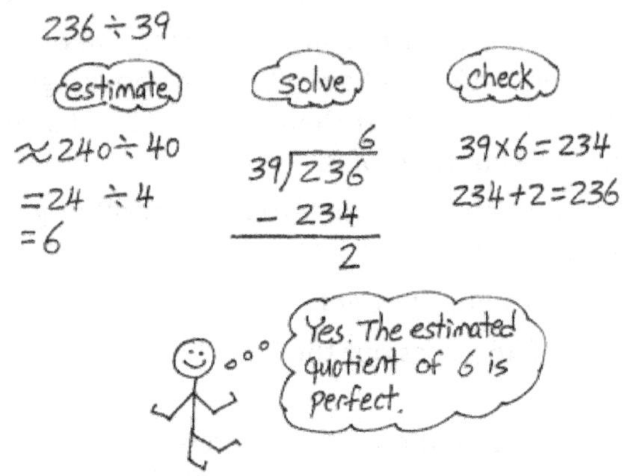


S: 240 ÷ 40 = 6.

T: What basic fact helped you to estimate?

S: 24 ÷ 6 = 4.

T: On your board, solve this problem with your partner using the standard algorithm. Partner A will divide using the standard algorithm, and Partner B will check the answer.

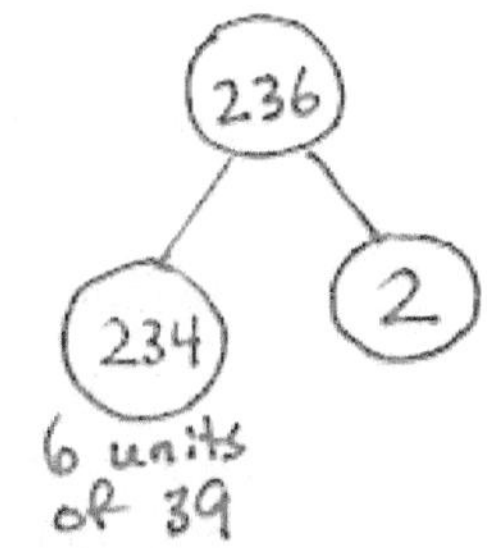


T: Let's go over the answer. Analyze why our estimate was perfect.

S: 39 is really close to our estimated divisor, 40. → The total was less than the rounded whole but by just a little bit. It was close. → 39 is one less than 40, so 6 groups of 39 will be 6 less than 240. The rounded quotient was 4 less than 240, so the difference is 2, our remainder!

T: What is 236 divided by 39?

S: The quotient is 6 with a remainder of 2.

T: Check it. How much is 39 × 6 + 2?

S: 236.

**NOTES ON MULTIPLE MEANS OF REPRESENTATION:**

Please note that there are multiple ways of recording the general method for division. Alternate ways of correcting underestimated quotients are included here. While the standard method for recording the algorithm is considered the goal in *A Story of Units,* some students may find these alternative recording methods more accessible.

**Problem 3: 369 ÷ 46**

T: (Write 369 ÷ 46 horizontally on the board.) How will you estimate the quotient?

S: 350 ÷ 50 = 7. → 400 ÷ 40 = 10. → 360 ÷ 40 = 9.

T: These are all reasonable estimates. Let's use 350 ÷ 50 = (350 ÷ 10) ÷ 5 = 35 ÷ 5 = 7. (Write the estimate below the problem.)

T: (Write the problem in a vertical algorithm, and record 7 in the ones column in the quotient.) How much is 46 × 7? You may solve on your board.

S: 322.

T: Subtract this from our whole. How many ones are remaining?

S: 47. (Record – 322 and 47 in the algorithm.)

MP.2

T: What do you notice about the remainder of 47 ones? Turn and discuss with your partner.

S: The remainder is larger than the group size, which means I have enough to make another group. → 47 is greater than the divisor of 46. We haven't made enough groups. We only made 7 groups of 46, but we can make 8. → Since 47 is bigger than 46, it means that the quotient of 7 is not big enough. We could try to use the quotient of 8.

T: We have 47 remaining. We agree that's enough to make another group of 46. We can record this several ways. (Write on the board.)

- Erase, start over, and use 8 as our quotient.
- Subtract one more group of 46, cross out the 7 at the top, and write in an 8.
- Subtract one more group of 46, and record a 1 above the 7 in our vertical algorithm.

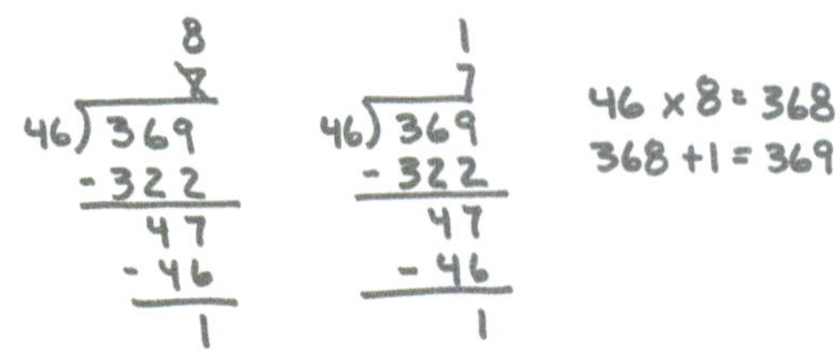


T: To state our final quotient, we will need to remember to add 7 and 1.

T: (Subtract one more unit of 46.) Now, how many are remaining?

S: 1.

T: (Record this in the algorithm.) Is that enough for another group of 46?

S: No.

T: How many forty-sixes are in 369?

S: 8 units of 46 with 1 one remaining.

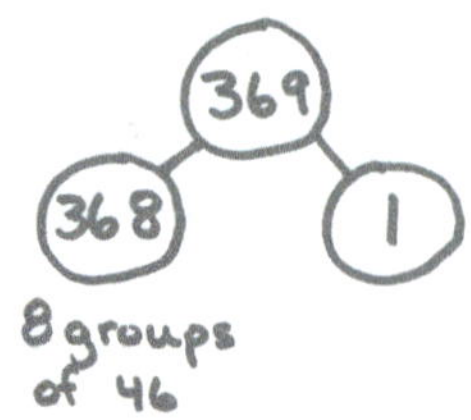


T: Check it. Remember that we have 8 units of 46. Solve 8 × 46 + 1. (Write the expression on the board.)

S: 369.

T: Let's go back and look at our original estimation. If you remember, I suggested 350 ÷ 50. Turn and talk to your partner about how we ended up with a quotient that was too small.

S: Our actual divisor was a lot smaller than the estimate. If the divisor is smaller, you can make more groups. → Also, our actual whole amount was bigger than our estimate. If the whole is larger, we can make more groups. → So, a smaller group size and larger whole meant our estimate was too small.

T: So, what can we say about estimating quotients?

S: Sometimes when we estimate a quotient, we need to be prepared to adjust it if necessary.

**Problem 4: 712 ÷ 94**

T: (Write 712 ÷ 94 horizontally on the board.) Use your personal white board. Talk with your partner, and estimate the quotient.

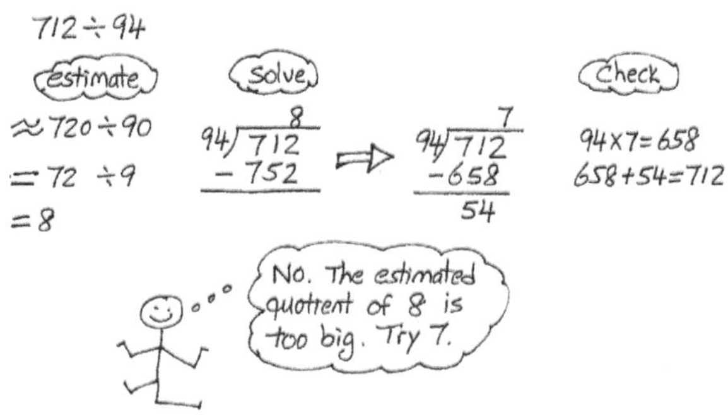


S: 700 ÷ 100 or just 7. → 720 ÷ 90, which is the same as 72 ÷ 9 = 8.

T: Both are reasonable estimates. Let's use the estimate that divides 720 by 90. That gives us an estimated quotient of 8. (Record this estimate on the board.) Talk with your partner about this estimate. What do you notice?

S: An estimate of 8 is too much because 8 groups of 90 is already more than 712, 8 × 90 = 720. We'll try 7 as our quotient.

T: What was your estimated quotient when you divided 700 by 10?

S: 7.

T: So, either estimate helped us get a starting place for our actual division. Even our imperfect estimate of 8 led us to the correct quotient. Now, finish the division, and check on your board. When you're finished, check it with a neighbor.

T: What's the answer for 712 divided by 94?

S: The quotient is 7 with a remainder of 54 ones.

T: Tell me the equations that you'd use to check your answer.

S: 94 × 7 = 658 and 658 + 54 = 712.

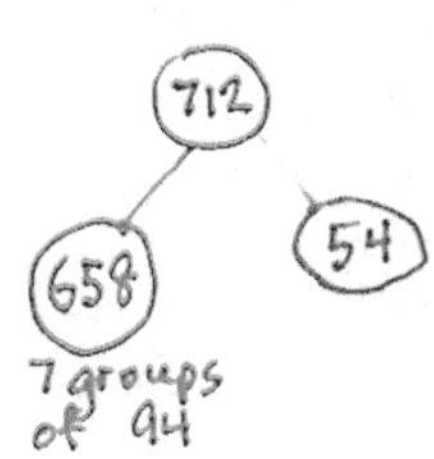

## Problem Set (10 minutes)

Students should do their personal best to complete the Problem Set within the allotted 10 minutes. For some classes, it may be appropriate to modify the assignment by specifying which problems they work on first. Some problems do not specify a method for solving. Students should solve these problems using the RDW approach used for Application Problems.

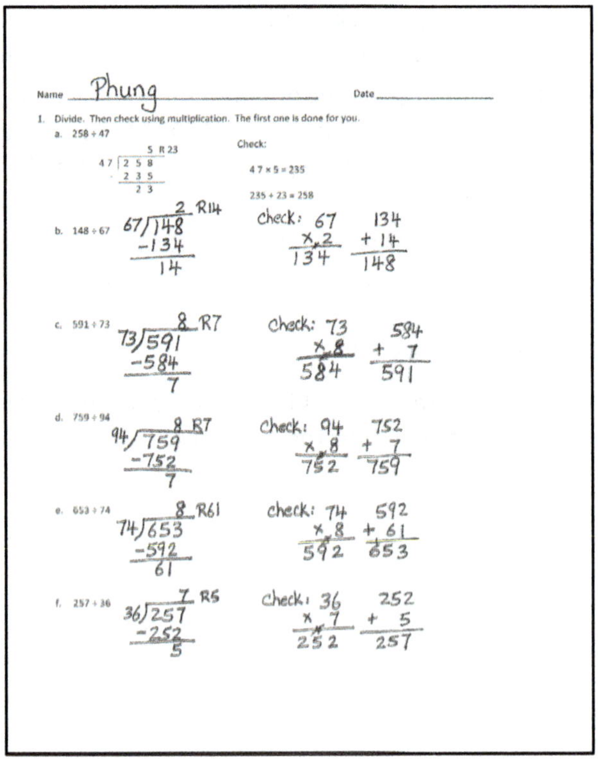

Name Phung Date

1. Divide. Then check using multiplication. The first one is done for you.

a. 258 ÷ 47 — 5 R 23; Check: 47 × 5 = 235; 235 + 23 = 258

b. 148 ÷ 67 — 2 R14; Check: 67 × 2 = 134; 134 + 14 = 148

c. 591 ÷ 73 — 8 R7; Check: 73 × 8 = 584; 584 + 7 = 591

d. 759 ÷ 94 — 8 R7; Check: 94 × 8 = 752; 752 + 7 = 759

e. 653 ÷ 74 — 8 R61; Check: 74 × 8 = 592; 592 + 61 = 653

f. 257 ÷ 36 — 7 R5; Check: 36 × 7 = 252; 252 + 5 = 257

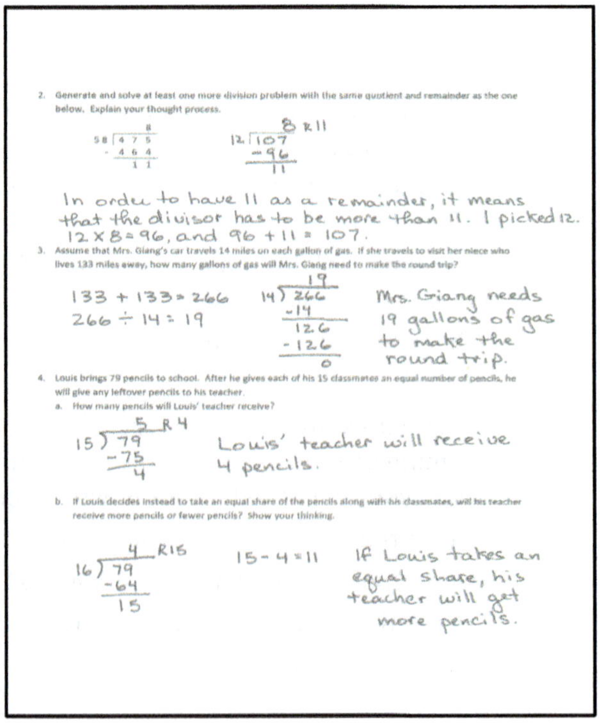

2. Generate and solve at least one more division problem with the same quotient and remainder as the one below. Explain your thought process.

58 ) 475 = 8, − 464, 11 12 ) 107 = 8 R11, − 96, 11

In order to have 11 as a remainder, it means that the divisor has to be more than 11. I picked 12. 12 × 8 = 96, and 96 + 11 = 107.

3. Assume that Mrs. Giang's car travels 14 miles on each gallon of gas. If she travels to visit her niece who lives 133 miles away, how many gallons of gas will Mrs. Giang need to make the round trip?

133 + 133 = 266; 266 ÷ 14 = 19; 14 ) 266 = 19, − 14, 126, − 126, 0

Mrs. Giang needs 19 gallons of gas to make the round trip.

4. Louis brings 79 pencils to school. After he gives each of his 15 classmates an equal number of pencils, he will give any leftover pencils to his teacher.

a. How many pencils will Louis' teacher receive?

15 ) 79 = 5 R 4, − 75, 4

Louis' teacher will receive 4 pencils.

b. If Louis decides instead to take an equal share of the pencils along with his classmates, will his teacher receive more pencils or fewer pencils? Show your thinking.

16 ) 79 = 4 R15, − 64, 15; 15 − 4 = 11

If Louis takes an equal share, his teacher will get more pencils.

## Student Debrief (10 minutes)

**Lesson Objective:** Divide two- and three-digit dividends by two-digit divisors with single-digit quotients, and make connections to a written method.

The Student Debrief is intended to invite reflection and active processing of the total lesson experience.

Invite students to review their solutions for the Problem Set. They should check work by comparing answers with a partner before going over answers as a class. Look for misconceptions or misunderstandings that can be addressed in the Debrief. Guide students in a conversation to debrief the Problem Set and process the lesson.

Any combination of the questions below may be used to lead the discussion.

- What similarity did you notice between Problems 1(c) and 1(d)? Since the quotient was 8 with remainder 7 for both problems, does that mean the two division expressions are equal to each other? Discuss the meaning of the quotient and remainder for both problems.
- In Problem 1, did your estimates need adjusting at times? When? What did you do in order to continue dividing?
- Share your thought process as you solved Problem 2. Can anyone share his or her solution? How many solutions might there be to this problem? Can you create another solution to it? How did your understanding of the check process help you answer this? Explain how the expression ($n \times 8$) + 11 might be used to solve this problem.

- What steps did you take as you solved Problem 3? Raise your hand if you doubled the distance (since 133 miles is just one way) *before* dividing. Try to find a classmate who solved this problem differently from you (one who doubled the quotient after dividing, perhaps). Compare your answers. What did you find?
- If the distance is doubled first, a quotient of 19 with no remainder is found. That is, Mrs. Giang only needs 19 gallons of gas. Solving the problem this way, however, results in a division problem with a two-digit quotient. Although finding a 2-digit quotient is taught in Lesson 22, many students are capable of the mathematics involved.
- If 133 (the one-way distance) is divided first, a quotient of 9 with 7 miles left to drive is found. Some students may interpret the remainder and conclude that 10 gallons is needed each way and double to arrive at a total of 20 gallons. (This amount of fuel would certainly allow Mrs. Giang to arrive at her destination with extra gas in her tank.) This is good reasoning.
- Students who divide first but are thinking more deeply may realize that if the quotient (9) is doubled, then the remainder (7 miles) must also be doubled. This yields 18 gallons of gas and 14 miles left over. This additional *left over* 14 miles requires 1 more gallon of gas, so Mrs. Giang needs at least 19 gallons of gas.
- Discuss thoroughly the remainders in Problem 4. It might be fruitful to allow students to make a prediction about the size of the remainder in Part (b) before computing. Many students may be surprised that the teacher receives *more* pencils even when *more* students are taking pencils. Discuss how this could be possible.
- Talk about how estimating makes the process of long division more efficient.
- The estimated quotient sometimes needs to be adjusted. Talk about why this may happen.

## Exit Ticket (3 minutes)

After the Student Debrief, instruct students to complete the Exit Ticket. A review of their work will help with assessing students' understanding of the concepts that were presented in today's lesson and planning more effectively for future lessons. The questions may be read aloud to the students.

Name ______________________________ Date ________________

1. Divide. Then, check using multiplication. The first one is done for you.

   a. 258 ÷ 47

$$\begin{array}{r} 5 \text{ R }23 \\ 47\overline{)258} \\ -\ 235 \\ \hline 23 \end{array}$$

*Check:*

47 × 5 = 235

235 + 23 = 258

   b. 148 ÷ 67

   c. 591 ÷ 73

   d. 759 ÷ 94





EUREKA MATH™

e. 653 ÷ 74

f. 257 ÷ 36

2. Generate and solve at least one more division problem with the same quotient and remainder as the one below. Explain your thought process.

$$\begin{array}{r} 8 \\ 58 \overline{) 475} \\ - \underline{464} \\ 11 \end{array}$$

3. Assume that Mrs. Giang's car travels 14 miles on each gallon of gas. If she travels to visit her niece who lives 133 miles away, how many gallons of gas will Mrs. Giang need to make the round trip?

4. Louis brings 79 pencils to school. After he gives each of his 15 classmates an equal number of pencils, he will give any leftover pencils to his teacher.

   a. How many pencils will Louis's teacher receive?

   b. If Louis decides instead to take an equal share of the pencils along with his classmates, will his teacher receive more pencils or fewer pencils? Show your thinking.

Name ______________________________ Date ______________

Divide. Then, check using multiplication.

a. 326 ÷ 53

b. 192 ÷ 38

Name ________________________________________ Date ____________________

1. Divide. Then, check using multiplication. The first one is done for you.

   a. $129 \div 21$

$$\begin{array}{r} 6 \text{ R}3 \\ 21\overline{)129} \\ -\ 126 \\ \hline 3 \end{array}$$

*Check:*

$21 \times 6 = 126$

$126 + 3 = 129$

   b. $158 \div 37$

   c. $261 \div 49$

   d. $574 \div 82$





EUREKA MATH™

e. $464 \div 58$

f. $640 \div 79$

2. It takes Juwan exactly 35 minutes by car to get to his grandmother's. The nearest parking area is a 4-minute walk from her apartment. One week, he realized that he spent 5 hours and 12 minutes traveling to her apartment and then back home. How many round trips did he make to visit his grandmother?

3. How many eighty-fours are in 672?

# Lesson 22

## Objective: Divide three- and four-digit dividends by two-digit divisors resulting in two- and three-digit quotients, reasoning about the decomposition of successive remainders in each place value.

### Suggested Lesson Structure

| | |
|---|---|
| Fluency Practice | (12 minutes) |
| Application Problem | (6 minutes) |
| Concept Development | (32 minutes) |
| Student Debrief | (10 minutes) |
| **Total Time** | **(60 minutes)** |

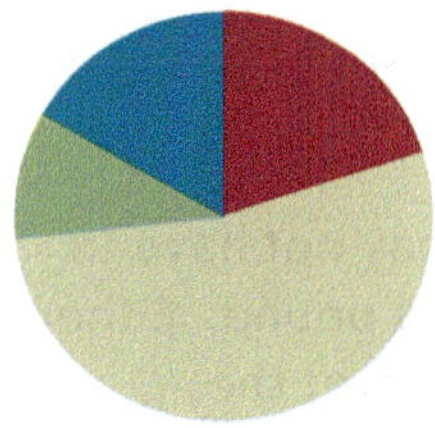

## Fluency Practice (12 minutes)

- Divide Decimals **5.NBT.7** (3 minutes)
- Group Count by Multi-Digit Numbers **5.NBT.6** (4 minutes)
- Divide by Two-Digit Numbers **5.NBT.6** (5 minutes)

### Divide Decimals (3 minutes)

Materials: (S) Personal white board

Note: This fluency activity prepares students for Lesson 24's Concept Development.

T: (Write 6 hundreds ÷ 2 = ____.) Say the division sentence in unit form.

S: 6 hundreds ÷ 2 = 3 hundreds.

Repeat the process with 6 tens ÷ 2, 6 ones ÷ 2, and 6 tenths ÷ 2.

T: On your personal white boards, write 6 tenths ÷ 2 in decimal form.

S: (Write 0.6 ÷ 2 = 0.3.)

Repeat the process for 6 hundredths ÷ 2, 8 thousands ÷ 2, 8 ones ÷ 2, 8 tenths ÷ 2, and 8 hundredths ÷ 2.

### Group Count by Multi-Digit Numbers (4 minutes)

Materials: (S) Personal white board

Note: This fluency activity prepares students for this lesson's Concept Development.

Repeat the process from Lesson 19 for 17 and 32.

**Divide by Two-Digit Numbers (5 minutes)**

Materials: (S) Personal white board

Note: This fluency exercise reviews Lesson 21 content.

Repeat the process from Lesson 21 for the following possible sequence: 208 ÷ 37, 128 ÷ 57, and 664 ÷ 83.

## Application Problem (6 minutes)

Zenin's baby sister weighed 132 ounces at birth. How much did his sister weigh in pounds and ounces?

Note: Depending on the class, you may or may not have to remind students that there are 16 ounces in a pound. Either way, it can be used as an opportunity to interpret the remainder (i.e., what does the remainder of 4 represent in this problem?).

$$\begin{array}{r} 8 \\ 16\overline{)132} \\ -128 \\ \hline 4 \end{array}$$

His baby sister weighed 8 pounds and 4 ounces.

## Concept Development (32 minutes)

Materials: (S) Personal white board

**Problem 1: 590 ÷ 17**

T: (Write 590 ÷ 17 in the algorithm on the board.) Can we divide 5 hundreds by 17?

S: Not without regrouping.

T: Let's work with 59 tens, then. We can divide 59 tens into 17 groups or groups of 17. Tell me how to estimate to divide 59 tens by 17.

S: 60 tens ÷ 20 = 3 tens

T: Record 3 tens, and find the remainder in the tens place. 3 tens times 17 is?

S: 51 tens.

MP.7

T: (Record 51 tens below 59 tens.) Remind me why we record here. (Point to the algorithm.)

S: We record the 5 in the hundreds place and the 1 is in the tens place because we know 51 tens is the same as 510.

T: How many tens are remaining?

S: 8 tens.

T: Can we divide 8 tens by 17?

S: Not without regrouping.

estimates

60 tens ÷ 20 = 3 tens

80 ones ÷ 20 = 4 ones

solution

$$\begin{array}{r} 34 \\ 17\overline{)590} \\ -51\phantom{0} \\ \hline 80 \\ -68 \\ \hline 12 \end{array}$$

check

$$\begin{array}{r} 34 \\ \times 17 \\ \hline 238 \\ +340 \\ \hline 578 \end{array} \qquad \begin{array}{r} 578 \\ +12 \\ \hline 590 \end{array}$$

**NOTES ON STANDARDS ALIGNMENT:**

The standards specifically require students to find quotients "using strategies based on place value" (**5.NBT.6**). When dividing, students are decomposing units just as they have done when subtracting since Grade 2: "I don't have enough tens to subtract, so I'll change 1 hundred for 10 tens." When dividing, they also change each larger unit that cannot be divided for smaller units: "I'll change 8 remaining tens for 80 ones."

T: We need to decompose these 8 tens into 80 ones. There are no ones in the whole to add in. (Point to the zero in the ones place of the whole.)

T: Now, we have 80 ones divided by 17. Tell me how to divide 80 by 17?

S: 80 ones ÷ 20 = 4 ones.

T: Record 4 ones in the quotient. What is 17 × 4 ones?

S: 68 ones.

T: What is 80 – 68? How many ones remain?

S: 12.

T: Could we make another group of 17?

S: No.

T: What is our quotient?

S: 34, remainder 12.

T: What is 34 units of 17 plus 12 ones?

S: 590.

**NOTES ON MULTIPLE MEANS OF ENGAGEMENT:**

At this point in the module on division, some students are ready for independent practice, while others clearly need more scaffolding and support. Teachers can allow some students to work independently while working with some students in a small group.

**Problem 2: 887 ÷ 27**

T: (Write 887 ÷ 27 in the algorithm on the board.) Let's divide together. Can we divide 8 hundreds by 27? (Point to the first digit of the dividend.)

S: No, we have to change the 8 hundreds to 80 tens. 80 tens and 8 tens is 88 tens. → We have to regroup to have 88 tens.

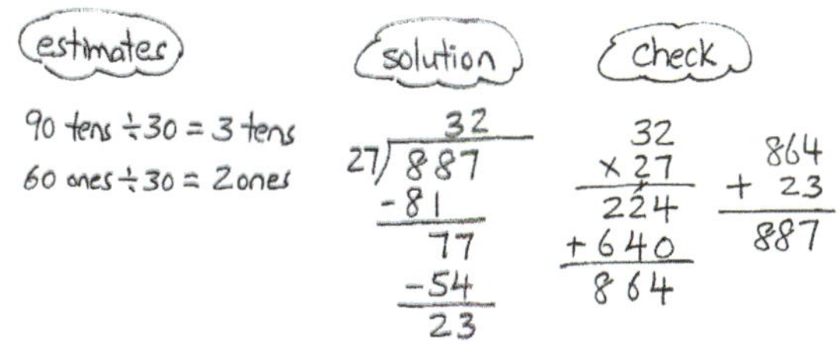


T: (Point to the first two digits of the dividend.) How would you estimate 88 tens divided by 27? Show me on your personal white board.

S: 90 tens ÷ 30 = 3 tens.

T: Record 3 tens in the quotient, and find the product of 3 tens and 27.

S: 81 tens.

T: How many tens are remaining?

S: 7 tens.

T: Can we divide 7 tens by 27, or must we regroup? Explain.

S: We need to regroup the 7 tens to 70 ones and combine them with the 7 ones in the whole to make 77 ones.

T: Now, we have 77 ones divided by 27. Show me how you'll estimate.

S: 60 ones ÷ 30 = 2 ones.

T: Record 2 ones in the quotient. What is 2 × 27?

S: 54 ones.

T: How many ones remain?

S: 23 ones.

T: Can we divide 23 ones by 27?
S: No, 23 is the remainder.
T: How many groups of 27 are in 887?
S: 32 groups.
T: With how many left over?
S: 23 remaining.
T: Complete the two-part check to make sure.

**Problem 3: 839 ÷ 41**

T: (Write 839 ÷ 41 in the algorithm on the board.) Solve this problem with a partner. As you finish each step, share your thinking with your partner.
S: (Work.)
T: Okay. Let's share your work. How did you first estimate to begin dividing?
S: 80 tens ÷ 40 = 2 tens.
T: 2 tens times 41 equals...?
S: 82 tens.
T: How many tens remain?
S: 1 ten.
T: What did you do next?
S: We regrouped the 1 ten and made 10 ones, and then we combined them with the 9 ones in the whole to make 19 ones.
T: What is 19 ones divided by 41?
S: Zero. It can't be divided.
T: What is the quotient, then?
S: 20, remainder 19.
T: Explain how you knew that the quotient was 20 with a remainder of 19 and not 2 with a remainder of 19. Turn and talk.
S: (Share.)
T: Did you check the answer? Was it correct?
S: Yes.

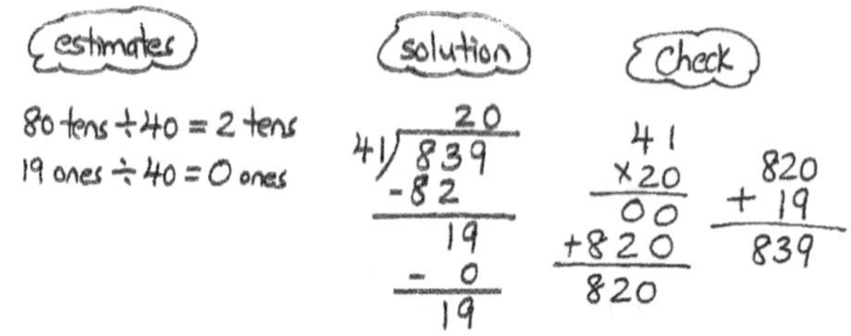


## Problem Set (10 minutes)

Students should do their personal best to complete the Problem Set within the allotted 10 minutes. For some classes, it may be appropriate to modify the assignment by specifying which problems they work on first. Some problems do not specify a method for solving. Students should solve these problems using the RDW approach used for Application Problems.





EUREKA MATH

## Student Debrief (10 minutes)

**Lesson Objective:** Divide three- and four-digit dividends by two-digit divisors resulting in two- and three-digit quotients, reasoning about the decomposition of successive remainders in each place value.

The Student Debrief is intended to invite reflection and active processing of the total lesson experience.

Invite students to review their solutions for the Problem Set. They should check work by comparing answers with a partner before going over answers as a class. Look for misconceptions or misunderstandings that can be addressed in the Debrief. Guide students in a conversation to debrief the Problem Set and process the lesson.

Any combination of the questions below may be used to lead the discussion.

- In Problem 1, was it ever necessary to adjust your quotients after estimating? If so, what did you do in order to continue dividing?
- While checking your work today, did anyone discover an error in his or her division? If so, how did you fix it? How did you know what to do?
- Explain your thought process as you solved Problem 1(f). What were you thinking as you recorded a digit in the ones place of your quotient and recorded the remainder? Was anyone tempted to say the answer was 4 with a remainder of 14?
- Talk to your partner about how you set up and solved Problem 2. What was your thinking like? How could you use your thinking to solve 660 ÷ 48 or 661 ÷ 48 or 662 ÷ 48, etc.? What would the total need to be in order to have a quotient of exactly 13?
- What did you have to do in order to solve Problem 3(b)? Talk with a neighbor.
- How did estimation help you to divide today?

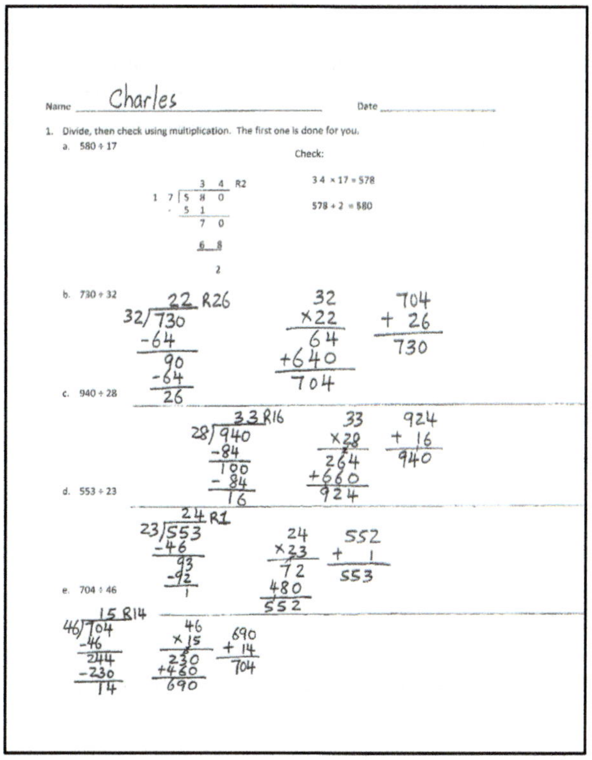

Name Charles Date

1. Divide, then check using multiplication. The first one is done for you.

a. 580 ÷ 17 = 34 R2 Check: 34 × 17 = 578; 578 + 2 = 580

b. 730 ÷ 32 = 22 R26 Check: 32 × 22 = 704; 704 + 26 = 730

c. 940 ÷ 28 = 33 R16 Check: 33 × 28 = 924; 924 + 16 = 940

d. 553 ÷ 23 = 24 R1 Check: 24 × 23 = 552; 552 + 1 = 553

e. 704 ÷ 46 = 15 R14 Check: 46 × 15 = 690; 690 + 14 = 704

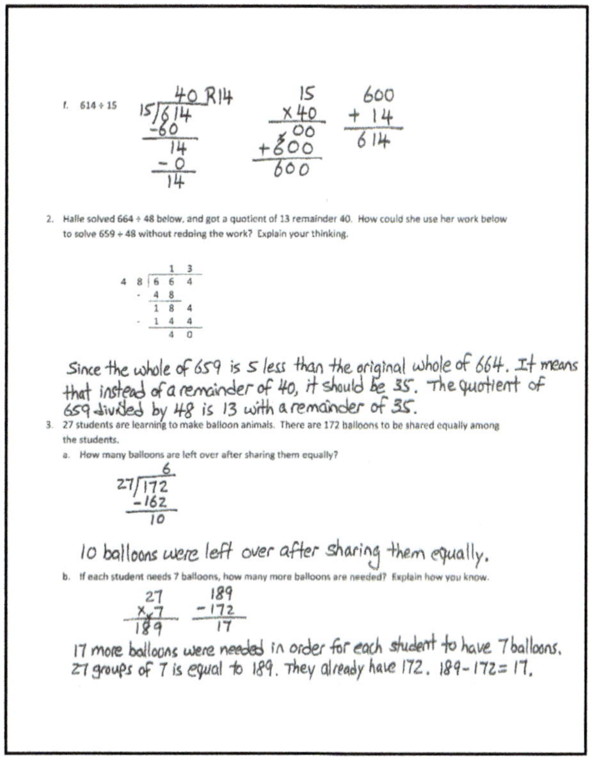

f. 614 ÷ 15 = 40 R14 Check: 15 × 40 = 600; 600 + 14 = 614

2. Halle solved 664 ÷ 48 below, and got a quotient of 13 remainder 40. How could she use her work below to solve 659 ÷ 48 without redoing the work? Explain your thinking.

Since the whole of 659 is 5 less than the original whole of 664. It means that instead of a remainder of 40, it should be 35. The quotient of 659 divided by 48 is 13 with a remainder of 35.

3. 27 students are learning to make balloon animals. There are 172 balloons to be shared equally among the students.

a. How many balloons are left over after sharing them equally?

10 balloons were left over after sharing them equally.

b. If each student needs 7 balloons, how many more balloons are needed? Explain how you know.

17 more balloons were needed in order for each student to have 7 balloons. 27 groups of 7 is equal to 189. They already have 172. 189 − 172 = 17.

## Exit Ticket (3 minutes)

After the Student Debrief, instruct students to complete the Exit Ticket. A review of their work will help with assessing students' understanding of the concepts that were presented in today's lesson and planning more effectively for future lessons. The questions may be read aloud to the students.

Name ____________________________________________ Date ____________________

1. Divide. Then, check using multiplication. The first one is done for you.

   a. 580 ÷ 17

```
      3 4 R2
17 | 5 8 0
   - 5 1
     -----
       7 0
     - 6 8
     -----
         2
```

*Check:*

34 × 17 = 578

578 + 2 = 580

   b. 730 ÷ 32

   c. 940 ÷ 28

   d. 553 ÷ 23

e. $704 \div 46$

f. $614 \div 15$

2. Halle solved $664 \div 48$ below. She got a quotient of 13 with a remainder of 40. How could she use her work below to solve $659 \div 48$ without redoing the work? Explain your thinking.

```
       1 3
48 | 6 6 4
  -  4 8
     1 8 4
   - 1 4 4
       4 0
```

3. 27 students are learning to make balloon animals. There are 172 balloons to be shared equally among the students.

   a. How many balloons are left over after sharing them equally?

   b. If each student needs 7 balloons, how many more balloons are needed? Explain how you know.

Name ______________________________ Date ______________

Divide. Then, check using multiplication.

a. $413 \div 19$

b. $708 \div 67$

Name ______________________________ Date ________________

1. Divide. Then, check using multiplication. The first one is done for you.

   a. 487 ÷ 21

$$\begin{array}{r} 23 \text{ R}4 \\ 21\overline{)487} \\ -\underline{42\phantom{7}} \\ 67 \\ -\underline{63} \\ 4 \end{array}$$

   *Check:*

   21 × 23 = 483

   483 + 4 = 487

   b. 485 ÷ 15

   c. 700 ÷ 21

   d. 399 ÷ 31

e. $820 \div 42$

f. $908 \div 56$

2. When dividing **878** by 31, a student finds a quotient of 28 with a remainder of 11. Check the student's work, and use the check to find the error in the solution.

3. A baker was going to arrange 432 desserts into rows of 28. The baker divides 432 by 28 and gets a quotient of 15 with remainder 12. Explain what the quotient and remainder represent.

# Lesson 23

## Objective: Divide three- and four-digit dividends by two-digit divisors resulting in two- and three-digit quotients, reasoning about the decomposition of successive remainders in each place value.

### Suggested Lesson Structure

| | |
|---|---|
| Fluency Practice | (12 minutes) |
| Application Problem | (5 minutes) |
| Concept Development | (33 minutes) |
| Student Debrief | (10 minutes) |
| **Total Time** | **(60 minutes)** |

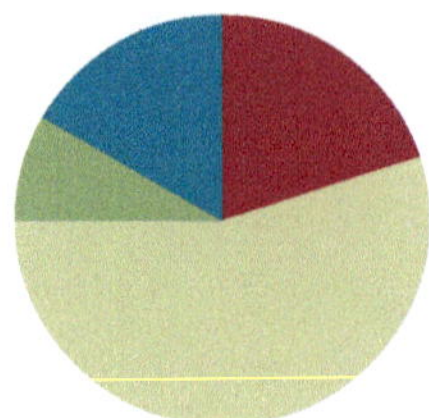

## Fluency Practice (12 minutes)

- Divide Decimals **5.NBT.7** (3 minutes)
- Rename Tenths and Hundredths **5.NBT.2** (4 minutes)
- Divide by Two-Digit Numbers **5.NBT.6** (5 minutes)

### Divide Decimals (3 minutes)

Materials: (S) Personal white board

Note: This fluency activity prepares students for the Concept Development in Lesson 24.

Repeat the process from Lesson 22 for the following possible sequence: 6 tens ÷ 3, 6 tenths ÷ 3, 6 hundredths ÷ 3, 9 thousands ÷ 3, 9 hundreds ÷ 3, 9 hundredths ÷ 3, and 9 tenths ÷ 3.

**NOTES ON MULTIPLE MEANS OF ENGAGEMENT:**

Today's lesson makes the transition from three-digit dividends to four-digit dividends. It is important to assess yesterday's Exit Ticket to determine if students are ready for this new complexity. It is not important that students master the skill yet. If the majority of the students are not yet showing an understanding of division concepts, the use of estimation, or displaying sound number sense, then consider doing an extra day of three-digit dividend work.

### Rename Tenths and Hundredths (4 minutes)

Materials: (S) Personal white board

Note: This exercise prepares students for estimating decimal quotients in Lesson 25.

T: I'll say a number, and you state it as you would write it. 1 tenth.

S: Zero point one.





EUREKA MATH™

Repeat the process for 2 tenths, 3 tenths, 8 tenths, and 9 tenths.

T: (Write 10 tenths =____.) Write the number.

S: (Write 1.)

Repeat the process for 11 tenths, 19 tenths, 20 tenths, 30 tenths, 80 tenths, 90 tenths, 100 tenths, and 200 tenths.

Repeat the process for 1 hundredth, 2 hundredths, 3 hundredths, 8 hundredths, 9 hundredths, 10 hundredths, 20 hundredths, 30 hundredths, 90 hundredths, 100 hundredths, 200 hundredths, 900 hundredths, 1,000 hundredths, and 2,000 hundredths.

### Divide by Two-Digit Numbers (5 minutes)

Materials: (S) Personal white board

Note: This exercise reviews Lesson 22 content.

Repeat the process from Lesson 21 for the following possible sequence: 650 ÷ 16, 740 ÷ 32, and 890 ÷ 27.

## Application Problem (5 minutes)

The rectangular room measures 224 square feet. One side of the room is 14 feet long. What is the perimeter of the room?

Note: This Application Problem builds on the previous day's lesson involving three-digit totals divided by two-digit divisors. It also provides a review of area and is a two-step problem.

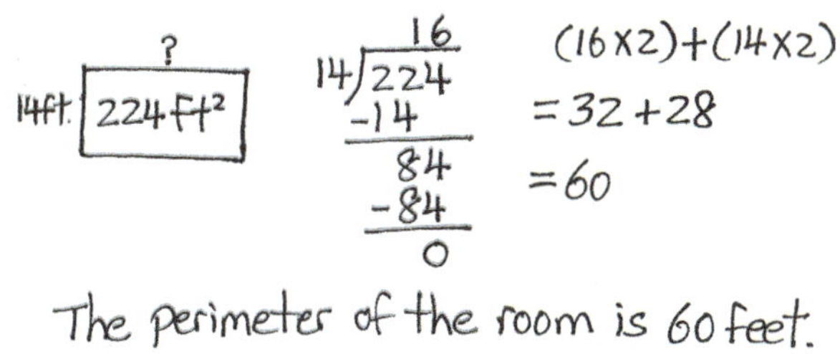


## Concept Development (33 minutes)

Materials: (S) Personal white board

**Problem 1: 6,247 ÷ 29**

T: (Write 6,247 ÷ 29 in the algorithm on the board.) Can we divide 6 thousands by 29?

S: Not without changing them to 62 hundreds.

T: Okay, then, work with 62 hundreds, which we can divide into 29 groups or groups of 29.

T: Divide 62 hundreds by 29. Show me how to estimate 62 hundreds divided by 29.

S: 60 hundreds ÷ 30 = 2 hundreds.

T: Record 2 in the hundreds place of the quotient.

T: What is 2 hundreds × 29? Solve on your personal white board.

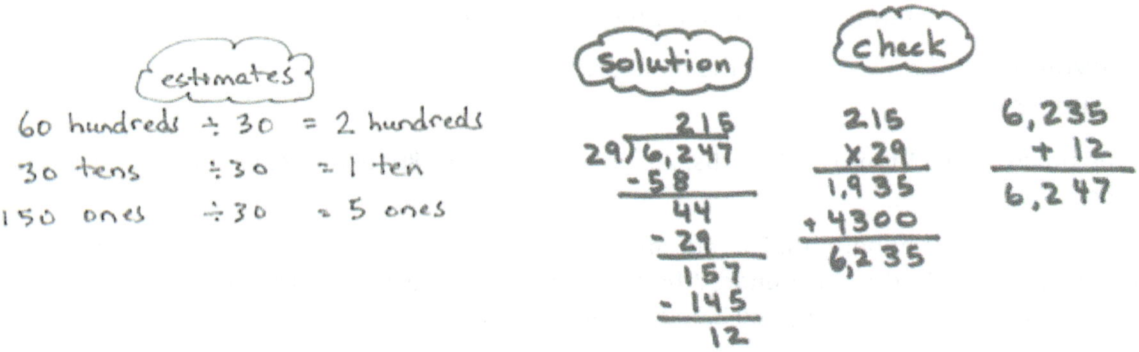


S: (Solve.) 58 hundreds.

T: Pay attention to place value as you carefully record this.

T: (Record in the algorithm.) How many hundreds are remaining?

S: 4 hundreds.

T: Decompose (regroup) those 4 hundreds into 40 tens plus the 4 tens in the whole. How many tens is that?

S: 44 tens.

T: Now, we must divide 44 tens by 29. Show me how you estimate 44 ÷ 29.

S: 30 tens ÷ 30 = 1 ten.

T: What is 1 ten × 29?

S: 29 tens.

T: 44 tens – 29 tens is ...?

S: 15 tens.

T: Can we divide again, or must we decompose? Explain.

S: We need to decompose 15 tens into 150 ones, plus the 7 ones in our whole, to make 157 ones. → We can't divide again because the remainder is less than the divisor.

T: Now, we have 157 ones divided by 29. Show me how you estimate 157 ÷ 29.

S: 150 ÷ 30 = 5.

T: What is 5 × 29?

S: 145.

T: How many are remaining?

S: 12.

T: What does that mean? Turn and talk.

S: When we divide 6,247 into twenty-nines we can make exactly 215 units of 29, with 12 left over. → Or, you could think of it as sharing 6,247 into 29 groups. There are 215 in each group with 12 left over.

**NOTES ON MULTIPLE MEANS OF ENGAGEMENT:**

While estimating, it is fair to assume that not all students in every class agree to round the dividend and divisor in the same way. For example, in Problem 1, some students may want to estimate 6,300 ÷ 30 = 210, while others may see 6,000 ÷ 25 = 240, and the majority probably want to estimate 6,000 ÷ 30. The intent here is not to rob students of their number sense or pigeonhole them into estimating one way but rather to cultivate their sense of how numbers relate to one another and be able to defend why they rounded how they did. In the end, however, in order to complete the problem as a group, the teacher must decide which approximation to use for the example being done on the board.

T: Let's check. Solve 215 × 29. (Wait for students to solve.)

S: 6,235.

T: 6,235 + 12?

S: 6,247.

**Problem 2: 4,289 ÷ 52**

T: (Write 4,289 ÷ 52 in the algorithm on the board.) Let's all complete this problem together. I'll work on the board. You work on your personal white boards.

S: (Work.)

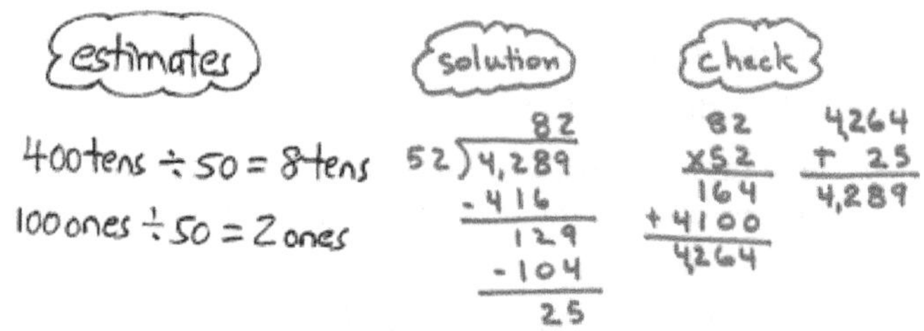


T: First, can we divide 4 thousands by 52?

S: No, we have to decompose.

T: Yes. How many hundreds do we have?

S: 42 hundreds.

T: Can we divide 42 hundreds by 52?

S: No. We have to decompose again.

T: Okay. How many tens do we have?

S: 428 tens.

T: Good. Now, we can divide 428 tens by 52. Show me how to estimate for 428 tens divided by 29.

S: 400 tens ÷ 50 = 8 tens.

T: Record 8 in the tens place of the quotient.

T: What is 8 tens × 52?

S: 416 tens.

T: Pay attention to place value as you carefully record this.

T: (Record in the algorithm.) How many tens are remaining?

S: 12 tens.

T: Decompose (regroup) those 12 tens into 120 ones, plus the 9 ones in the whole. How many ones is that?

S: 129 ones.

T: Now, we divide 129 ones by 52? Show me how to estimate 129 ÷ 52.

S: 100 ones ÷ 50 = 2 ones.

T: What is 2 × 52?

S: 104 ones.

T: 129 ones – 104 ones gives a remainder of…?

S: 25 ones.

T: Are we finished, or do we continue to decompose and divide? Explain.

S: We are finished. 25 is our remainder, and we don't need to continue to decompose to the tenths place.

T: Did you check your answer? Was it correct?

S: Yes.

**Problem 3: 6,649 ÷ 63**

T: (Write 6,649 ÷ 63 in the algorithm on the board.) Solve this problem with a partner. As you finish each step, share your thinking with your partner.

S: (Work while the teacher circulates and assists where necessary.)

T: Okay. Let's share your work. How did you first estimate to begin dividing?

S: 60 hundreds ÷ 60 = 1 hundred.

T: 1 hundred times 63 equals...?

S: 63 hundreds.

T: How many hundreds remain?

S: 3 hundreds.

T: What did you do next?

S: We regrouped the 3 hundreds and made 30 tens. Then, we combined the 30 tens with the 4 tens in the whole to make 34 tens.

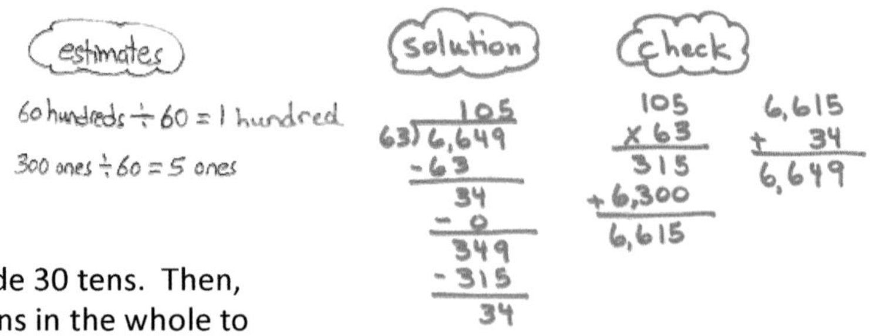


T: Can we divide 34 tens by 63?

S: No. We have to decompose.

T: Yes. Record 0 in the tens place of the quotient. Now, we decompose; what's 340 ones plus 9 ones?

S: 349 ones.

T: How did you estimate 349 divided by 63?

S: 300 ÷ 60 = 5.

T: What's 5 × 63?

S: 315.

T: What's the remainder?

S: 34.

T: Did you check the answer? Was it correct?

S: Yes.

**Problem 4: 3,164 ÷ 45**

T: (Write 3,164 ÷ 45 in the algorithm on the board.) Solve this problem independently. Do all three steps independently: estimate, solve, and check. After you finish each step, check your answer with a partner before moving on.

Follow the questioning sequence from above. Allow students to discuss the recording of 0 ones thoroughly.

## Problem Set (10 minutes)

Students should do their personal best to complete the Problem Set within the allotted 10 minutes.
For some classes, it may be appropriate to modify the assignment by specifying which problems they work on first. Some problems do not specify a method for solving. Students should solve these problems using the RDW approach used for Application Problems.

## Student Debrief (10 minutes)

**Lesson Objective:** Divide three- and four-digit dividends by two-digit divisors resulting in two- and three-digit quotients, reasoning about the decomposition of successive remainders in each place value.

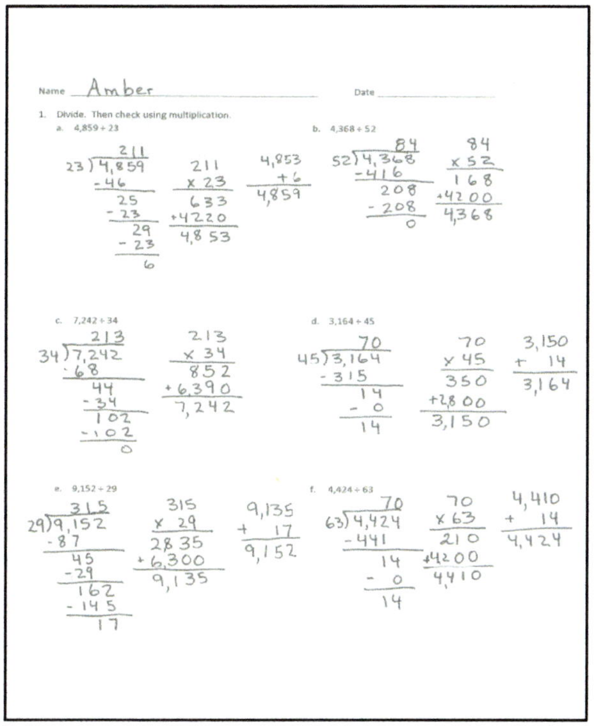


The Student Debrief is intended to invite reflection and active processing of the total lesson experience.

Invite students to review their solutions for the Problem Set. They should check work by comparing answers with a partner before going over answers as a class. Look for misconceptions or misunderstandings that can be addressed in the Debrief. Guide students in a conversation to debrief the Problem Set and process the lesson.

Any combination of the questions below may be used to lead the discussion.

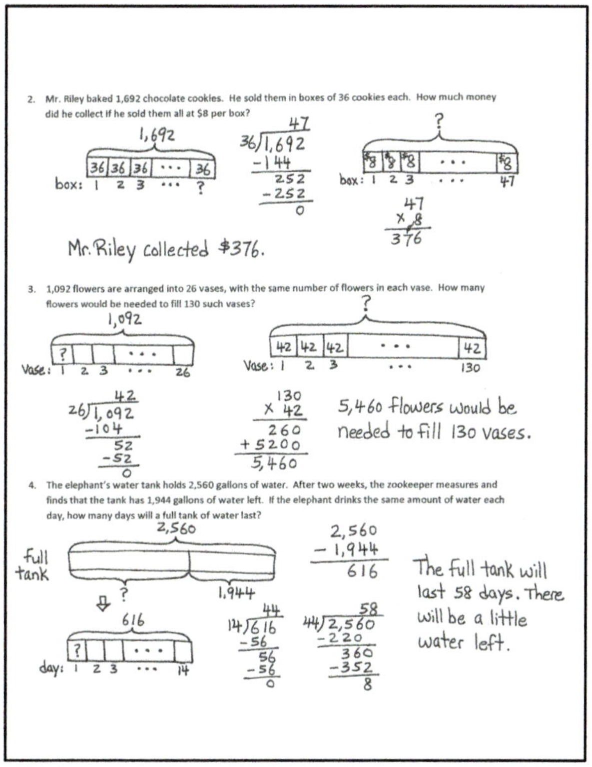


- What pattern did you notice between Problems 1(d) and 1(f)? Since the quotient was 70 with a remainder of 14 for both problems, does that mean these division expressions are equal? Discuss the meaning of the remainder for both problems. Does the remainder of 14 represent the same thing? Does the quotient of 70 represent the same thing? Are the 70 units in Problem 1(d) equal to 70 units in Problem 1(f)? (The quotient in Problem 1(d) means 70 groups of 45, with 14 remaining. → The quotient in Problem 1(f) means 70 groups of 63, with 14 remaining.)
- When dividing, did your estimate need to be adjusted at times? When? What did you do in order to continue dividing?
- Compare your quotients in Problem 1. What did you notice in Problems 1(a), (b), and (c)? Will a four-digit total divided by a two-digit divisor always result in a three-digit quotient? How does the relationship between the divisor and the whole impact the number of digits in the quotient? Can you create a problem that will result in a two-digit quotient? A three-digit quotient?
- Discuss student approaches to finding the number of days the full tank will last in Problem 4. Various interpretations of the remainders will engender different answers between 56 and 57 days.

## Exit Ticket (3 minutes)

After the Student Debrief, instruct students to complete the Exit Ticket. A review of their work will help with assessing students' understanding of the concepts that were presented in today's lesson and planning more effectively for future lessons. The questions may be read aloud to the students.

Name ____________________________________________ Date ________________________

1. Divide. Then, check using multiplication.

a. 4,859 ÷ 23

b. 4,368 ÷ 52

c. 7,242 ÷ 34

d. 3,164 ÷ 45

e. 9,152 ÷ 29

f. 4,424 ÷ 63

2. Mr. Riley baked 1,692 chocolate cookies. He sold them in boxes of 36 cookies each. How much money did he collect if he sold them all at $8 per box?

3. 1,092 flowers are arranged into 26 vases, with the same number of flowers in each vase. How many flowers would be needed to fill 130 such vases?

4. The elephant's water tank holds 2,560 gallons of water. After two weeks, the zookeeper measures and finds that the tank has 1,944 gallons of water left. If the elephant drinks the same amount of water each day, how many days will a full tank of water last?

Name ____________________________________ Date ____________________

Divide. Then, check using multiplication.

a. 8,283 ÷ 19

b. 1,056 ÷ 37

Name ___________________________________ Date ____________________

1. Divide. Then, check using multiplication.

a. 9,962 ÷ 41

b. 1,495 ÷ 45

c. 6,691 ÷ 28

d. 2,625 ÷ 32

e. 2,409 ÷ 19

f. 5,821 ÷ 62

2. A political gathering in South America was attended by 7,910 people. Each of South America's 14 countries was equally represented. How many representatives attended from each country?

3. A candy company packages caramel into containers that hold 32 fluid ounces. In the last batch, 1,848 fluid ounces of caramel were made. How many containers were needed for this batch?

Topic G

# Partial Quotients and Multi-Digit Decimal Division

**5.NBT.2, 5.NBT.7**

| | | |
|---|---|---|
| **Focus Standards:** | 5.NBT.2 | Explain patterns in the number of zeros of the product when multiplying a number by powers of 10, and explain patterns in the placement of the decimal point when a decimal is multiplied or divided by a power of 10. Use whole-number exponents to denote power of 10. |
| | 5.NBT.7 | Add, subtract, multiply, and divide decimals to hundredths, using concrete models or drawings and strategies based on place value, properties of operations, and/or the relationship between addition and subtraction; relate the strategy to a written method and explain the reasoning used. |
| **Instructional Days:** | 4 | |
| **Coherence -Links from:** | G5–M1 | Place Value and Decimal Fractions |
| **-Links to:** | G6–M2 | Arithmetic Operations Including Division of Fractions |

Topic G uses the knowledge students have accumulated about whole number division with double-digit divisors and extends it to division of decimals by double-digit divisors (**5.NBT.7**). Parallels between sharing or grouping whole number units and sharing or grouping decimal units are the emphasis of Topic G. Students quickly surmise that the concepts of division remain the same regardless of the size of the units being shared or grouped. Placement of the decimal point in quotients is based on students' reasoning about when wholes are being shared or grouped and when the part being shared or grouped transitions into fractional parts. Students reason about remainders in a deeper way than in previous grades. Students consider cases in which remainders expressed as whole numbers appear to be equivalent; however, equivalence is disproven when such remainders are decomposed as decimal units and shared or grouped.



EUREKA MATH

## A Teaching Sequence Toward Mastery of Partial Quotients and Multi-Digit Decimal Division

**Objective 1:** **Divide decimal dividends by multiples of 10, reasoning about the placement of the decimal point and making connections to a written method.**
**(Lesson 24)**

**Objective 2:** **Use basic facts to approximate decimal quotients with two-digit divisors, reasoning about the placement of the decimal point.**
**(Lesson 25)**

**Objective 3:** **Divide decimal dividends by two-digit divisors, estimating quotients, reasoning about the placement of the decimal point, and making connections to a written method.**
**(Lessons 26–27)**

# Lesson 24

## Objective: Divide decimal dividends by multiples of 10, reasoning about the placement of the decimal point and making connections to a written method.

### Suggested Lesson Structure

| | |
|---|---|
| Fluency Practice | (12 minutes) |
| Application Problem | (7 minutes) |
| Concept Development | (31 minutes) |
| Student Debrief | (10 minutes) |
| **Total Time** | **(60 minutes)** |

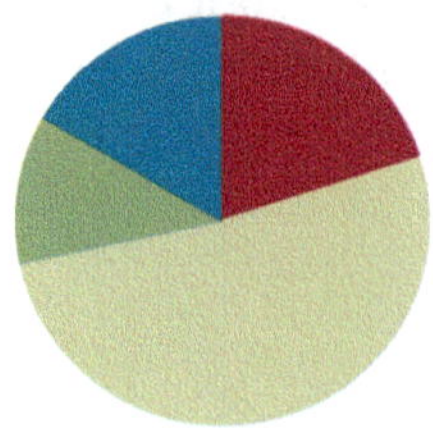

## Fluency Practice (12 minutes)

- Rename Tenths and Hundredths **5.NBT.2** (4 minutes)
- Divide Decimals **5.NBT.7** (3 minutes)
- Divide by Two-Digit Numbers **5.NBT.6** (5 minutes)

### Rename Tenths and Hundredths (4 minutes)

Materials: (S) Personal white board

Note: This fluency activity prepares students for estimating decimal quotients in Lesson 25.

Repeat the process from Lesson 23 using the following possible sequence: 9 tenths, 10 tenths, 20 tenths, 90 tenths, 95 tenths, 100 tenths, 200 tenths, 600 tenths, 650 tenths, 657 tenths, 832 tenths, 9 hundredths, 10 hundredths, 20 hundredths, 90 hundredths, 95 hundredths, 100 hundredths, 200 hundredths, 900 hundredths, 950 hundredths, 1,000 hundredths, 2,000 hundredths, 5,000 hundredths, 5,800 hundredths, 5,830 hundredths, 5,834 hundredths, and 2,834 hundredths.

### Divide Decimals (3 minutes)

Materials: (S) Personal white board

Note: This fluency activity prepares students for the Concept Development in today's lesson.

Repeat the process from Lesson 22 using the following possible sequence:
15 ones ÷ 5, 15 tenths ÷ 5, 15 hundredths ÷ 5, 12 tens ÷ 3, 12 tenths ÷ 3, 24 hundreds ÷ 6, and 24 hundredths ÷ 6.

### Divide by Two-Digit Numbers (5 minutes)

Materials: (S) Personal white board

Note: This exercise reviews Lesson 23 content.

Repeat the process from Lesson 21 using the following possible sequence: 5,349 ÷ 21, 6,816 ÷ 32, and 4,378 ÷ 51.

## Application Problem (7 minutes)

A long-time runner compiled her training distances in the following chart. Fill in the missing values.

**Runner's Log**

| Total Number of Miles Run | Number of Days | Miles Run Each Day |
|---|---|---|
| 420 | | 12 |
| 14.5 | 5 | |
| 38.0 | 10 | |
| | 17 | 16.5 |

Note: This Application Problem serves as a quick review of multi-digit multiplication and division with double-digit divisors. Students must determine which operation is needed, and the review of whole number division can serve as an anticipatory set for today's continuation into decimal divisors.

**NOTES ON MULTIPLE MEANS OF ENGAGEMENT:**

It may be challenging for some students to articulate their ideas without a moment to prepare. One strategy that can help struggling students is to ask them to restate what they hear the teacher or another student say. For example, the teacher might say, "When I've finished explaining this problem, I'm going to ask you to restate my explanation in your own words."

| Total Miles | # of Days | Miles Run |
|---|---|---|
| 420 | 35 | 12 |
| 14.5 | 5 | 2.9 |
| 38.0 | 10 | 3.8 |
| 280.5 | 17 | 16.5 |

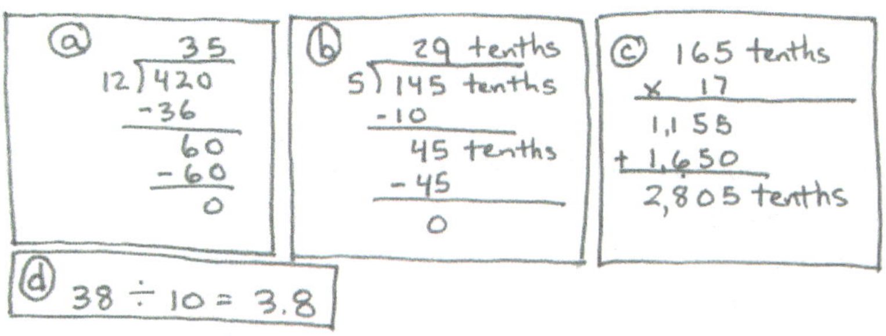

## Concept Development (31 minutes)

Materials: (S) Millions to thousandths place value chart (Lesson 1 Template), personal white board

**Problems 1–3**

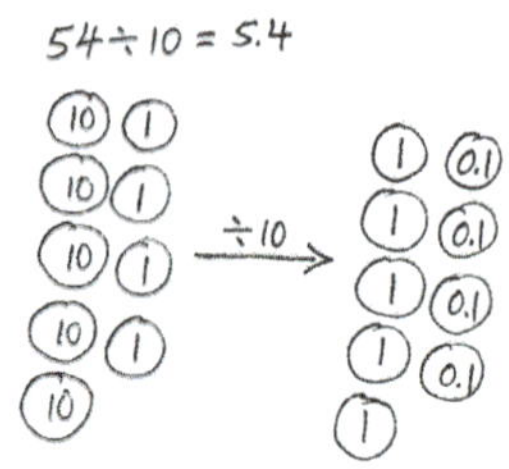


54 ÷ 10

5.4 ÷ 10

0.54 ÷ 10

MP.2

T: (Write 54 ÷ 10 horizontally on the board.) Let's solve this problem using place value disks. Draw 5 tens disks and 4 ones disks on your personal white board.

Student and teacher draw 5 tens disks and 4 ones disks, as shown to the right.

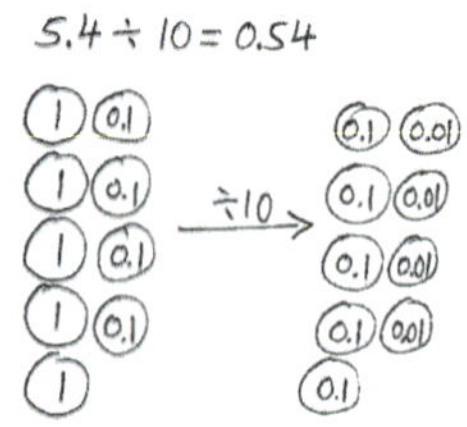


T: Say this in unit form.
S: 5 tens 4 ones.
T: What is 1 ten divided by 10?
S: 1 one.
T: If 1 ten divided by 10 is 1 one, what is 5 tens divided by 10?
S: 5 ones.

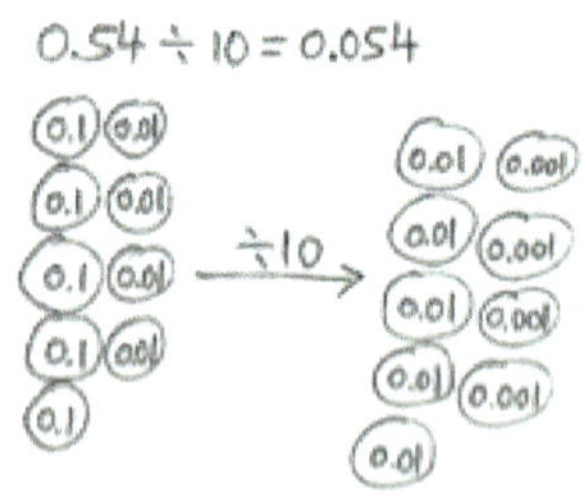


T: I'll show that division with my place value disks. You do the same. (Draw an arrow showing ÷ 10 and 5 ones disks.)
T: What is 1 one divided by 10?
S: 1 tenth.
T: If 1 one divided by 10 is 1 tenth, what is 4 ones divided by 10?
S: 4 tenths.
T: Show that division with place value disks.
T: (Point to the original problem.) Read the equation with the solution.
S: 54 ÷ 10 = 5.4.
T: (Write 5.4 ÷ 10 on the board.) Compare this problem to 54 divided by 10. Turn and talk.
S: The whole is less than the first one, but we are still dividing by 10. → 5.4 is 1 tenth as large as 54. → The quotient from our first problem is now the whole. → The first whole is 10 times as large, so its quotient should also be 10 times larger than the quotient of 5.4 ÷ 10.

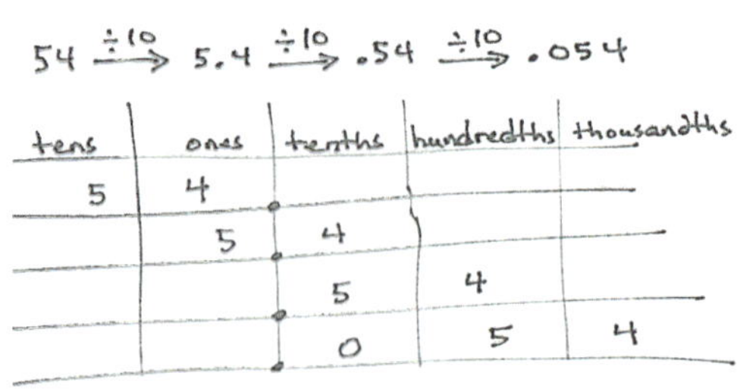


T: Imagine what this number would look like on a place value chart. When we divide, what will happen to the digits and why?
S: They will move to the right one place value because they are being divided into smaller units.

T: What pattern do you notice in the placement of the decimal? Turn and talk.

S: (Share.)

Follow a similar sequence for this problem and the others in this Problem Set. Use Module 1 knowledge of the place value chart to support division with the disks. Please refer to the graphics for examples of student work.

**Problems 4–8**

$54 \div 90$

$5.4 \div 90$

$0.54 \div 90$

$54 \div 900$

$5.4 \div 900$

T: (Write $54 \div 90$ horizontally on the board.) How is this problem different than the others we've solved? Turn and talk.

S: I know 54 divided by 9 equals 6. → We're still dividing with tens, but there are 9 tens rather than 1 ten.

T: Our divisor this time is 90. Can you decompose 90 with 10 as a factor?

S: Yes, $10 \times 9 = 90$.

T: I'll rewrite the problem to reflect our thinking. (Write $54 \div 90 = 54 \div 10 \div 9$.) Turn and tell your neighbor the quotient of 54 divided by 10. If necessary, you may use your place value disks, chart, or visualize what happens when dividing by 10.

T: What is 54 divided by 10?

S: 5.4.

T: Are we finished?

S: No, we still need to divide by 9.

T: Say the division equation we now have to solve.

S: Five and four tenths divided by 9.

T: Read this equation naming 5.4 as tenths.

S: 54 tenths divided by 9.

T: Solve it on your personal white boards.

T: Say the original division equation with the quotient.

S: 54 divided by 90 equals 6 tenths.

T: When we factored our divisor as $10 \times 9$, we first divided by 10. Then, we divided by 9. Would our quotient be affected if we divided by 9 and then by 10? Why or why not? Turn and talk.

**NOTES ON MULTIPLE MEANS OF ENGAGEMENT:**

Many students may benefit if teachers think aloud as they solve a problem. This strategy is often referred to as *self-talk,* wherein a teacher doesn't ask any questions as the problem is solved. Instead, the teacher talks through each step, verbalizing why each decision is made, as if talking out loud to his or herself.

This strategy is beneficial for students who do not have enough background knowledge or vocabulary to answer questions.

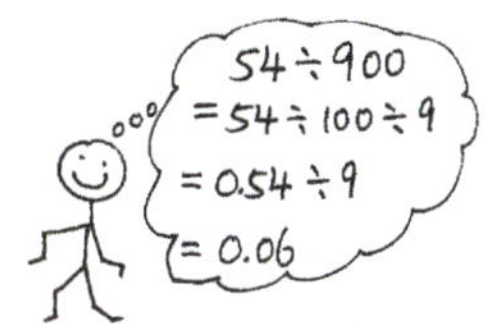

S: No. It wouldn't matter because we are still dividing by 90 either way. → $9 \times 10$ and $10 \times 9$ are both equal to 90. $54 \div (10 \times 9) = 54 \div (9 \times 10)$. Our divisor wasn't changed, so the quotient wouldn't change. → $(54 \div 10) \div 9 = (54 \div 9) \div 10$.

Repeat this sequence with the other problems in the set. Please refer to the graphics for student work.

### Problem Set (10 minutes)

Students should do their personal best to complete the Problem Set within the allotted 10 minutes.
For some classes, it may be appropriate to modify the assignment by specifying which problems they work on first. Some problems do not specify a method for solving. Students should solve these problems using the RDW approach used for Application Problems.

## Student Debrief (10 minutes)

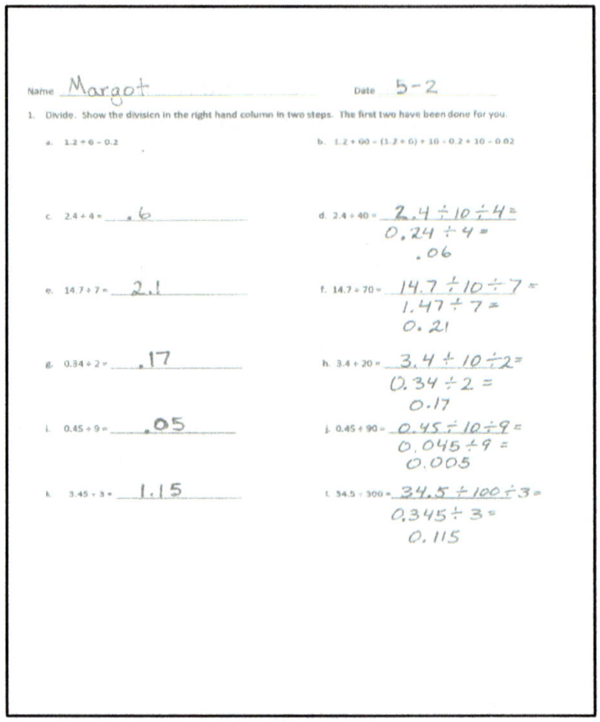
Name Margot Date 5-2

1. Divide. Show the division in the right hand column in two steps. The first two have been done for you.

a. $1.2 \div 6 = 0.2$

b. $1.2 \div 60 = (1.2 \div 6) \div 10 = 0.2 \div 10 = 0.02$

c. $2.4 \div 4 =$ .6

d. $2.4 \div 40 =$ $2.4 \div 10 \div 4 =$ $0.24 \div 4 =$ .06

e. $14.7 \div 7 =$ 2.1

f. $14.7 \div 70 =$ $14.7 \div 10 \div 7 =$ $1.47 \div 7 =$ 0.21

g. $0.34 \div 2 =$ .17

h. $3.4 \div 20 =$ $3.4 \div 10 \div 2 =$ $0.34 \div 2 =$ 0.17

i. $0.45 \div 9 =$ .05

j. $0.45 \div 90 =$ $0.45 \div 10 \div 9 =$ $0.045 \div 9 =$ 0.005

k. $3.45 \div 3 =$ 1.15

l. $34.5 \div 300 =$ $34.5 \div 100 \div 3 =$ $0.345 \div 3 =$ 0.115

**Lesson Objective:** Divide decimal dividends by multiples of 10, reasoning about the placement of the decimal point and making connections to a written method.

The Student Debrief is intended to invite reflection and active processing of the total lesson experience.

Invite students to review their solutions for the Problem Set. They should check work by comparing answers with a partner before going over answers as a class. Look for misconceptions or misunderstandings that can be addressed in the Debrief. Guide students in a conversation to debrief the Problem Set and process the lesson.

Any combination of the questions below may be used to lead the discussion.

- Describe the pattern that you noticed in our lesson and Problem Set when a smaller number is divided by a greater number.
- Discuss Problems 1(g) and 1(h). Ask, "The divisors and wholes are different in these problems, yet the quotients are the same. How is this possible?"
- In Problem 1(j), by which factor of 90 did you divide first? Find someone who divided the same way you did. Now, find someone who did it differently. Compare your approach and quotients.
- Challenge students to generate another pair of problems similar to Problems 1(g) and 1(h).

## Exit Ticket (3 minutes)

After the Student Debrief, instruct students to complete the Exit Ticket. A review of their work will help with assessing students' understanding of the concepts that were presented in today's lesson and planning more effectively for future lessons. The questions may be read aloud to the students.

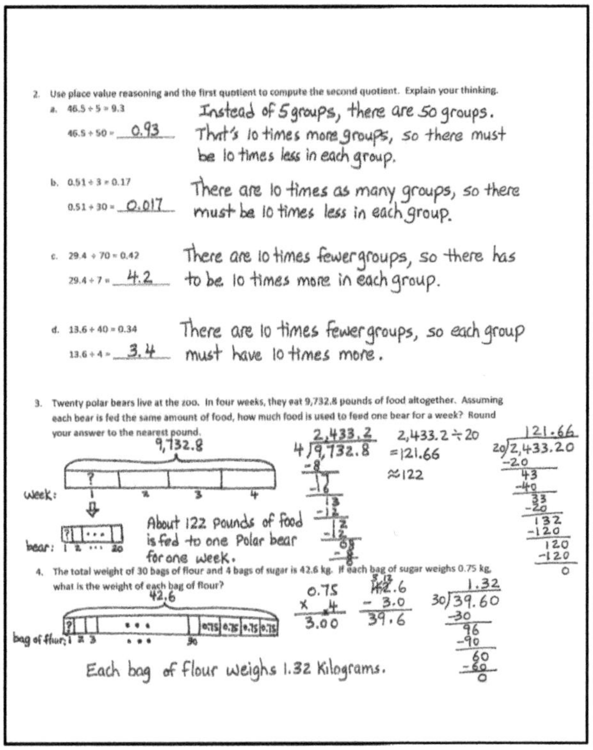

2. Use place value reasoning and the first quotient to compute the second quotient. Explain your thinking.

a. 46.5 ÷ 5 = 9.3
46.5 ÷ 50 = 0.93
Instead of 5 groups, there are 50 groups. That's 10 times more groups, so there must be 10 times less in each group.

b. 0.51 ÷ 3 = 0.17
0.51 ÷ 30 = 0.017
There are 10 times as many groups, so there must be 10 times less in each group.

c. 29.4 ÷ 70 = 0.42
29.4 ÷ 7 = 4.2
There are 10 times fewer groups, so there has to be 10 times more in each group.

d. 13.6 ÷ 40 = 0.34
13.6 ÷ 4 = 3.4
There are 10 times fewer groups, so each group must have 10 times more.

3. Twenty polar bears live at the zoo. In four weeks, they eat 9,732.8 pounds of food altogether. Assuming each bear is fed the same amount of food, how much food is used to feed one bear for a week? Round your answer to the nearest pound.

9,732.8 ÷ 4 = 2,433.2
2,433.2 ÷ 20 = 121.66 ≈ 122

About 122 pounds of food is fed to one polar bear for one week.

4. The total weight of 30 bags of flour and 4 bags of sugar is 42.6 kg. If each bag of sugar weighs 0.75 kg, what is the weight of each bag of flour?

0.75 × 4 = 3.00
42.6 − 3.0 = 39.6
39.60 ÷ 30 = 1.32

Each bag of flour weighs 1.32 kilograms.

Name ________________________________ Date ____________

1. Divide. Show the division in the right-hand column in two steps. The first two have been done for you.

a. $1.2 \div 6 = 0.2$

b. $1.2 \div 60 = (1.2 \div 6) \div 10 = 0.2 \div 10 = 0.02$

c. $2.4 \div 4 =$ ____________

d. $2.4 \div 40 =$ ____________

e. $14.7 \div 7 =$ ____________

f. $14.7 \div 70 =$ ____________

g. $0.34 \div 2 =$ ____________

h. $3.4 \div 20 =$ ____________

i. $0.45 \div 9 =$ ____________

j. $0.45 \div 90 =$ ____________

k. $3.45 \div 3 =$ ____________

l. $34.5 \div 300 =$ ____________





EUREKA MATH™

2. Use place value reasoning and the first quotient to compute the second quotient. Explain your thinking.

   a. 46.5 ÷ 5 = 9.3

      46.5 ÷ 50 = ___________

   b. 0.51 ÷ 3 = 0.17

      0.51 ÷ 30 = ___________

   c. 29.4 ÷ 70 = 0.42

      29.4 ÷ 7 = ___________

   d. 13.6 ÷ 40 = 0.34

      13.6 ÷ 4 = ___________

3. Twenty polar bears live at the zoo. In four weeks, they eat 9,732.8 pounds of food altogether. Assuming each bear is fed the same amount of food, how much food is used to feed one bear for a week? Round your answer to the nearest pound.

4. The total weight of 30 bags of flour and 4 bags of sugar is 42.6 kg. If each bag of sugar weighs 0.75 kg, what is the weight of each bag of flour?




G5-M2-TE-B2-1.3.1-01.2016

Name ______________________________ Date ______________

1. Divide.

   a. $27.3 \div 3$

   b. $2.73 \div 30$

   c. $273 \div 300$

2. If $7.29 \div 9 = 0.81$, then the quotient of $7.29 \div 90$ is ______________. Use place value reasoning to explain the placement of the decimal point.

Name ______________________________ Date ______________

1. Divide. Show every other division sentence in two steps. The first two have been done for you.

a. 1.8 ÷ 6 = 0.3

b. 1.8 ÷ 60 = (1.8 ÷ 6) ÷ 10 = 0.3 ÷ 10 = 0.03

c. 2.4 ÷ 8 = ______________________

d. 2.4 ÷ 80 = ______________________

e. 14.6 ÷ 2 = ______________________

f. 14.6 ÷ 20 = ______________________

g. 0.8 ÷ 4 = ______________________

h. 80 ÷ 400 = ______________________

i. 0.56 ÷ 7 = ______________________

j. 0.56 ÷ 70 = ______________________

k. 9.45 ÷ 9 = ______________________

l. 9.45 ÷ 900 = ______________________

2. Use place value reasoning and the first quotient to compute the second quotient. Use place value to explain how you placed the decimal point.

a. 65.6 ÷ 80 = 0.82

65.6 ÷ 8 = __________

b. 2.5 ÷ 50 = 0.05

2.5 ÷ 5 = __________

c. 19.2 ÷ 40 = 0.48

19.2 ÷ 4 = __________

d. 39.6 ÷ 6 = 6.6

39.6 ÷ 60 = __________

3. Chris rode his bike along the same route every day for 60 days. He logged that he had gone exactly 127.8 miles.

   a. How many miles did he bike each day? Show your work to explain how you know.

   b. How many miles did he bike over the course of two weeks?

4. 2.1 liters of coffee were equally distributed to 30 cups. How many milliliters of coffee were in each cup?

# Lesson 25

## Objective: Use basic facts to approximate decimal quotients with two-digit divisors, reasoning about the placement of the decimal point.

### Suggested Lesson Structure

| | |
|---|---|
| Fluency Practice | (12 minutes) |
| Application Problem | (7 minutes) |
| Concept Development | (31 minutes) |
| Student Debrief | (10 minutes) |
| **Total Time** | **(60 minutes)** |

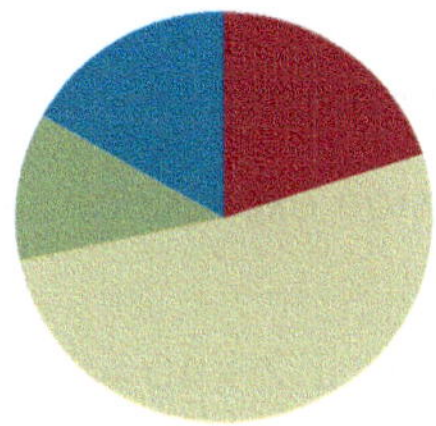

## Fluency Practice (12 minutes)

- Rename Tenths and Hundredths **5.NBT.2** (4 minutes)
- Divide Decimals by 10 **5.NBT.7** (4 minutes)
- Divide Decimals by Multiples of 10 **5.NBT.7** (4 minutes)

### Rename Tenths and Hundredths (4 minutes)

Materials: (S) Personal white board

Note: This fluency activity prepares students for estimating decimal quotients during the Concept Development.

Repeat the process from Lesson 23 for the following possible sequence: 10 tenths, 90 tenths, 94 tenths, 100 tenths, 700 tenths, 783 tenths, 372 tenths, 9 hundredths, 10 hundredths, 90 hundredths, 98 hundredths, 100 hundredths, 900 hundredths, 980 hundredths, 1,000 hundredths, 7,000 hundredths, 7,400 hundredths, 7,418 hundredths, and 4,835 hundredths.

### Divide Decimals by 10 (4 minutes)

Materials: (S) Personal white board

Note: This fluency activity reviews Lesson 24 content.

T: (Project 3,800 on a place value chart. To the side, write 3,800 ÷ 10.) Say the division sentence.

S: 3,800 ÷ 10 = 380.

T: (Cross out each digit, and draw arrows one place value to the right. Write 380 in the place value chart.) When dividing by 10, digits shift how many places to the right?

S: One.

T: (Project 380 on a place value chart. To the side, write 380 ÷ 10.) On your personal white board, write the division sentence and answer.

S: (Write 380 ÷ 10 = 38.)

Repeat the process for 38 ÷ 10, 3.8 ÷ 10, 270 ÷ 10, 2.7 ÷ 10, 4,900 ÷ 10, 49 ÷ 10, and 0.49 ÷ 10.

### Divide Decimals by Multiples of 10 (4 minutes)

Materials: (T) Millions to thousandths place value chart (Lesson 1 Template) (S) Personal white board

Note: This fluency activity reviews Lesson 24 content.

T: (Write 1.2 ÷ 4 = ____.) Solve the division expression saying the whole in tenths.

S: 12 tenths ÷ 4 = 3 tenths.

T: (Write 1.2 ÷ 4 = 0.3. To the right, write 1.2 ÷ 40 = ____.) On your personal white board, write 12 tenths ÷ 40 as a three-step division sentence, taking out the ten.

S: (Write (1.2 ÷ 10) ÷ 4 = 0.12 ÷ 4 = 0.03.)

Repeat the process for 2.4 ÷ 2, 2.4 ÷ 20, 8 ÷ 2, 8 ÷ 20, 0.35 ÷ 5, and 0.35 ÷ 50.

## Application Problem (7 minutes)

Ms. Heinz spent 12 dollars on 30 bus tokens for the field trip. What was the cost of 12 tokens?

Note: This Application Problem is based on Lesson 24, where students divided decimals by 10 and multiples of 10. This also asks students to multiply decimals to find the answer, which is a review of the first half of Module 2.

\$12.00 ÷ 30
= \$12.00 ÷ (3 × 10)
= (\$12.00 ÷ 3) ÷ 10
= \$4.00 ÷ 10
= \$0.40

1 unit = \$0.40
12 units = \$0.40 × 12
= \$4.80

$$\begin{array}{r} 0.40 \\ \times\ 12 \\ \hline 80 \\ +\ 400 \\ \hline 4.80 \end{array}$$

The cost of 12 tokens was \$4.80.

## Concept Development (31 minutes)

Materials: (S) Personal white board

**Problem 1**

39.1 ÷ 17

3.91 ÷ 17

T: In Module 1, we rounded our decimal factors to estimate the product. We will estimate quotients now by rounding the whole and the divisor. (Write 39.1 ÷ 19 horizontally on the board.)

T: Just as we did before, round the divisor first. What is 17 rounded to the nearest ten?

S: 20.

T: Let's record our estimation. (Under the original problem, write ≈ ____ ÷ 20 on the board.) We need to round our whole, 39.1, to a number that can easily be divided by 20. Turn and share your ideas with your partner.

T: What could we round 39.1 to?

S: I can round 39.1 to 40. → I can use mental math to divide 4 tens by 2 tens.

T: (Fill in the blank to get ≈ 40 ÷ 20.) Show me how to find the estimated quotient of 39.1 ÷ 17.

S: 40 ÷ 20 = 4 ÷ 2 = 2.

T: So, 39.1 ÷ 17 ≈ 2.

T: (Write 3.91 ÷ 17 on the board.) Think about the size of this related quotient based on the estimation we just made. Turn and talk.

> **NOTES ON MULTIPLE MEANS OF ACTION AND EXPRESSION:**
>
> Allow students to express the rounded whole number dividends in unit form because these representations may help students *see* the division of a smaller number by a larger number more easily. For example, students who struggle seeing 4 as a number that can be divided by 20 may have more success if it is written as 40 tenths.

S: The whole has the same digits, but it is 1 tenth the size of the first one. → I think the quotient will also be 1 tenth the size of the first one. → The quotient will probably be around 2 tenths because that is 1 tenth as large as 2.

T: Let's estimate the quotient. Because our divisor is the same, let's use the same estimate of 20. Can you think of a multiple of 2 that would be close to 3.91?

S: 4.

T: (Write 4 ÷ 20 ≈ ____.) Show me how to find the estimated quotient. Talk to your partner about your thinking.

S: 4 ÷ 20 = 4 ÷ 10 ÷ 2 = 0.4 ÷ 2 = 0.2. → 4 ones divided by 10 is 4 tenths, and 4 tenths divided by 2 is 2 tenths. → 4 ÷ 2 = 2 and 2 ÷ 10 = 0.2.

T: Show me the equation to find the estimated quotient.

S: 4 ÷ 20 = 0.2.

T: I noticed that you all factored 20 into 2 × 10, but some of you divided by 2 first, and others divided by 10 first. How did this affect your quotient?

S: It didn't affect it, because 2 × 10 and 10 × 2 are both 20. As long as our divisor is still 20, the order doesn't matter.

T: Why is estimating useful?

S: It helps give us a starting place when we need to find the actual quotient, just like with whole numbers. → I'm unsure of the value of 2.42 ÷ 12, but an estimate can help me think about the value of the quotient when a smaller number is divided by a larger number. → 24 tenths divided by 12 is easy: 2 tenths.

**Problem 2**

63.6 ÷ 73

6.36 ÷ 73

T: 63.6 pounds of rice were put into 73 bags. About how many pounds of rice were in each bag? (Write the problem horizontally on the board.) Thinking about this story problem, will the number of pounds in each bag be more than 1 pound or less than 1 pound? How do you know?

S: It should be less than 1 pound because there are 73 bags and only 63 pounds. There's not enough to put 1 pound in each bag. → Less than 1 pound. To put 1 pound in each bag, you'd need 73 pounds, and we don't have that much.

MP.1

T: Let's estimate this quotient and test that thinking. Turn and talk.

S: We can estimate the divisor as 70. I can see a 63 in the whole, and that is a multiple of 7. → I will round 73 to 70 and think of a close multiple of 7. 63 is close to 63.6.

T: What is the estimation expression in standard form?

S: 63 ÷ 70. → 63.0 ÷ 70.

T: Show your division in two steps on your personal white board.

S: 63 ÷ 10 = 6.3 and 6.3 ÷ 7 = 0.9. → 63 ÷ 7 = 9 and 9 ÷ 10 = 0.9.

Repeat this sequence with 6.36 ÷ 73. Have students reason about how large the quotient would be and how it relates to the first quotient.

**NOTES ON MULTIPLE MEANS OF ACTION AND EXPRESSION:**

In this lesson, students use easily identifiable multiples to find an estimated quotient. Remind students about the relationship between multiplication and division (the inverse property), so they can think of the following division sentence as a multiplication equation:

1,200 ÷ 40 = ___ → 40 × ___ = 1200.

**Problem 3**

11.72 ÷ 42

T: (Write 11.72 ÷ 42 horizontally on the board.) Read the division expression in word form.

S: 11 and 72 hundredths divided by 42.

T: Estimate the divisor. Turn and talk.

S: 42 is close to 40.

T: Whisper to your partner a multiple of 4 that is close to 11.72. Then, find the estimated quotient.

T: Show how to find the estimated quotient.

S: 1,200 hundredths ÷ 40. → 12 ÷ 40. → 12.00 ÷ 40.

T: (Write the problem horizontally on board.) 11.72 ÷ 42 ≈ 12 ÷ 40. What's your estimate?

S: 30 hundredths. → 0.30. → 3 tenths. → 0.3.

T: Explain how you found your answer.

S: (Record student thinking on the board.) I divided 1,200 by 10 to get 120. Then, I divided 120 by 4 and got 30.

T: When you are working on your Problem Set, I want you to remember to estimate using easily identifiable multiples.

## Problem Set (10 minutes)

Students should do their personal best to complete the Problem Set within the allotted 10 minutes. For some classes, it may be appropriate to modify the assignment by specifying which problems they work on first. Some problems do not specify a method for solving. Students should solve these problems using the RDW approach used for Application Problems.

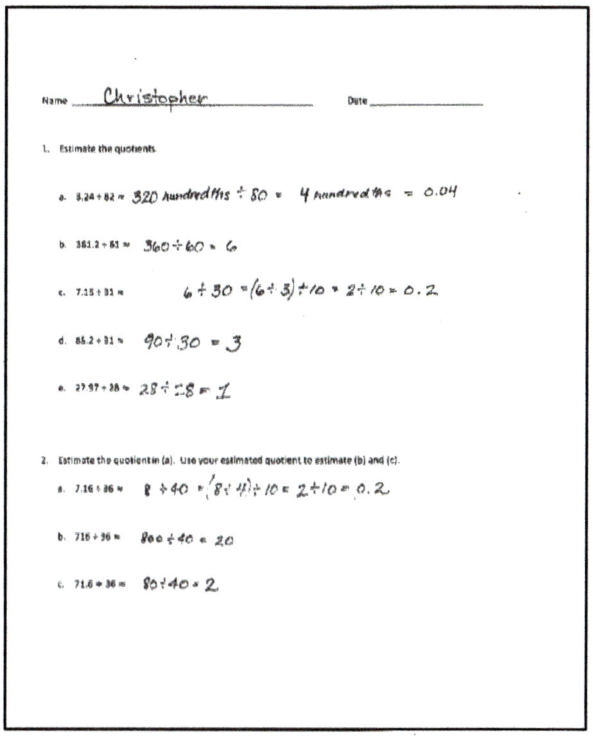
Name Christopher Date

1. Estimate the quotients.

a. 3.24 ÷ 82 ≈ 320 hundredths ÷ 80 = 4 hundredths = 0.04

b. 361.2 ÷ 61 ≈ 360 ÷ 60 = 6

c. 7.15 ÷ 31 ≈ 6 ÷ 30 = (6 ÷ 3) ÷ 10 = 2 ÷ 10 = 0.2

d. 85.2 ÷ 31 ≈ 90 ÷ 30 = 3

e. 27.97 ÷ 28 ≈ 28 ÷ 28 = 1

2. Estimate the quotient in (a). Use your estimated quotient to estimate (b) and (c).

a. 7.16 ÷ 36 ≈ 8 ÷ 40 = (8 ÷ 4) ÷ 10 = 2 ÷ 10 = 0.2

b. 716 ÷ 36 ≈ 800 ÷ 40 = 20

c. 71.6 ÷ 36 ≈ 80 ÷ 40 = 2

## Student Debrief (10 minutes)

**Lesson Objective:** Use basic facts to approximate decimal quotients with two-digit divisors, reasoning about the placement of the decimal point.

The Student Debrief is intended to invite reflection and active processing of the total lesson experience.

Invite students to review their solutions for the Problem Set. They should check work by comparing answers with a partner before going over answers as a class.

Look for misconceptions or misunderstandings that can be addressed in the Debrief. Guide students in a conversation to debrief the Problem Set and process the lesson.

Any combination of the questions below may be used to lead the discussion.

- Before students begin the Problem Set, have them predict and sort the tasks in Problem 1 into those with quotients more than 1 and quotients less than 1. Have them justify their thinking as they sort.
- Have students compare estimates for Problems 1(c) and 1(e) and defend their choices.
- Could your answer to Problem 1(c) help you find the answer to Problem 1(d) without having to make another estimate? (Although the divisors round to the same number, another estimate is needed.)
- How is Problem 4 like Problem 1(e)? (Divisors can be left and dividends estimated for both.) Are there other problems where this method of estimating the quotient makes sense?

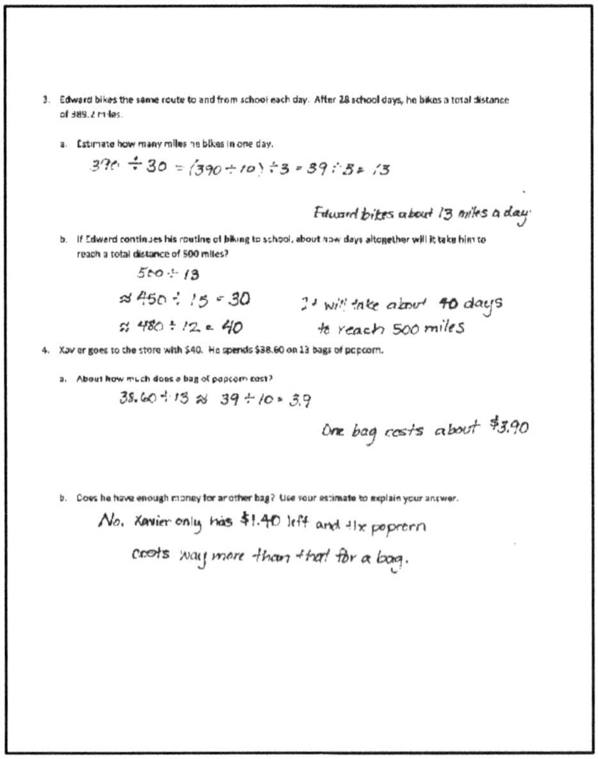
3. Edward bikes the same route to and from school each day. After 28 school days, he bikes a total distance of 389.2 miles.

a. Estimate how many miles he bikes in one day.

390 ÷ 30 = (390 ÷ 10) ÷ 3 = 39 ÷ 3 = 13

Edward bikes about 13 miles a day.

b. If Edward continues his routine of biking to school, about how many days altogether will it take him to reach a total distance of 500 miles?

500 ÷ 13
≈ 450 ÷ 15 = 30
≈ 480 ÷ 12 = 40

It will take about 40 days to reach 500 miles

4. Xavier goes to the store with $40. He spends $38.60 on 13 bags of popcorn.

a. About how much does a bag of popcorn cost?

38.60 ÷ 13 ≈ 39 ÷ 10 = 3.9

One bag costs about $3.90

b. Does he have enough money for another bag? Use your estimate to explain your answer.

No. Xavier only has $1.40 left and the popcorn costs way more than that for a bag.

## Exit Ticket (3 minutes)

After the Student Debrief, instruct students to complete the Exit Ticket. A review of their work will help with assessing students' understanding of the concepts that were presented in today's lesson and planning more effectively for future lessons. The questions may be read aloud to the students.

Name ______________________________ Date ________________

1. Estimate the quotients.

   a. $3.24 \div 82 \approx$

   b. $361.2 \div 61 \approx$

   c. $7.15 \div 31 \approx$

   d. $85.2 \div 31 \approx$

   e. $27.97 \div 28 \approx$

2. Estimate the quotient in (a). Use your estimated quotient to estimate (b) and (c).

   a. $7.16 \div 36 \approx$

   b. $716 \div 36 \approx$

   c. $71.6 \div 36 \approx$

3. Edward bikes the same route to and from school each day. After 28 school days, he bikes a total distance of 389.2 miles.

   a. Estimate how many miles he bikes in one day.

   b. If Edward continues his routine of biking to school, about how many days altogether will it take him to reach a total distance of 500 miles?

4. Xavier goes to the store with $40. He spends $38.60 on 13 bags of popcorn.

   a. About how much does one bag of popcorn cost?

   b. Does he have enough money for another bag? Use your estimate to explain your answer.

Name ______________________________ Date ________________

Estimate the quotients.

a. $1.64 \div 22 \approx$

b. $123.8 \div 62 \approx$

c. $6.15 \div 31 \approx$

Name ______________________________ Date ________________

1. Estimate the quotients.

   a. $3.53 \div 51 \approx$

   b. $24.2 \div 42 \approx$

   c. $9.13 \div 23 \approx$

   d. $79.2 \div 39 \approx$

   e. $7.19 \div 58 \approx$

2. Estimate the quotient in (a). Use your estimated quotient to estimate (b) and (c).

   a. $9.13 \div 42 \approx$

   b. $913 \div 42 \approx$

   c. $91.3 \div 42 \approx$

3. Mrs. Huynh bought a bag of 3 dozen toy animals as party favors for her son's birthday party. The bag of toy animals cost $28.97. Estimate the price of each toy animal.

4. Carter drank 15.75 gallons of water in 4 weeks. He drank the same amount of water each day.

   a. Estimate how many gallons he drank in one day.

   b. Estimate how many gallons he drank in one week.

   c. About how many days altogether will it take him to drink 20 gallons?

# Lesson 26

## Objective: Divide decimal dividends by two-digit divisors, estimating quotients, reasoning about the placement of the decimal point, and making connections to a written method.

### Suggested Lesson Structure

| | |
|---|---|
| ■ Fluency Practice | (12 minutes) |
| ■ Concept Development | (31 minutes) |
| ■ Application Problem | (7 minutes) |
| ■ Student Debrief | (10 minutes) |
| **Total Time** | **(60 minutes)** |

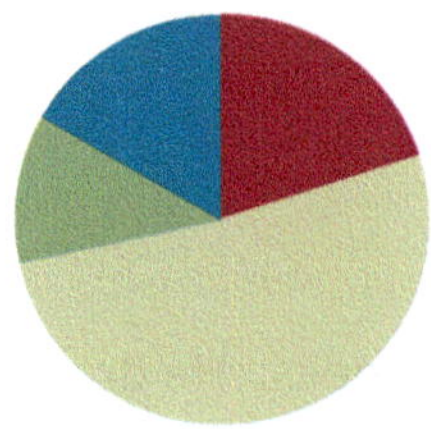

## Fluency Practice (12 minutes)

- Rename Tenths and Hundredths **5.NBT.2** (4 minutes)
- Divide Decimals by Multiples of 10 **5.NBT.7** (4 minutes)
- Estimate the Quotient **5.NBT.7** (4 minutes)

### Rename Tenths and Hundredths (4 minutes)

Materials: (S) Personal white board

Note: This fluency activity prepares students for estimating decimal quotients during the Concept Development.

Repeat the process from Lesson 23 using the following possible sequence: 10 tenths, 90 tenths, 93 tenths, 100 tenths, 800 tenths, 483 tenths, 9 hundredths, 10 hundredths, 90 hundredths, 97 hundredths, 100 hundredths, 900 hundredths, 970 hundredths, 1,000 hundredths, 8,000 hundredths, 8,417 hundredths, and 5,946 hundredths.

**NOTES ON MULTIPLE MEANS OF REPRESENTATION:**

Because unit form can be used, the only moment the decimal point is necessary is when representing the quotient. This allows students to use their work with whole-number division to support them as they transition into decimal division.

Fluencies, such as Rename Tenths and Hundredths, support the smooth movement between unit form and standard form.

### Divide Decimals by Multiples of 10 (4 minutes)

Materials: (S) Personal white board

Note: This fluency activity reviews Lesson 24 content.

Repeat the process from Lesson 25 using the following possible sequence: 1.2 ÷ 3, 1.2 ÷ 30, 9.6 ÷ 3, 9.6 ÷ 30, 8 ÷ 4, 8 ÷ 40, 0.45 ÷ 5, and 0.45 ÷ 50.

## Estimate the Quotient (4 minutes)

Materials: (S) Personal white board

Note: This fluency activity reviews Lesson 25 content.

T: (Write 15.4 ÷ 32.) Say the division expression in unit form.
S: 154 tenths ÷ 32.
T: Say the divisor rounded to the nearest ten.
S: 30.
T: Name the multiple of 30 that's closest to 154.
S: 150.
T: On your personal white board, write a division equation to find the estimated quotient.
S: (Write 150 tenths ÷ 30 = 5 tenths = 0.5.)

Repeat process using the following possible sequence: 2.53 ÷ 43 and 29.8 ÷ 31.

## Concept Development (31 minutes)

Materials: (S) Personal white board

**Problems 1–2**

904 ÷ 32

456 ÷ 16

Note: In preparation for this set, teachers should have the algorithm for both problems completed beforehand but solved only to the point of the solution being 28 with a remainder of 8 ones.

**1.**

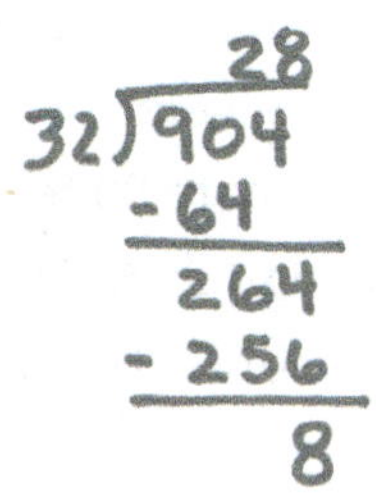


**2.**

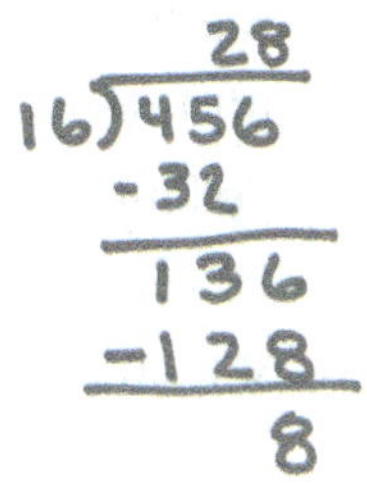


T: (Write 1: 904 ÷ 32 and 2: 456 ÷ 16 horizontally on the board.) Partner A should work on Problem 1, and Partner B should work on Problem 2. Estimate, solve, and check, but share your work with your partner after each step. (Allow time for students to solve.)
T: Say the quotient and remainder for Problem 1.
S: 28 remainder 8.
T: (Write an equal sign next to Problem 1, and record the quotient with the remainder.) Say the quotient and remainder for Problem 2.
S: 28 remainder 8.
T: (Write an equal sign next to Problem 2, and record the quotient with the remainder.) What do you notice about these quotients and remainders?
S: They are the same. → The quotients are the same, and the remainders are the same.

T: Because the quotients and remainders are the same, does that necessarily mean the two division expressions are equivalent? Is 904 ÷ 32 = 456 ÷ 16 a true equation? Turn and talk.

S: If the answer is the same, the expression must be equal, too. → Yes, the answer is the same like 2 + 2 = 4 and 3 + 1 = 4, so 2 + 2 = 3 + 1. → I'm not really sure. Since the divisor is different, I'm wondering if the remainder means something different.

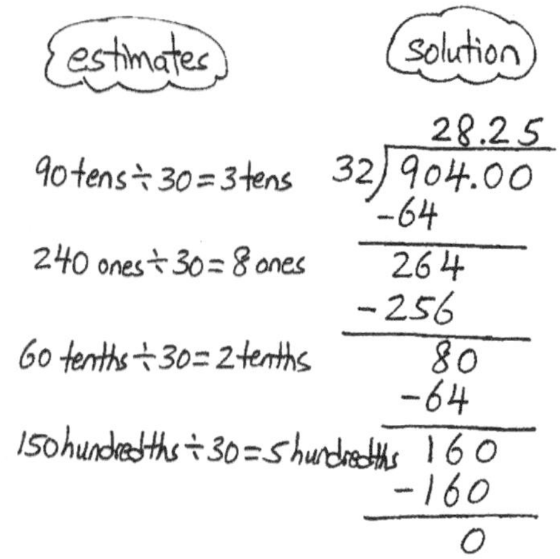


T: (Show students the completed algorithms.) Let's go back to Problem 1. We stopped dividing when we had 8 ones. Can we decompose 8 ones into a smaller unit that would allow us to continue to divide?

S: 8 ones is equal to 80 tenths.

T: This is true. Is there a digit in the tenths place of 904? (Point to the empty area next to the ones place.)

S: No, there are no tenths.

T: True, but I can name 904 as 9,040 tenths. Let's put a decimal point next to the 4 ones and a zero in the tenths place. (Place the decimal and zero in the dividend.) Can you see the 9,040 tenths now?

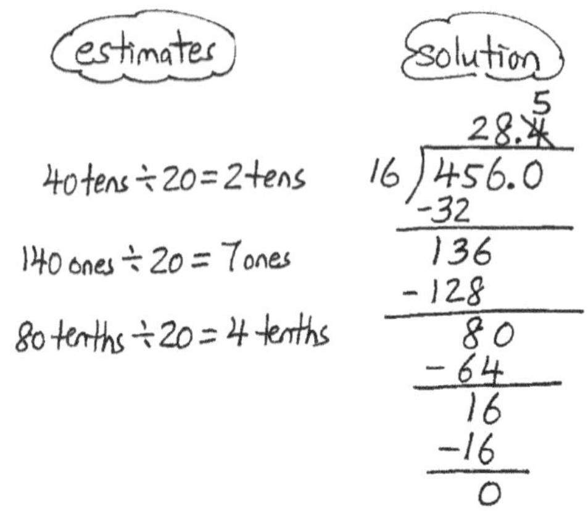


S: Yes.

T: Did I change the value of 904?

S: No.

T: (Point to the zero in the dividend, then write 80 tenths in the algorithm.) So, now we will rename our 8 ones as 80 tenths. Is this enough to divide by 32, or must we regroup again?

S: It's enough. We can divide.

T: Tell me how you estimate 80 tenths ÷ 32.

S: 60 tenths ÷ 30 = 2 tenths. → 80 tenths ÷ 40.

T: Watch where we record 2 tenths in the quotient. Why was it necessary to include the decimal in the quotient?

S: These are tenths. Without the decimal, we won't know the value of the 2. → If you leave out the decimal, it looks like the 2 means 2 ones instead of 2 tenths.

T: What is 2 tenths times 32?

S: 64 tenths.

T: 80 tenths – 64 tenths equals...?

S: 16 tenths.

T: Can we make another group of 32, or must we decompose?

S: We need to decompose 16 tenths into 160 hundredths.

**NOTES ON MULTIPLE MEANS OF REPRESENTATION:**

Allow students to use their own estimates in their quotient. In the given example, the initial under-estimate of 4 hundredths meant recording one more group above the first estimate. The goal is developing the strategic thinking to make the successive approximations in the division process and the steps necessary to adjust those approximations when needed.

Please see Lesson 21 for alternate methods of recording the extra group.

T: What digit is in the hundredths place of 904?

S: There is no digit there.

T: (Write 0 in the hundredths place.) Does this zero change the amount in our whole?

S: No.

T: Now, we can divide 160 hundredths by 32. Tell me how you'll estimate 160 hundredths ÷ 32.

S: 150 hundredths ÷ 30 = 5 hundredths. → 160 ÷ 40 = 4 hundredths.

T: Good estimates. I estimated using 150 ÷ 30, so I'm going to record a 5 in the hundredths place. What is 5 hundredths times 32?

S: 160 hundredths.

T: How many hundredths remain?

S: Zero hundredths.

T: So, what is the quotient?

S: 28.25.

T: Let's use the same process to divide in Problem 2. Similar to Problem 1, we stopped dividing with 8 ones remaining. What can we do to continue to divide?

S: We can decompose the remaining 8 ones into 80 tenths just like before.

T: Yes! We record a zero in the tenths place of the whole. What is 80 tenths divided by 16? Tell how you'll estimate.

S: 80 tenths ÷ 20 = 4 tenths. → 16 is close to the midpoint, so it could be 5.

T: I'll record a 4 in the tenths place of the quotient. What is 4 tenths × 16?

S: 64 tenths.

T: 80 tenths – 64 tenths equals...?

S: 16 tenths.

T: Can we make another group of 16, or must we decompose to make smaller units?

S: We have enough tenths to make another group of 16. → We can make one more group, so we don't need to decompose yet.

T: What is one tenth more than the 4 we have?

S: 5 tenths.

T: Let's adjust our quotient. Cross out the 4 in the tenths place of the quotient, and write a 5. What is 1 tenth × 16?

S: 16 tenths.

T: How many tenths remain?

S: Zero tenths.

T: What is the quotient?

S: 28.5.

T: Talk to your neighbor about what you notice about the quotients of Problems 1 and 2 now.

S: They aren't equal. 28.5 is more than 28.25. → Since the quotients are different, the division expressions are not equal to each other. → It's like two different fractions. 8 sixteenths is greater than 8 thirty-secondths.

T: The remainder is 5 tenths in one problem and 25 hundredths in the other.

T: We can write: 904 ÷ 32 ≠ 456 ÷ 16.

**Problem 3**

834.6 ÷ 26

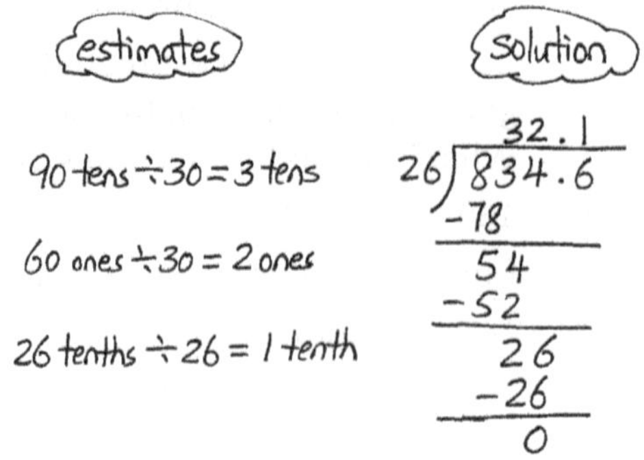


T: (Write 834.6 ÷ 26.) I'll work on the board. You work on your personal white board. Can we divide 8 hundreds by 26 without regrouping?

S: No, we have to decompose 8 hundreds as tens.

T: How many tens do we have?

S: 83 tens.

T: Tell how you'll estimate 83 tens ÷ 26.

S: 90 tens ÷ 30 = 3 tens.

T: (Record 3 in the tens place of the quotient.) What is 3 tens × 26?

S: 78 tens.

T: (Record 78 tens in the algorithm.) How many tens remain?

S: 5 tens.

T: (Record the difference in the algorithm.) Divide or decompose into smaller units?

S: Decompose 5 tens into 50 ones.

T: Plus the 4 ones in the whole, is how many ones?

S: 54 ones.

T: (Record this in the algorithm.) Tell how you'll estimate 54 ones ÷ 26.

S: 60 ones ÷ 30 = 2 ones.

T: (Record 2 ones in the quotient.) What is 2 ones × 26?

S: 52 ones.

T: (Record in the algorithm.) How many ones remain?

S: 2 ones.

T: (Record the difference in the algorithm.) Can we divide again, or must we decompose?

S: We need to decompose 2 ones to 20 tenths.

T: Plus the 6 tenths in our whole makes how many tenths?

S: 26 tenths.

T: (Record this in the algorithm.) What is 26 tenths ÷ 26?

S: One tenth.

T: How will I show 1 tenth in the quotient? Turn and talk.

S: Put a decimal point next to the 2 in the ones place. Then, put the 1 in the tenths place.

T: (Record this in the quotient.) Yes! What is 1 tenth × 26?

S: 26 tenths.

T: (Record this in the algorithm.) How many tenths remain?

S: Zero tenths.

T: (Record in the algorithm.) What is our quotient?

S: 32.1.

T: How many tenths is that?

S: 321 tenths.

T: Work with a partner to check with multiplication.

T: What is 321 tenths times 26?

S: 834.6.

**Problem 4**

48.36 ÷ 39

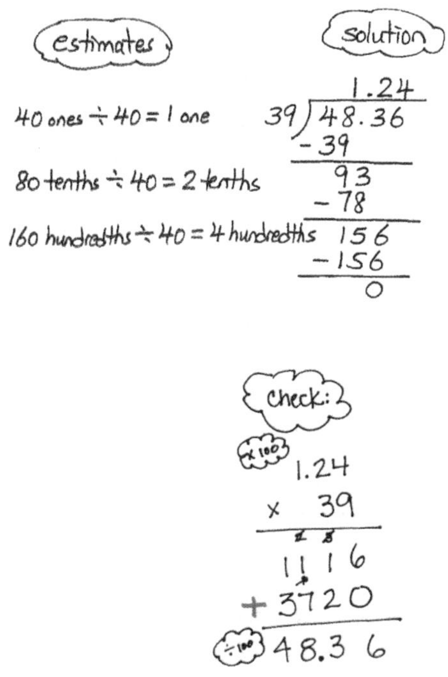


T: (Write 48.36 ÷ 39 on the board.) Before dividing, let's reason about what our quotient might be. Show me how you'll estimate 48.36 ÷ 39.

S: 40 ones ÷ 40 = 1 one.

T: Is 1 a reasonable estimate? It looks like our whole has four digits. How could the quotient be only 1? Turn and talk.

S: Yes, it's reasonable. There is only 1 group of 39 in 48.
→ Yes, because it's basically 48 ÷ 39. That will be like 1 group of 39, and a little bit more. But, it's not enough for 2 groups of 39.

T: Work with a partner to solve. As you finish each step of the division process, share your thinking with a partner. Check your final answer with multiplication.

T: Say the complete division sentence with the quotient.

S: 48.36 ÷ 39 = 1.24.

T: Is the actual quotient reasonable considering the estimating you did previously?

S: Yes, our estimated quotient was 1.

T: Did you check your work?

**Problem 5**

8.61 ÷ 41

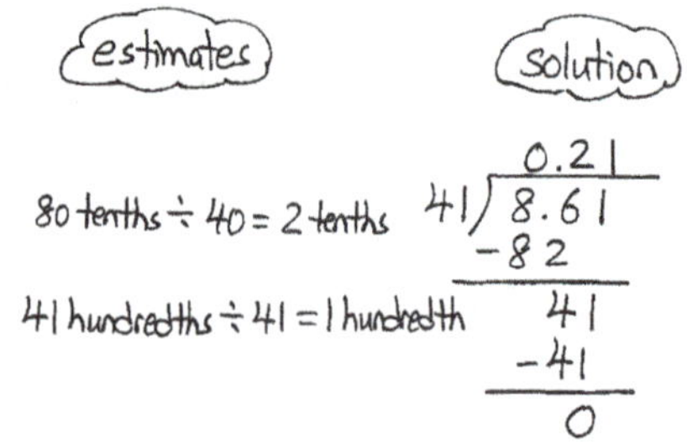


T: (Write 8.61 ÷ 41 on the board.) Before dividing, let's reason about what our quotient might be. Will our quotient be more than 1 or less than 1? How do you know?

S: Less than one. There are only 8 ones, and that is not nearly enough to make a group of 41. → It should be a lot less than one. We can't divide 8 by 41 without regrouping. Our first digit will be tenths, and that's less than 1.

T: Because there will be no ones in our quotient, what will we record in the ones place?

S: A zero.

T: Keep that in mind as you work independently to solve. Continue to reason as you work through the division process. Share your work with a neighbor after each digit you record in the quotient.

S: (Solve while the teacher circulates and supports where necessary.)

T: Say the complete division sentence with the quotient.

S: 8.61 ÷ 41 = 0.21.

T: Check your work with multiplication.

T: Let's talk for a moment about the placement of our decimal in the quotient. Does the placement of the decimal in the quotient make sense? Why or why not?

S: It does make sense. It couldn't be put between the 2 and the 1 because we said our answer had to be less than 1 when we started. → It couldn't be after the 1 because that would be way too big. It wouldn't make sense at all. No way could you have 21 groups of 41 made from 8 ones. → If the divisor was 4, the quotient would be around 2, but it's 10 times larger than that. We'd need to divide again by 10, which makes 2 tenths. Our quotient is very reasonable, and the decimal could only go where we put it. → When we check, we can also see if our decimal was placed correctly. If it's not, the product in our check won't be the same as our whole.

## Application Problem (7 minutes)

Find the whole number quotient and remainder of the following two expressions:

201 ÷ 12          729 ÷ 45

Use >, <, or = to complete this sentence:
201 ÷ 12 ____ 729 ÷ 45, and justify your answer using decimal quotients.

Note: This Application Problem provides another opportunity for students to explore the idea that whole numbers with remainder quotients, with the same digits, are not necessarily equivalent. Finding decimal quotients is one way to obtain a more precise comparison.

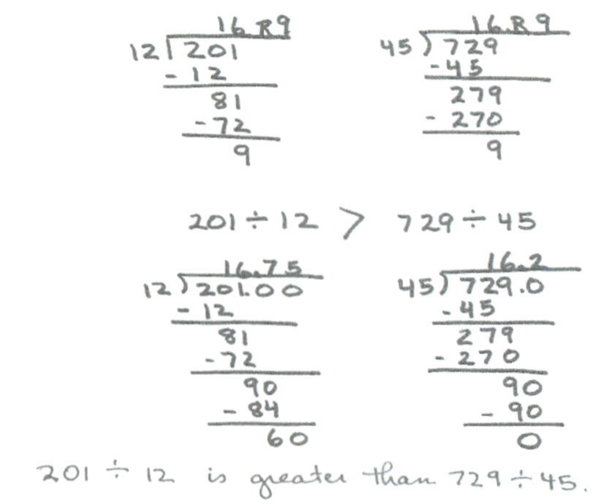






EUREKA MATH

## Problem Set (10 minutes)

Students should do their personal best to complete the Problem Set within the allotted 10 minutes. For some classes, it may be appropriate to modify the assignment by specifying which problems they work on first. Some problems do not specify a method for solving. Students should solve these problems using the RDW approach used for Application Problems.

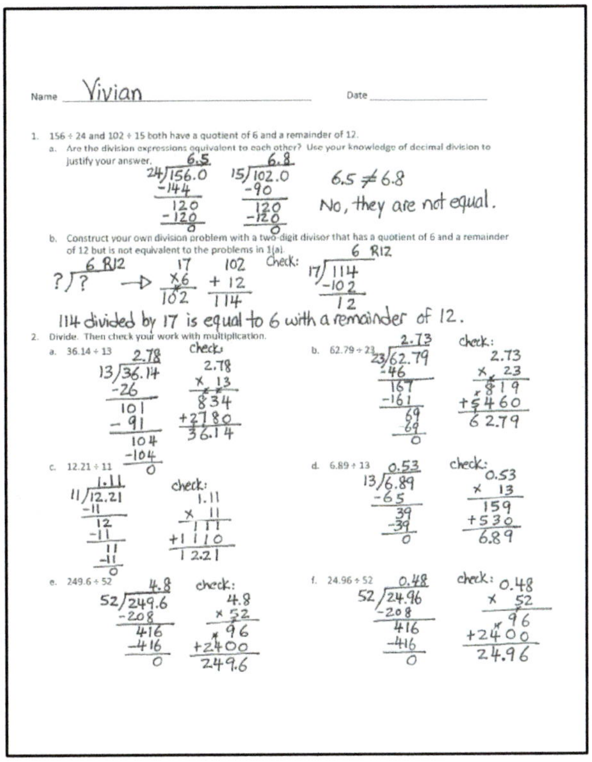

Name Vivian Date

1. 156 ÷ 24 and 102 ÷ 15 both have a quotient of 6 and a remainder of 12.
   a. Are the division expressions equivalent to each other? Use your knowledge of decimal division to justify your answer.
   6.5 ≠ 6.8 No, they are not equal.
   b. Construct your own division problem with a two-digit divisor that has a quotient of 6 and a remainder of 12 but is not equivalent to the problems in 1(a).
   114 divided by 17 is equal to 6 with a remainder of 12.
2. Divide. Then check your work with multiplication.
   a. 36.14 ÷ 13 = 2.78
   b. 62.79 ÷ 23 = 2.73
   c. 12.21 ÷ 11 = 1.11
   d. 6.89 ÷ 13 = 0.53
   e. 249.6 ÷ 52 = 4.8
   f. 24.96 ÷ 52 = 0.48

## Student Debrief (10 minutes)

**Lesson Objective:** Divide decimal dividends by two-digit divisors, estimating quotients, reasoning about the placement of the decimal point, and making connections to a written method.

The Student Debrief is intended to invite reflection and active processing of the total lesson experience.

Invite students to review their solutions for the Problem Set. They should check work by comparing answers with a partner before going over answers as a class. Look for misconceptions or misunderstandings that can be addressed in the Debrief. Guide students in a conversation to debrief the Problem Set and process the lesson.

Any combination of the questions below may be used to lead the discussion.

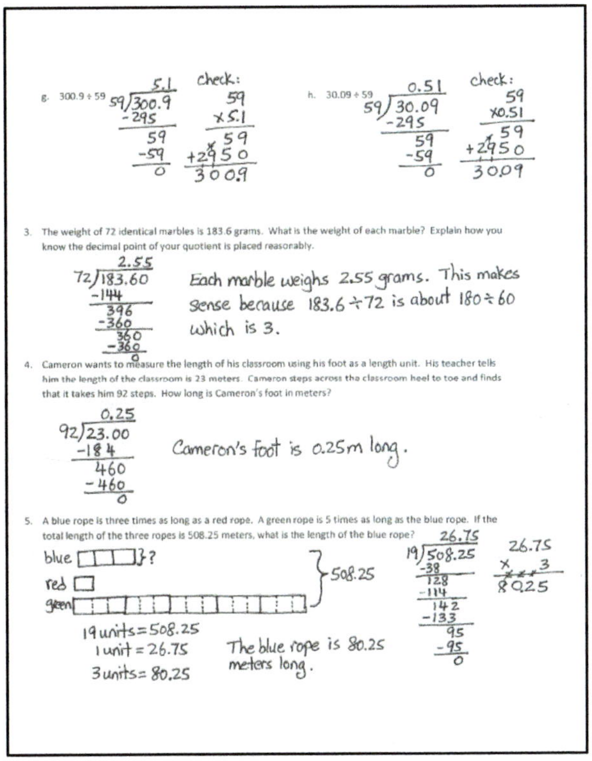

g. 300.9 ÷ 59 = 5.1
h. 30.09 ÷ 59 = 0.51

3. The weight of 72 identical marbles is 183.6 grams. What is the weight of each marble? Explain how you know the decimal point of your quotient is placed reasonably.
   Each marble weighs 2.55 grams. This makes sense because 183.6 ÷ 72 is about 180 ÷ 60 which is 3.
4. Cameron wants to measure the length of his classroom using his foot as a length unit. His teacher tells him the length of the classroom is 23 meters. Cameron steps across the classroom heel to toe and finds that it takes him 92 steps. How long is Cameron's foot in meters?
   Cameron's foot is 0.25m long.
5. A blue rope is three times as long as a red rope. A green rope is 5 times as long as the blue rope. If the total length of the three ropes is 508.25 meters, what is the length of the blue rope?
   19 units = 508.25
   1 unit = 26.75
   3 units = 80.25
   The blue rope is 80.25 meters long.

- Explain how you could prove whether two division expressions with the same whole number quotients and remainders are equivalent.
- Turn to a partner, and compare your work and thinking for Problem 1(b). (Take the necessary time here for students to compare approaches. Possibly, give the students the following challenge: Is it possible to create a pair of division problems whose quotient and whole number remainder look equal and actually are equal when decimal division is used?)
- Explain how you check to see if your quotient's decimal point is placed reasonably.
- How did the Application Problem connect to today's lesson?
- How does your knowledge of multiplication facts help you find a reasonable estimate?

## Exit Ticket (3 minutes)

After the Student Debrief, instruct students to complete the Exit Ticket. A review of their work will help with assessing students' understanding of the concepts that were presented in today's lesson and planning more effectively for future lessons. The questions may be read aloud to the students.

Name ______________________________ Date ________________

1. $156 \div 24$ and $102 \div 15$ both have a quotient of 6 and a remainder of 12.

   a. Are the division expressions equivalent to each other? Use your knowledge of decimal division to justify your answer.

   b. Construct your own division problem with a two-digit divisor that has a quotient of 6 and a remainder of 12 but is not equivalent to the problems in 1(a).

2. Divide. Then, check your work with multiplication.

   a. $36.14 \div 13$

   b. $62.79 \div 23$

   c. $12.21 \div 11$

   d. $6.89 \div 13$

e. $249.6 \div 52$

f. $24.96 \div 52$

g. $300.9 \div 59$

h. $30.09 \div 59$

3. The weight of 72 identical marbles is 183.6 grams. What is the weight of each marble? Explain how you know the decimal point of your quotient is placed reasonably.

4. Cameron wants to measure the length of his classroom using his foot as a length unit. His teacher tells him the length of the classroom is 23 meters. Cameron steps across the classroom heel to toe and finds that it takes him 92 steps. How long is Cameron's foot in meters?

5. A blue rope is three times as long as a red rope. A green rope is 5 times as long as the blue rope. If the total length of the three ropes is 508.25 meters, what is the length of the blue rope?

Name ______________________________ Date ______________

1. Estimate. Then, divide using the standard algorithm and check.

   a. $45.15 \div 21$

   b. $14.95 \div 65$

2. We learned today that division expressions that have the same quotient and remainders are not necessarily equal to each other. Explain how this is possible.

Name ______________________________ Date ______________

1. Create two whole number division problems that have a quotient of 9 and a remainder of 5. Justify which is greater using decimal division.

2. Divide. Then, check your work with multiplication.

a. 75.9 ÷ 22

b. 97.28 ÷ 19

c. 77.14 ÷ 38

d. 12.18 ÷ 29

3. Divide.

   a. 97.58 ÷ 34

   b. 55.35 ÷ 45

4. Use the equations on the left to solve the problems on the right. Explain how you decided where to place the decimal in the quotient.

   a. 520.3 ÷ 43 = 12.1 52.03 ÷ 43 = ____________________

   b. 19.08 ÷ 36 = 0.53 190.8 ÷ 36 = ____________________

5. You can look up information on the world's tallest buildings at http://www.infoplease.com/ipa/A0001338.html.

   a. The Aon Centre in Chicago, Illinois, is one of the world's tallest buildings. Built in 1973, it is 1,136 feet high and has 80 stories. If each story is of equal height, how tall is each story?

   b. Burj al Arab Hotel, another one of the world's tallest buildings, was finished in 1999. Located in Dubai, it is 1,053 feet high with 60 stories. If each floor is the same height, how much taller or shorter is each floor than the height of the floors in the Aon Center?

# Lesson 27

## Objective: Divide decimal dividends by two-digit divisors, estimating quotients, reasoning about the placement of the decimal point, and making connections to a written method.

### Suggested Lesson Structure

| | |
|---|---|
| Fluency Practice | (12 minutes) |
| Application Problem | (5 minutes) |
| Concept Development | (33 minutes) |
| Student Debrief | (10 minutes) |
| **Total Time** | **(60 minutes)** |

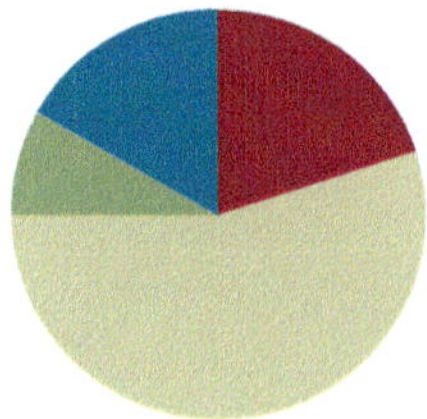

## Fluency Practice (12 minutes)

- Divide Decimals by Multiples of 10 **5.NBT.7** (3 minutes)
- Unit Conversions **5.MD.1** (4 minutes)
- Divide Decimals by Two-Digit Numbers **5.NBT.7** (5 minutes)

### Divide Decimals by Multiples of 10 (3 minutes)

Materials: (S) Personal white board

Note: This fluency activity reviews Lesson 24 content.

Repeat the process from Lesson 25 using the following possible sequence: 1.2 ÷ 6, 1.2 ÷ 60, 8.4 ÷ 4, 8.4 ÷ 40, 6 ÷ 3, 6 ÷ 30, 0.32 ÷ 4, and 0.32 ÷ 40.

### Unit Conversions (4 minutes)

Materials: (S) Personal white board

Note: This fluency activity reviews unit conversions and prepares students for problem solving in Lessons 28 and 29.

T: (Write 1 liter = ____ mL.) How many milliliters are in 1 liter?
S: 1,000 milliliters.

Repeat the process for 1 ft = ____ in, 1 kg = ____ g, and 1 lb = ____ oz.

T: (Write 0.732 liters = ____ mL.) On your personal white board, write an equation to solve, and then show how many milliliters are in 0.732 liters.

S: (Write 0.732 × 1,000 = 732 and 0.732 liters = 732 mL.)

Repeat the process using the following possible sequence: 0.037 liters = ____ mL, 0.537 kg = ____ g, and 0.04 kg = ____ g.

### Divide Decimals by Two-Digit Numbers (5 minutes)

Materials: (S) Personal white board

Note: This fluency activity reviews Lesson 26 content.

T: (Write 83.03 ÷ 23.) On your personal white board, write a division equation to estimate the quotient.

S: (Write 80 ÷ 20 = 4.)

T: Use the algorithm to solve.

Repeat the process using the following possible sequence: 6.76 ÷ 13, 12.43 ÷ 11, and 65.94 ÷ 21.

## Application Problem (5 minutes)

Michael has 567 pennies, Jorge has 464 pennies, and Jaime has 661 pennies. If the pennies are shared equally by the 3 boys and 33 of their classmates, how much money will each classmate receive? Express your final answer in dollars.

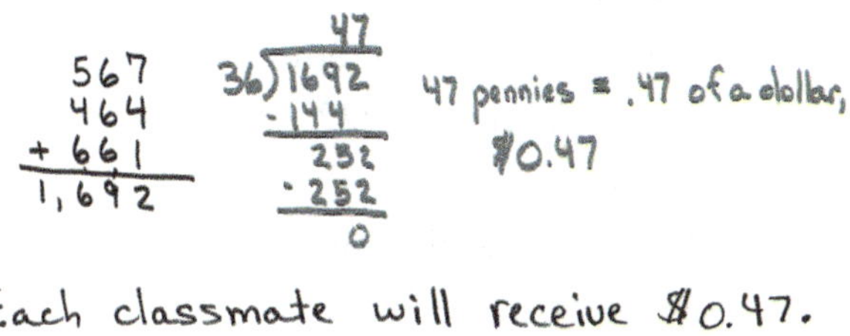


Note: This problem invites different ways of working with the quantities, either as decimals or whole numbers, at different stages of the problem. Students might place the decimal point at the very end of their work or as they add from the beginning. Have them share their approach and express their dollar amounts as decimal units, too.

## Concept Development (33 minutes)

Materials: (S) Personal white board

**Problem 1**

In a 77-kilometer relay race, each of 22 team members will run an equal distance. How many kilometers will each team member run?

77 ÷ 22

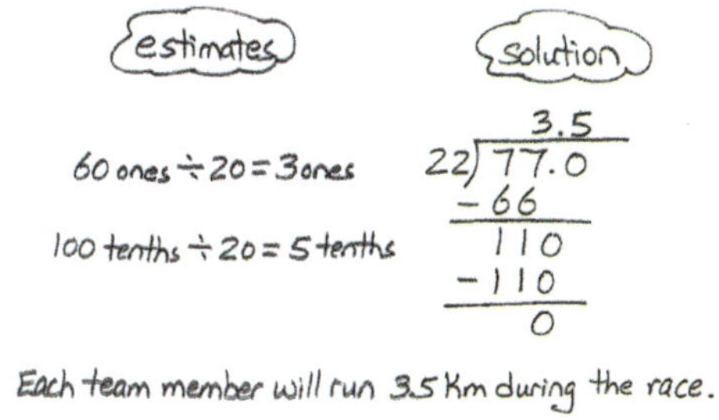


T: Write a division expression to solve for the number of kilometers run by each team member.

S: (Work.) 77 ÷ 22.

T: (Using the standard algorithm, write 77 ÷ 22 on the board.) Let's solve together. Tell how you'll estimate.

S: 60 ones ÷ 20 = 3 ones.

T: (Record each step in the algorithm.) What is 3 ones times 22?

S: 66 ones.

T: How many ones remain?

S: 11 ones.

T: Decompose 11 ones into tenths. How many tenths is that?

S: 110 tenths.

T: Tell your neighbor how you'll show the zero tenths in the whole.

S: I'll write a decimal point and a zero in the tenths place next to the 7 ones in 77.

T: Now, divide 110 tenths by 22. Tell me how you'll estimate.

S: 100 tenths ÷ 20 = 5 tenths.

T: What is 5 tenths × 22?

S: 110 tenths.

T: How many tenths remain?

S: Zero tenths.

T: What is our quotient?

S: 3.5.

T: So, how many kilometers will each team member run during the race?

S: Each team member will run 3 and 5 tenths kilometers.

T: Is your answer reasonable? What is 5 tenths kilometer as meters?

S: 500 meters!

T: What fraction of a kilometer is 500 meters?

S: Half! → So, each runner ran 3 and a half kilometers. → 3.5 kilometers is the same as 3 kilometers and 500 meters, or 3 and a half kilometers.

**NOTES ON MULTIPLE MEANS OF ACTION AND EXPRESSION:**

Students should continue to be encouraged to interpret remainders and decimal portions of quotients. Challenge students with the following questions:

- How many kilometers would need to be added to the race for each runner to run a 4-kilometer distance? A 4.5-kilometer distance? A 5-kilometer distance?
- Change the context of the problem so that the interpretation of the remainder must change.
  - 77 students need to board buses. The buses have 22 seats. How many buses are needed?
  - 22 students will share 77 t-shirts. How many students could receive more than 1 shirt?

**Problem 2**

A vial contains 14.7 mL of serum that is then split equally into 21 tiny containers. How much serum is in each new container?

14.7 ÷ 21

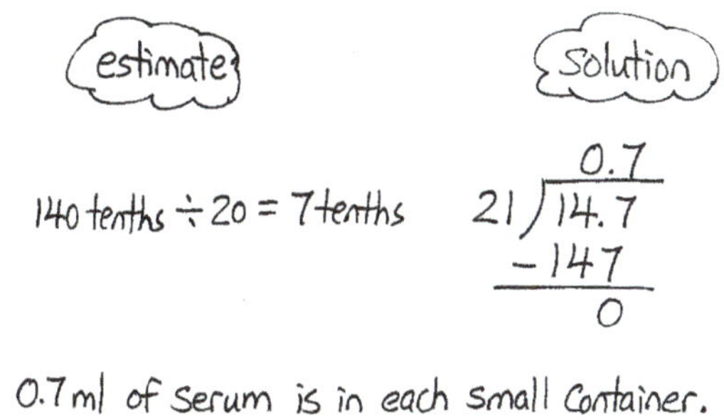


T: Work with a partner to write a division expression that matches this story problem.

S: (Work.) 14.7 ÷ 21.

T: (Using the standard algorithm, write 14.7 ÷ 21 on the board.) Before dividing, let's reason about what our quotient might be. Will there be more than 1 mL in each container or less than 1 mL? Justify your thinking.

S: It will have to be less than 1 mL because there are more containers than mL of serum. → To have 1 mL in each container, there would have to be 21 mL of serum. We only have about 15.

T: Great reasoning. Now, tell me how you will estimate 14.7 ÷ 21 numerically.

S: 140 tenths ÷ 20 = 7 tenths.

T: Work with a partner to solve.

T: What is the quotient?

S: 0.7.

T: Is our actual quotient reasonable? Does the placement of the decimal make sense?

S: Yes, it's the exact same as the estimated quotient. → We said we should have less than 1 mL, and we do. If the decimal was behind the 7, it wouldn't make sense because that would be 7 mL in each container. → It couldn't have been 7 hundredths. If the divisor had been 2, then the answer would be 7. We had 21, which is about 10 times as large, so we had to divide by 10, which is 7 tenths, not 7 hundredths.

T: Did you check your work?

S: Yes.

T: Answer the question using a complete sentence.

S: Zero and 7 tenths milliliters of serum is in each tiny container.

**NOTES ON MULTIPLE MEANS OF ACTION AND EXPRESSION:**

Many lessons, including this one, require students to understand and use precise vocabulary. Teachers can help students gain familiarity with new words by displaying them on posters. There are several ways to make posters more effective:

- Be judicious when deciding how many posters to display. If there are too many, students tend to not see them.
- Place the posters in odd places, such as on the floor in the doorway or in the washroom. This may capture attention.
- Try to use pictures or graphics instead of wordy definitions.
- When a student is struggling for a precise word, point to a poster. This shows students that posters can be support materials.

The same context may be repeated for the following: 22.47 ÷ 21. This problem requires the recording of a zero in the quotient.

**Problem 3**

The surface area of a rectangular piece of construction paper is 140.25 square inches. If the paper's length is 17 inches, what is the width?

140.25 ÷ 17

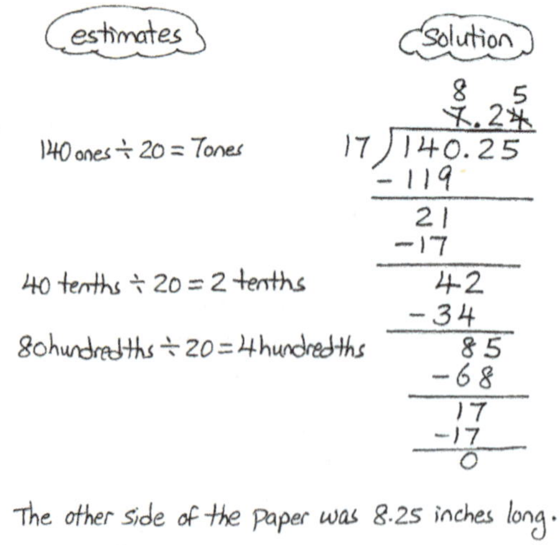


T: What expression would you use to solve this problem?

S: (Work.) 140.25 ÷ 17.

T: (Using the standard algorithm, write 140.25 ÷ 17 on the board.) Before dividing, let's reason about what our quotient might be. Tell me how you'll estimate 140.25 ÷ 17.

S: 140 ones ÷ 20 = 7 ones.

T: Work independently to solve this problem. Share your work with a neighbor after each step in the division process.

T: What is the quotient?

S: 8.25.

T: Answer the question using a complete sentence.

S: The width of the paper was 8 and 25 hundredths inches long.

## Problem Set (10 minutes)

Students should do their personal best to complete the Problem Set within the allotted 10 minutes. For some classes, it may be appropriate to modify the assignment by specifying which problems they work on first. Some problems do not specify a method for solving. Students should solve these problems using the RDW approach used for Application Problems.

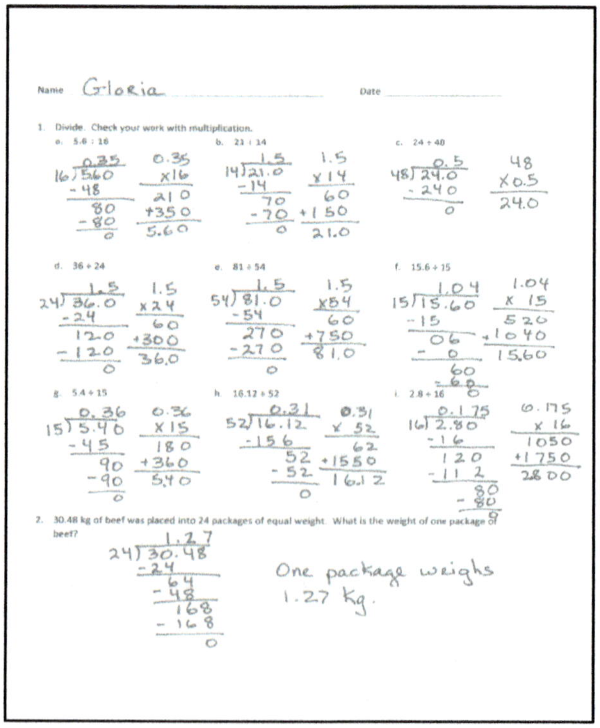


## Student Debrief (10 minutes)

**Lesson Objective:** Divide decimal dividends by two-digit divisors, estimating quotients, reasoning about the placement of the decimal point, and making connections to a written method.

The Student Debrief is intended to invite reflection and active processing of the total lesson experience.

Invite students to review their solutions for the Problem Set. They should check work by comparing answers with a partner before going over answers as a class.

Look for misconceptions or misunderstandings that can be addressed in the Debrief. Guide students in a conversation to debrief the Problem Set and process the lesson.





EUREKA MATH™

Any combination of the questions below may be used to lead the discussion.

- When dividing a decimal by a two-digit divisor, when is it useful to think of decimals in various units? (During estimation, it can be easier to think of a whole number as an equivalent amount of smaller units.)
- Discuss the multi-step problems in the Problem Set. Ask students to explain how they knew their placement of the decimal point was reasonable, how they knew their quotient was reasonable, and how to interpret the decimal portion of the quotient.
- The quotients for Problems 1(d) and (e) are the same. Divide them again. This time, do not go beyond the ones place. Compare the whole number remainders. What do you notice? Are the division equations equal to each other? Why or why not?
- We expressed our remainders today using decimals. Does it always make sense to do this? Give an example of a situation where a whole number remainder makes more sense. Do you notice a pattern to these examples?

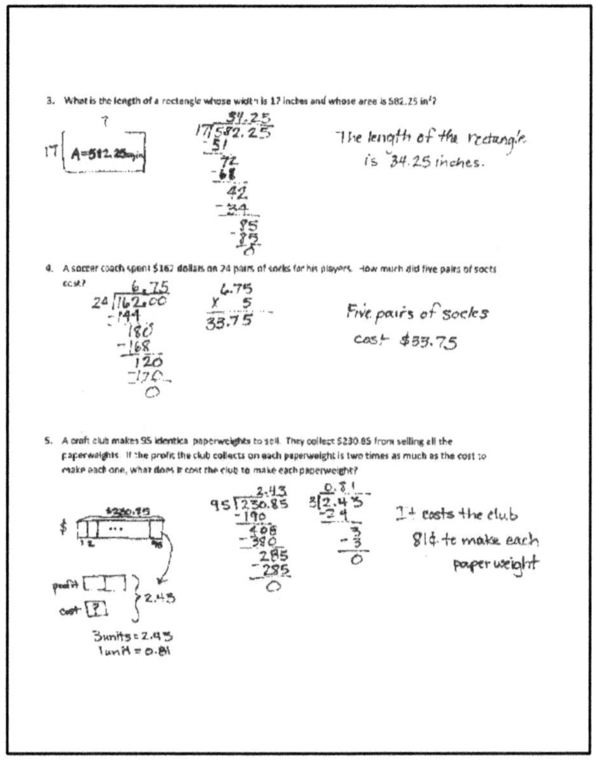


## Exit Ticket (3 minutes)

After the Student Debrief, instruct students to complete the Exit Ticket. A review of their work will help with assessing students' understanding of the concepts that were presented in today's lesson and planning more effectively for future lessons. The questions may be read aloud to the students.

Name ______________________________ Date ______________

1. Divide. Check your work with multiplication.

a. $5.6 \div 16$

b. $21 \div 14$

c. $24 \div 48$

d. $36 \div 24$

e. $81 \div 54$

f. $15.6 \div 15$

g. $5.4 \div 15$

h. $16.12 \div 52$

i. $2.8 \div 16$

2. 30.48 kg of beef was placed into 24 packages of equal weight. What is the weight of one package of beef?

3. What is the length of a rectangle whose width is 17 inches and whose area is 582.25 $in^2$?

4. A soccer coach spent $162 dollars on 24 pairs of socks for his players. How much did five pairs of socks cost?

5. A craft club makes 95 identical paperweights to sell. They collect $230.85 from selling all the paperweights. If the profit the club collects on each paperweight is two times as much as the cost to make each one, what does it cost the club to make each paperweight?

Name ____________________________ Date ______________

Divide.

a. $28 \div 32$

b. $68.25 \div 65$

Name ______________________________ Date ______________

1. Divide. Check your work with multiplication.

a. $7 \div 28$

b. $51 \div 25$

c. $6.5 \div 13$

d. $132.16 \div 16$

e. $561.68 \div 28$

f. $604.8 \div 36$

2. In a science class, students water a plant with the same amount of water each day for 28 consecutive days. If the students use a total of 23.8 liters of water over the 28 days, how many liters of water did they use each day? How many milliliters did they use each day?





EUREKA MATH™

3. A seamstress has a piece of cloth that is 3 yards long. She cuts it into shorter lengths of 16 inches each. How many of the shorter pieces can she cut?

4. Jenny filled 12 pitchers with an equal amount of lemonade in each. The total amount of lemonade in the 12 pitchers was 41.4 liters. How many liters of lemonade would be in 7 pitchers?

## Topic H

# Measurement Word Problems with Multi-Digit Division

**5.NBT.6, 5.NBT.7**

| | | | |
|---|---|---|---|
| **Focus Standards:** | | 5.NBT.6 | Find whole-number quotients of whole numbers with up to four-digit dividends and two-digit divisors, using strategies based on place value, the properties of operations, and/or the relationship between multiplication and division. Illustrate and explain the calculation by using equations, rectangular arrays, and/or area models. |
| | | 5.NBT.7 | Add, subtract, multiply, and divide decimals to hundredths, using concrete models or drawings and strategies based on place value, properties of operations, and/or the relationship between addition and subtraction; relate the strategy to a written method and explain the reasoning used. |
| **Instructional Days:** | | 2 | |
| **Coherence** | **-Links from:** | G4–M3 | Multi-Digit Multiplication and Division |
| | **-Links to:** | G6–M2 | Arithmetic Operations Including Division of Fractions |

In Topic H, students apply the work of the module to solve multi-step word problems using multi-digit division (**5.NBT.6**). Cases include unknowns representing either the group size or number of groups. In this topic, an emphasis on checking the reasonableness of their solutions draws on skills learned throughout the module, which includes using knowledge of place value, rounding, and estimation. Students relate calculations to reasoning about division through a variety of strategies including place value, properties of operations, equations, and area models.

**A Teaching Sequence Toward Mastery of Measurement Word Problems with Multi-Digit Division**

**Objective 1: Solve division word problems involving multi-digit division with group size unknown and the number of groups unknown.**
**(Lessons 28–29)**

# Lesson 28

## Objective: Solve division word problems involving multi-digit division with group size unknown and the number of groups unknown.

### Suggested Lesson Structure

| | |
|---|---|
| Fluency Practice | (12 minutes) |
| Concept Development | (38 minutes) |
| Student Debrief | (10 minutes) |
| **Total Time** | **(60 minutes)** |

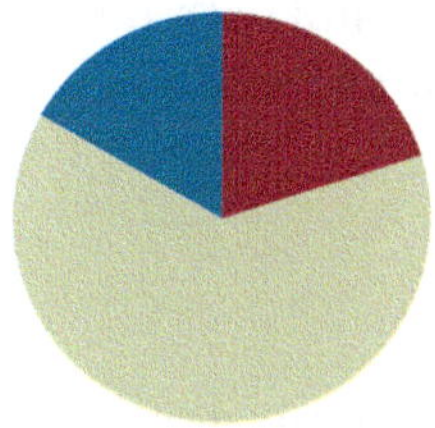

## Fluency Practice (12 minutes)

- Sprint: Divide Decimals by Multiples of 10 **5. NBT.7** (9 minutes)
- Unit Conversions **5.MD.1** (3 minutes)

### Sprint: Divide Decimals by Multiples of 10 (9 minutes)

Materials: (S) Divide Decimals by Multiples of 10 Sprint

Note: This Sprint builds automaticity of Lesson 24 content.

### Unit Conversions (3 minutes)

Materials: (S) Personal white board

Note: This fluency activity reviews unit conversions and prepares students for problem solving in this lesson's Concept Development.

**NOTES ON MULTIPLE MEANS OF ENGAGEMENT:**

Some students may still need support while dividing in the standard algorithm. Refer to Lessons 21–27 to help guide students through the algorithm using place value language.

Repeat the process from Lesson 27 for each unit conversion, using the following possible sequence: 1 m = ___ cm, 1 L = ___ mL, 1 ft = ___ in, 0.37 L = ___ mL, 0.152 kg = ___ g, and 0.08 m = ___ cm.

## Concept Development (38 minutes)

Materials: (S) Problem Set

### Suggested Delivery of Instruction for Solving Topic H Word Problems

**1. Model the problem.**

Have two pairs of students, who can be successful with modeling the problem, work at the board while the others work independently or in pairs at their seats. Review the following questions before beginning the first problem:

- Can you draw something?
- What can you draw?
- What conclusions can you make from your drawing?

As students work, circulate. Reiterate the questions above. After two minutes, have the two pairs of students share only their labeled diagrams. For about one minute, have the demonstrating students receive and respond to feedback and questions from their peers.

**2. Calculate to solve, and write a statement.**

Give everyone two minutes to finish work on that question, sharing their work and thinking with a peer. All should then write their equations and statements of the answer.

**3. Assess the solution for reasonability.**

Give students one to two minutes to assess and explain the reasonableness of their solution.

Note: In Lessons 28–29, the Problem Set comprises the word problems from the Concept Development.

**Problem 1**

Ava is saving for a new computer that costs $1,218. She has already saved half of the money. Ava earns $14.00 per hour. How many hours must Ava work to save the rest of the money?

This two-step *equal groups with number of groups unknown* problem is a step forward for students as they divide the total in half and use their decimal division skills to divide 609 by 14 to find the number of hours Ava needs to work. In this case, the divisor represents the size of the unit. While circulating, use professional judgment to look for other alternate modeling strategies that can be quickly mentioned or explored more deeply.

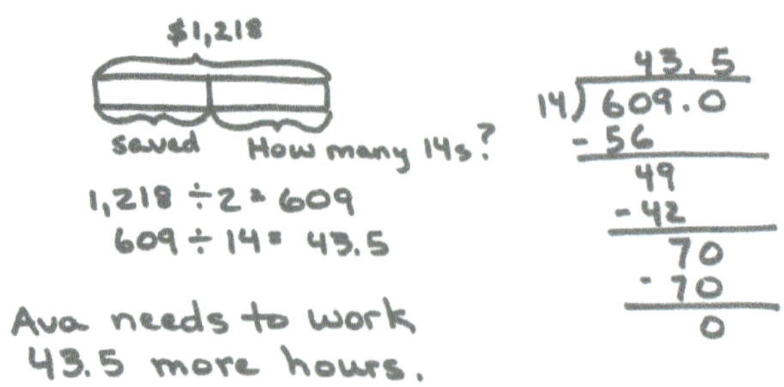


After the students have solved the problem, ask them to check their answer for reasonableness.

T: How can you know if 43.5 is a reasonable answer? Discuss with your partner.



EUREKA MATH

**Problem 2**

Michael has a collection of 1,404 sports cards. He hopes to sell the collection in packs of 36 cards and make $633.75 when all the packs are sold. If each pack is priced the same, how much should Michael charge per pack?

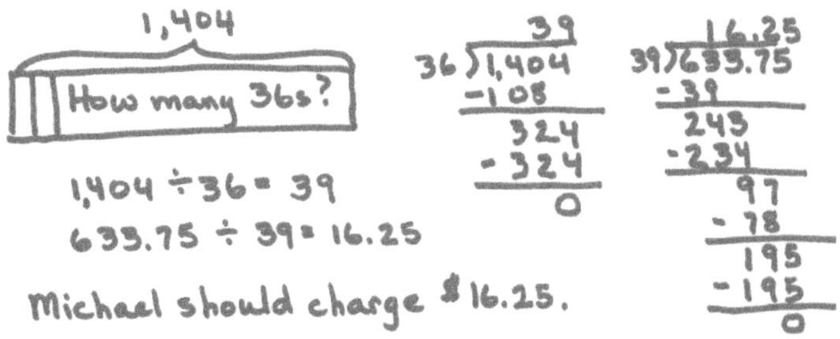


This two-step *equal groups with number of groups unknown* problem involves both whole number and decimal division. Students must first find the number of packs of cards and then find the price per each pack. In the whole number division, the divisor represents the size of the unit: How many groups of 36 cards are there? While in the decimal division, the 39 packs of cards are "sharing" the total amount of money: How much money is in each group? Because the accuracy of the second quotient is determined by the accuracy of the first, students may wish to check the first division problem before moving to the second.

After students have solved the problem, ask them to check their answers for reasonability.

T: How can you know your answer of $16.25 is reasonable?

S: I thought about the money as $640 and the number of packs as 40. That's like 64 ÷ 4, which is 16. My estimate of the number of packs was 1 more than the actual, so it made sense that each pack would cost more money. $16.25 is really close to $16.

T: Did you check the first division problem before moving on to the second? Why or why not?

S: I did check to be sure I had the right number of thirty-sixes. I knew if I didn't have the right number of packs, my price for each would be off. → I didn't check until the end, but I did check both my division problems.

T: Compare the meaning of the divisors for these two different division equations.

**Problem 3**

Jim Nasium is building a tree house for his two daughters. He cuts 12 pieces of wood from a board that is 128 inches long. He cuts 5 pieces that measure 15.75 inches each and 7 pieces evenly cut from what is left. Jim calculates that, due to the width of his cutting blade, he will lose a total of 2 inches of wood after making all the cuts. What is the length of each of the seven pieces?

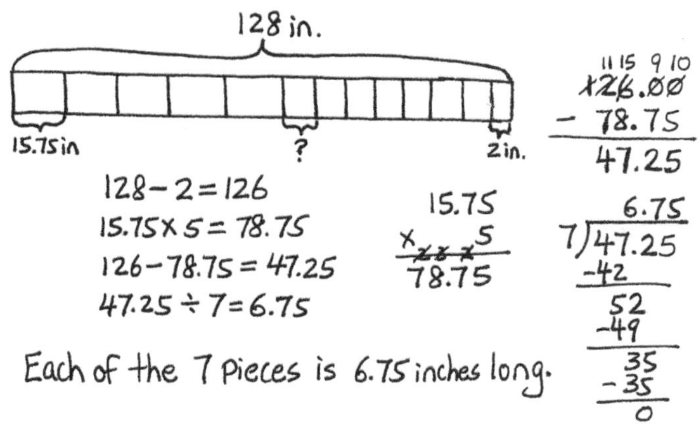

Careful drawing is essential for success in this multi-step *equal groups with group size unknown* problem because it requires students to first subtract the 2 inches lost to the blade's kerf. Then, students must subtract the total from the 5 larger pieces cut. This remaining wood is then divided into 7 parts, and the length is found for each. The divisor represents the number of units.

T: How can you be sure your final answer is reasonable?

T: How did you organize your work so that you could keep track of all the different steps? Compare your organization with that of your partner.

**Problem 4**

A load of bricks is twice as heavy as a load of sticks. The total weight of 4 loads of bricks and 4 loads of sticks is 771 kilograms. What is the total weight of 1 load of bricks and 3 loads of sticks?

The new complexity of this *equal groups with group size unknown* problem is that students must consider the number of units that must be used to represent the weight of the bricks and then consider those units when choosing the number of units to multiply by 64.25. Alternatively, after identifying the value of the base unit, in the final step, students might calculate the weight of a single load of bricks and a single load of sticks, multiply the bricks by 3, and then add. Also, the division of two whole numbers results in a decimal. Students must rename ones as tenths and tenths as hundredths, placing additional zeros in the dividend. In this situation, the divisor represents the number of units. After solving and assessing the solution for reasonability, consider the following questions:

T: What was the first thing that you drew? What did one unit represent in your model?

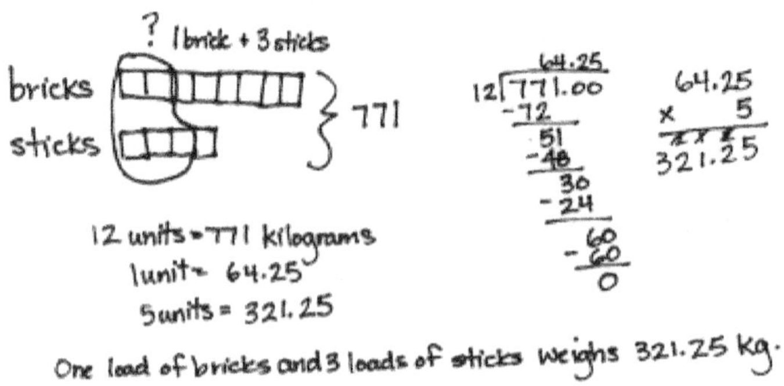


S: I drew 1 unit for the load of sticks and 2 units for the load of bricks. Then, I drew the other boxes as I counted out the rest of the loads of bricks and sticks. → I knew that the brick units would be twice as many as the stick units because the bricks were two times heavier than the sticks. I just drew 4 units for the loads of sticks and then doubled the units for the loads of bricks.

T: Compare your approach to finding the total weight of 3 loads of bricks and 1 load of sticks to your partner's.

## Problem Set (10 minutes)

Students should do their personal best to complete the Problem Set within the allotted 10 minutes. For some classes, it may be appropriate to modify the assignment by specifying which problems they work on first. Some problems do not specify a method for solving. Students should solve these problems using the RDW approach used for Application Problems.

## Student Debrief (10 minutes)

**Lesson Objective:** Solve division word problems involving multi-digit division with group size unknown and the number of groups unknown.

The Student Debrief is intended to invite reflection and active processing of the total lesson experience.

Invite students to review their solutions for the Problem Set. They should check work by comparing answers with a partner before going over answers as a class. Look for misconceptions or misunderstandings that can be addressed in the Debrief. Guide students in a conversation to debrief the Problem Set and process the lesson.

Any combination of the questions below may be used to lead the discussion.

- How are the problems alike? How are they different?
- How was your solution the same and different from those that were demonstrated?
- Did you see other solutions that surprised you or made you see the problem differently?
- Why should we assess reasonability after solving?
- Sort the problems into those in which the group size was unknown, and those in which the number of groups was unknown. There may be problems that must be placed into both categories.

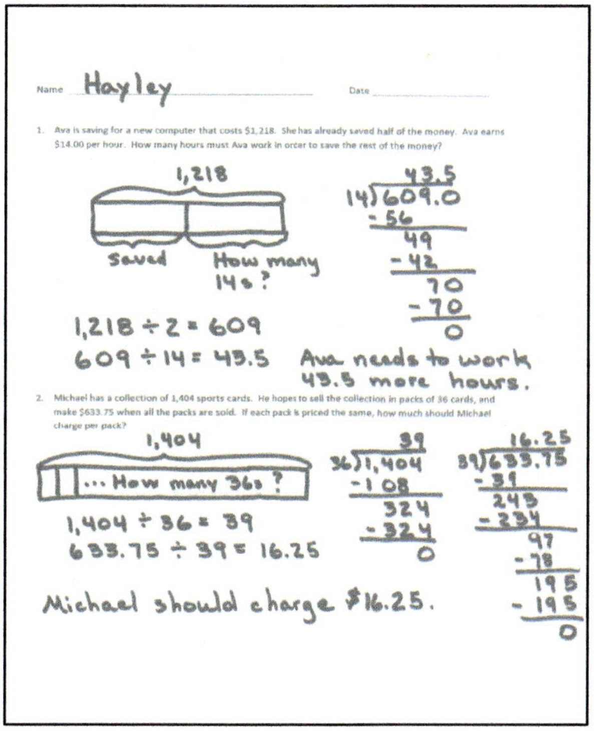
Name Hayley Date

1. Ava is saving for a new computer that costs $1,218. She has already saved half of the money. Ava earns $14.00 per hour. How many hours must Ava work in order to save the rest of the money?



2. Michael has a collection of 1,404 sports cards. He hopes to sell the collection in packs of 36 cards, and make $633.75 when all the packs are sold. If each pack is priced the same, how much should Michael charge per pack?



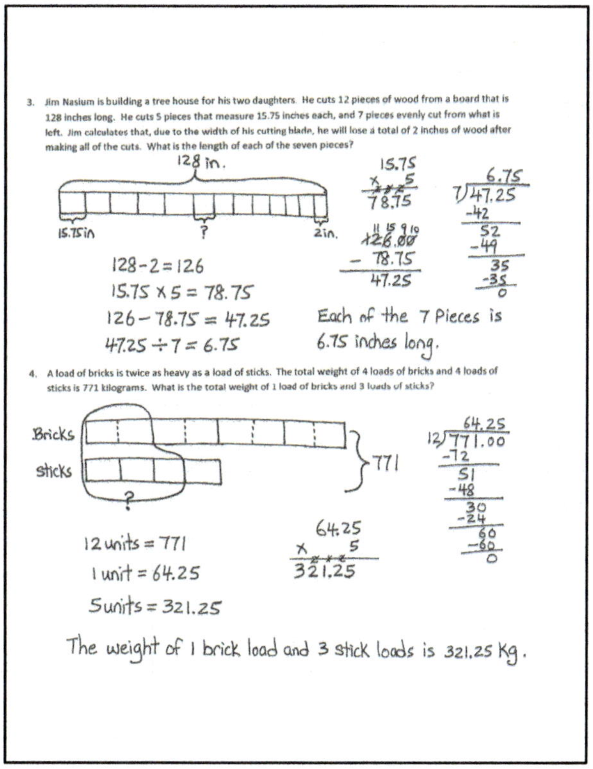
3. Jim Nasium is building a tree house for his two daughters. He cuts 12 pieces of wood from a board that is 128 inches long. He cuts 5 pieces that measure 15.75 inches each, and 7 pieces evenly cut from what is left. Jim calculates that, due to the width of his cutting blade, he will lose a total of 2 inches of wood after making all of the cuts. What is the length of each of the seven pieces?



4. A load of bricks is twice as heavy as a load of sticks. The total weight of 4 loads of bricks and 4 loads of sticks is 771 kilograms. What is the total weight of 1 load of bricks and 3 loads of sticks?



### Exit Ticket (3 minutes)

After the Student Debrief, instruct students to complete the Exit Ticket. A review of their work will help with assessing students' understanding of the concepts that were presented in today's lesson and planning more effectively for future lessons. The questions may be read aloud to the students.

# A

Number Correct: _______

Divide Decimals by Multiples of 10

| | | | | | |
|---|---|---|---|---|---|
| 1. | 6 ÷ 10 = | . | 23. | 25 ÷ 50 = | . |
| 2. | 6 ÷ 20 = | . | 24. | 2.5 ÷ 50 = | . |
| 3. | 6 ÷ 60 = | . | 25. | 4.5 ÷ 50 = | . |
| 4. | 8 ÷ 10 = | . | 26. | 4.5 ÷ 90 = | . |
| 5. | 8 ÷ 40 = | . | 27. | 0.45 ÷ 90 = | . |
| 6. | 8 ÷ 20 = | . | 28. | 0.45 ÷ 50 = | . |
| 7. | 4 ÷ 10 = | . | 29. | 0.24 ÷ 60 = | . |
| 8. | 4 ÷ 20 = | . | 30. | 0.63 ÷ 90 = | . |
| 9. | 4 ÷ 40 = | . | 31. | 0.48 ÷ 80 = | . |
| 10. | 9 ÷ 3 = | . | 32. | 0.49 ÷ 70 = | . |
| 11. | 9 ÷ 30 = | . | 33. | 6 ÷ 30 = | . |
| 12. | 12 ÷ 3 = | . | 34. | 14 ÷ 70 = | . |
| 13. | 12 ÷ 30 = | . | 35. | 72 ÷ 90 = | . |
| 14. | 12 ÷ 40 = | . | 36. | 6.4 ÷ 80 = | . |
| 15. | 12 ÷ 60 = | . | 37. | 0.48 ÷ 40 = | . |
| 16. | 12 ÷ 20 = | . | 38. | 0.36 ÷ 30 = | . |
| 17. | 15 ÷ 3 = | . | 39. | 0.55 ÷ 50 = | . |
| 18. | 15 ÷ 30 = | . | 40. | 1.36 ÷ 40 = | . |
| 19. | 15 ÷ 50 = | . | 41. | 2.04 ÷ 60 = | . |
| 20. | 18 ÷ 30 = | . | 42. | 4.48 ÷ 70 = | . |
| 21. | 24 ÷ 30 = | . | 43. | 6.16 ÷ 80 = | . |
| 22. | 16 ÷ 40 = | . | 44. | 5.22 ÷ 90 = | . |

# B

Number Correct: _______

Improvement: _______

Divide Decimals by Multiples of 10

| | | | | | |
|---|---|---|---|---|---|
| 1. | 4 ÷ 10 = | . | 23. | 25 ÷ 50 = | . |
| 2. | 4 ÷ 20 = | . | 24. | 2.5 ÷ 50 = | . |
| 3. | 4 ÷ 40 = | . | 25. | 3.5 ÷ 50 = | . |
| 4. | 8 ÷ 10 = | . | 26. | 3.5 ÷ 70 = | . |
| 5. | 8 ÷ 20 = | . | 27. | 0.35 ÷ 70 = | . |
| 6. | 8 ÷ 40 = | . | 28. | 0.35 ÷ 50 = | . |
| 7. | 9 ÷ 10 = | . | 29. | 0.42 ÷ 60 = | . |
| 8. | 9 ÷ 30 = | . | 30. | 0.54 ÷ 90 = | . |
| 9. | 9 ÷ 90 = | . | 31. | 0.56 ÷ 80 = | . |
| 10. | 6 ÷ 2= | . | 32. | 0.63 ÷ 70 = | . |
| 11. | 6 ÷ 20 = | . | 33. | 6 ÷ 30 = | . |
| 12. | 12 ÷ 2 = | . | 34. | 18 ÷ 90 = | . |
| 13. | 12 ÷ 20 = | . | 35. | 72 ÷ 80 = | . |
| 14. | 12 ÷ 30 = | . | 36. | 4.8 ÷ 80 = | . |
| 15. | 12 ÷ 40 = | . | 37. | 0.36 ÷ 30 = | . |
| 16. | 12 ÷ 60 = | . | 38. | 0.48 ÷ 40 = | . |
| 17. | 15 ÷ 5 = | . | 39. | 0.65 ÷ 50 = | . |
| 18. | 15 ÷ 50 = | . | 40. | 1.38 ÷ 30 = | . |
| 19. | 15 ÷ 30 = | . | 41. | 2.64 ÷ 60 = | . |
| 20. | 21 ÷ 30 = | . | 42. | 5.18 ÷ 70 = | . |
| 21. | 27 ÷ 30 = | . | 43. | 6.96 ÷ 80 = | . |
| 22. | 36 ÷ 60 = | . | 44. | 6.12 ÷ 90 = | . |

Name ______________________________ Date ________________

1. Ava is saving for a new computer that costs $1,218. She has already saved half of the money. Ava earns $14.00 per hour. How many hours must Ava work in order to save the rest of the money?

2. Michael has a collection of 1,404 sports cards. He hopes to sell the collection in packs of 36 cards and make $633.75 when all the packs are sold. If each pack is priced the same, how much should Michael charge per pack?





EUREKA MATH

3. Jim Nasium is building a tree house for his two daughters. He cuts 12 pieces of wood from a board that is 128 inches long. He cuts 5 pieces that measure 15.75 inches each and 7 pieces evenly cut from what is left. Jim calculates that, due to the width of his cutting blade, he will lose a total of 2 inches of wood after making all of the cuts. What is the length of each of the seven pieces?

4. A load of bricks is twice as heavy as a load of sticks. The total weight of 4 loads of bricks and 4 loads of sticks is 771 kilograms. What is the total weight of 1 load of bricks and 3 loads of sticks?

Name ____________________________________________ Date ____________________

Solve this problem, and show all of your work.

Kenny is ordering uniforms for both the girls' and boys' tennis clubs. He is ordering shirts for 43 players and two coaches at a total cost of \$658.35. Additionally, he is ordering visors for each player at a total cost of \$368.51. How much will each player pay for the shirt and visor?

Name ________________________________________ Date ____________________

1. Mr. Rice needs to replace the 166.25 ft of edging on the flower beds in his backyard. The edging is sold in lengths of 19 ft each. How many lengths of edging will Mr. Rice need to purchase?

2. Olivia is making granola bars. She will use 17.9 ounces of pistachios, 12.6 ounces of almonds, 12.5 ounces of walnuts, and 12.5 ounces of cashews. This amount makes 25 bars. How many ounces of nuts are in each granola bar?

3. Adam has 16.45 kg of flour, and he uses 6.4 kg to make hot cross buns. The remaining flour is exactly enough to make 15 batches of scones. How much flour, in kg, will be in each batch of scones?

4. There are 90 fifth-grade students going on a field trip. Each student gives the teacher $9.25 to cover admission to the theater and for lunch. Admission for all of the students will cost $315, and each student will get an equal amount to spend on lunch. How much will each fifth grader get to spend on lunch?

5. Ben is making math manipulatives to sell. He wants to make at least $450. Each manipulative costs $18 to make. He is selling them for $30 each. What is the minimum number he can sell to reach his goal?

# Lesson 29

## Objective: Solve division word problems involving multi-digit division with group size unknown and the number of groups unknown.

### Suggested Lesson Structure

| | |
|---|---|
| Fluency Practice | (10 minutes) |
| Application Problem | (8 minutes) |
| Concept Development | (32 minutes) |
| Student Debrief | (10 minutes) |
| **Total Time** | **(60 minutes)** |

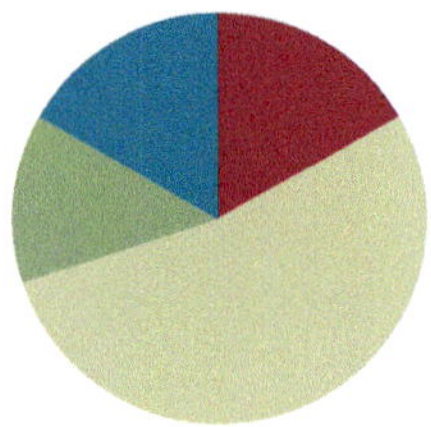

## Fluency Practice (10 minutes)

- Unit Conversions **5.MD.1** (3 minutes)
- Divide Decimals by Two-Digit Numbers **5.NBT.7** (7 minutes)

### Unit Conversions (3 minutes)

Materials: (S) Personal white board

Note: This fluency activity reviews unit conversions and prepares students for problem solving in the Concept Development.

Repeat the process from Lesson 27 for each unit conversion, using the following possible sequence: 1 kg = ___ g, 1 lb = ___ oz, 1 ft = ___ in, 1 L = ___ mL, 0.42 L = ___ mL, 0.678 kg = ___ g, and 0.953 m = ___ cm.

### Divide Decimals by Two-Digit Numbers (7 minutes)

Materials: (S) Personal white board

Note: This fluency activity reviews Lesson 27 content.

Repeat the process from Lesson 27, using the following possible sequence: 8.61 ÷ 21, 4.9 ÷ 14, and 24 ÷ 16.

## Application Problem (8 minutes)

A one-year (52-week) subscription to a weekly magazine is $39.95. Greg calculates that he would save $219.53 if he subscribed to the magazine instead of purchasing it each week at the store. What is the price of the individual magazine at the store?

Note: This Application Problem uses concepts from Grade 5 Module 1 in the first step of the problem and division of decimals with group size unknown from Module 2 in the second step of the problem. A tape diagram or place value chart can be used to add the decimals. A tape diagram is the ideal strategy to represent the division.

$259.48 ÷ 52 is best solved through estimation because the dividend can be estimated as an easily identifiable multiple of 50. However, if more time is needed for Concept Development, the Application Problem may be used for homework or journal entry.

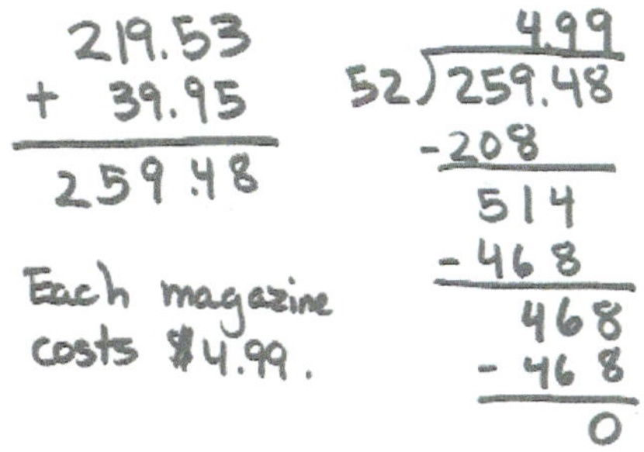


## Concept Development (32 minutes)

Materials: (S) Problem Set

Note: This lesson is a continuation of the problem solving from Lesson 28. It is recommended that delivery of today's lesson follow that of Lesson 28. It is acceptable to allow students as much independence in solving as is appropriate for specific student populations.

**Problem 1**

Lamar has 1,354.5 kilograms of potatoes to deliver in equal amounts to 18 stores. Twelve of the stores are in the Bronx. How many kilograms of potatoes will be delivered to stores in the Bronx?

Before solving:

T: Will the amount delivered to the stores in the Bronx be more or less than half of the total amount of potatoes delivered? How do you know?

S: More than half because more than half of the stores are in the Bronx.

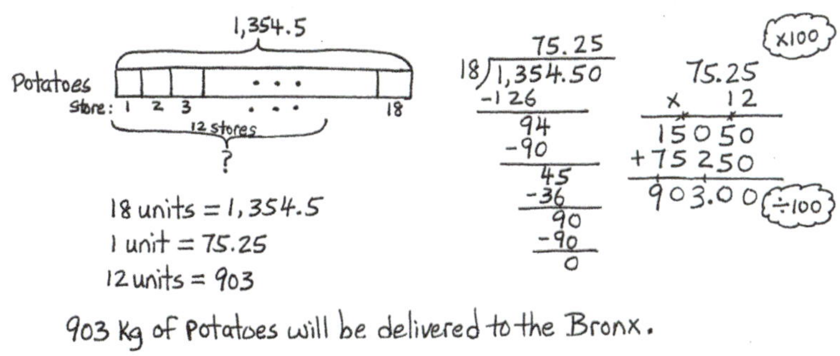


This two-step *equal groups with group size unknown problem* requires first dividing to find the value of one unit and then multiplying to find the value of 12 of those units.

T: How can you know that your final answer is reasonable? Was the amount delivered to the Bronx stores more than half of the total?

T: How did you determine if your decimal was placed reasonably in your product?

S: I was multiplying by 12. I knew that my answer needed to be more than 750 but less than 7,500. The only place that made sense to put the decimal made the answer 903, not 90.3 or 9,030. → I mentally multiplied 75.25 by 100 to make it 7,525 hundredths before I multiplied by 12. I knew I needed to adjust my product by dividing by 100 at the end.

**Problem 2**

MP.2

Valerie uses 12 fluid oz of detergent each week for her laundry. If there are 75 fluid oz of detergent in the bottle, in how many weeks will she need to buy a new bottle of detergent? Explain how you know.

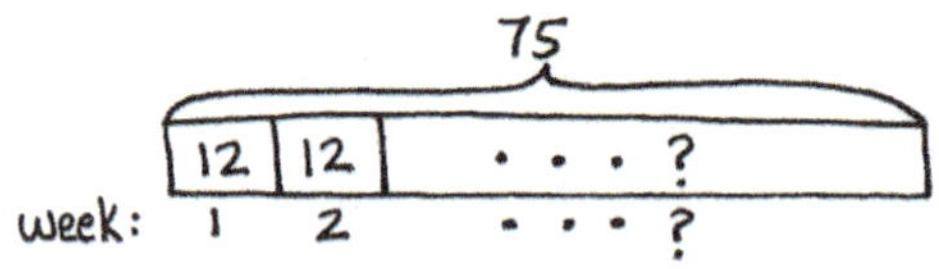


$12\overline{)75.00} = 6.25$; $-72$, $30$, $-24$, $60$, $-60$, $0$

Valerie will need to buy a new bottle of detergent after 6 weeks. She will have a little left over after 6 weeks, but not enough to do all her laundry in the 7th week.

The interpretation of the remainder in this single-step *equal groups with number of groups unknown* problem requires that students recognize the need to buy the detergent in 6 weeks. Although there will be a small amount of detergent left after the sixth week, there is not enough to do a seventh week of laundry.

After solving and assessing reasonability:

T: The quotient was more than 6. Why can't Valerie wait another week before buying detergent?

S: The quotient is the number of weeks that the detergent will last. It will last a little more than 6 weeks, but that means she won't have enough for all the laundry in the seventh week. → To have enough for 7 weeks, the detergent bottle would need to hold 7 × 12 fluid oz, which is 84 fluid oz. It's less than that, so she has to buy another bottle before the seventh week.

**Problems 3–4**

*Problem 3: The area of a rectangle is 56.96 m². If the length is 16 m, what is its perimeter?*

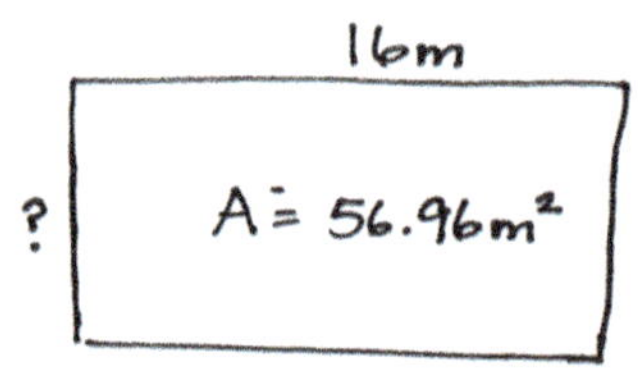


$16\overline{)56.96} = 3.56$; $-48$, $89$, $-80$, $96$, $-96$, $0$

A = l × w

56.96 = 16 × ?

P = 16 + 16 + ? + ?

P = 32 + (3.56 × 2)

= 32 + 7.12

= 39.12

The perimeter is 39.12 m.





EUREKA MATH

*Problem 4: A city block is 3 times as long as it is wide. If the total distance around the block is 0.48 kilometers, what is the area of the block in square meters?*

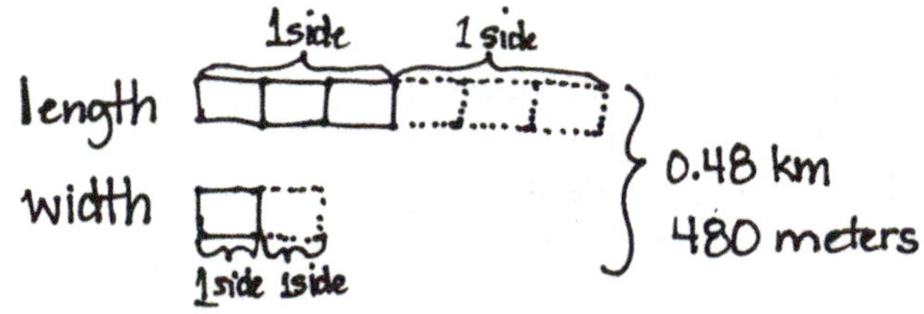


8 units = 480 meters
1 unit = 60 meters
length = 3 units = 180 m
width = 1 unit = 60 m

A = 60 × 180
= 6 × 18 × 100
= 108 × 100
= 10800

The area of the block is 10,800 m$^2$.

Problems 3 and 4 require students to apply their knowledge of area and perimeter to find missing sides using division and then use that information to answer the question. In Problem 3, area information must be used to find the perimeter, and in Problem 4, the perimeter must be used to find the area. In both cases, students must consider the existence of 2 pairs of equal sides in their calculations. In Problem 3, students may find it more helpful to draw a rectangle rather than a tape diagram. However, the *3 times as long* relationship in Problem 4 might be better modeled using a tape diagram. An added complexity of Problem 4 is the need to convert between kilometers and meters.

**NOTES ON MULTIPLE MEANS OF REPRESENTATION:**

When using a tape diagram that is divided into more than 10 equal parts, encourage students to use *dot, dot, dot* to indicate the uniformity of the equal parts in the tape diagram. This saves time and space. For students who are having difficulty with the tape diagram or calculations, it is better to work with smaller numbers that allow for a greater understanding of the concept when modeled.

After solving and assessing reasonability:

T: Find someone whose drawing looks different than yours for Problem 3 or 4. Compare your approaches.

T: How are these two problems alike, and how are they different?

S: Both are about rectangles with missing information. → One asks for area, and the other asks for perimeter. → You have to remember how to find area and perimeter. You have to find the missing side before you can answer the question.

## Problem Set (10 minutes)

Students should do their personal best to complete the Problem Set within the allotted 10 minutes. For some classes, it may be appropriate to modify the assignment by specifying which problems they work on first. Some problems do not specify a method for solving. Students should solve these problems using the RDW approach used for Application Problems.

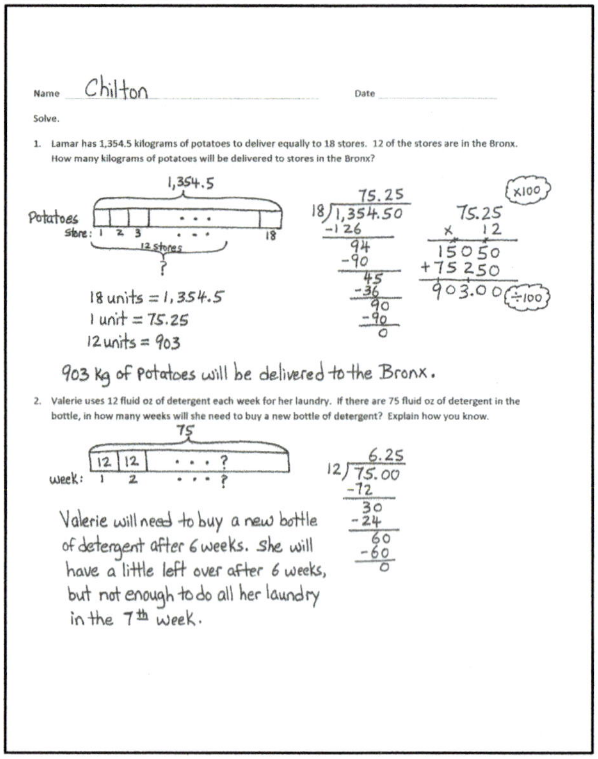
Name Chilton Date

Solve.

1. Lamar has 1,354.5 kilograms of potatoes to deliver equally to 18 stores. 12 of the stores are in the Bronx. How many kilograms of potatoes will be delivered to stores in the Bronx?

18 units = 1,354.5
1 unit = 75.25
12 units = 903

903 kg of potatoes will be delivered to the Bronx.

2. Valerie uses 12 fluid oz of detergent each week for her laundry. If there are 75 fluid oz of detergent in the bottle, in how many weeks will she need to buy a new bottle of detergent? Explain how you know.

Valerie will need to buy a new bottle of detergent after 6 weeks. She will have a little left over after 6 weeks, but not enough to do all her laundry in the 7th week.

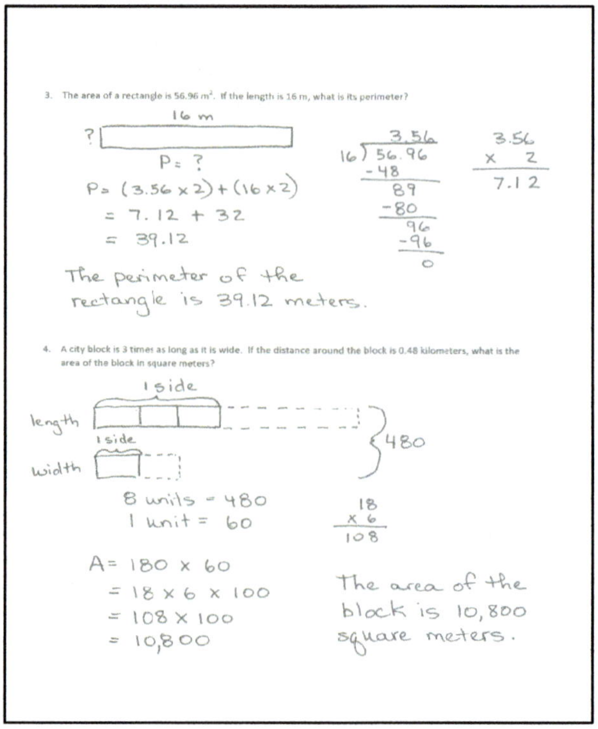
3. The area of a rectangle is 56.96 m². If the length is 16 m, what is its perimeter?

P = (3.56 × 2) + (16 × 2)
= 7.12 + 32
= 39.12

The perimeter of the rectangle is 39.12 meters.

4. A city block is 3 times as long as it is wide. If the distance around the block is 0.48 kilometers, what is the area of the block in square meters?

8 units = 480
1 unit = 60

A = 180 × 60
= 18 × 6 × 100
= 108 × 100
= 10,800

The area of the block is 10,800 square meters.

## Student Debrief (10 minutes)

**Lesson Objective:** Solve division word problems involving multi-digit division with group size unknown and the number of groups unknown.

The Student Debrief is intended to invite reflection and active processing of the total lesson experience.

Invite students to review their solutions for the Problem Set. They should check work by comparing answers with a partner before going over answers as a class. Look for misconceptions or misunderstandings that can be addressed in the Debrief. Guide students in a conversation to debrief the Problem Set and process the lesson.

Any combination of the questions below may be used to lead the discussion. As there is so much time given to debriefing each individual problem in the set, the culminating questions for today's lesson are brief.

- Compare Problems 3 and 4 and Problems 1 and 2. Students may note the following:
  - A tape diagram is not as helpful compared to a picture of the rectangle in Problem 3.
  - In Problems 3 and 4, it is harder to say if the divisor is the number of groups or the size of the group.
  - All four problems involve measurement.
- What did the divisor represent in each equation? What did the unknown represent for each? How did that change the model you drew? Which is easier to draw?

## Exit Ticket (3 minutes)

After the Student Debrief, instruct students to complete the Exit Ticket. A review of their work will help with assessing students' understanding of the concepts that were presented in today's lesson and planning more effectively for future lessons. The questions may be read aloud to the students.

Name ______________________________ Date ______________

Solve.

1. Lamar has 1,354.5 kilograms of potatoes to deliver equally to 18 stores. 12 of the stores are in the Bronx. How many kilograms of potatoes will be delivered to stores in the Bronx?

2. Valerie uses 12 fluid oz of detergent each week for her laundry. If there are 75 fluid oz of detergent in the bottle, in how many weeks will she need to buy a new bottle of detergent? Explain how you know.

3. The area of a rectangle is 56.96 $m^2$. If the length is 16 m, what is its perimeter?

4. A city block is 3 times as long as it is wide. If the distance around the block is 0.48 kilometers, what is the area of the block in square meters?

Name ______________________________________ Date ____________________

Solve.

Hayley borrowed $1,854 from her parents. She agreed to repay them in equal installments throughout the next 18 months. How much will Hayley still owe her parents after a year?

Name ______________________________ Date ______________

Solve.

1. Michelle wants to save $150 for a trip to the Six Flags amusement park. If she saves $12 each week, how many weeks will it take her to save enough money for the trip?

2. Karen works for 85 hours throughout a two-week period. She earns $1,891.25 throughout this period. How much does Karen earn for 8 hours of work?

3. The area of a rectangle is 256.5 $m^2$. If the length is 18 m, what is the perimeter of the rectangle?

4. Tyler baked 702 cookies. He sold them in boxes of 18. After selling all of the boxes of cookies for the same amount each, he earned $136.50. What was the cost of one box of cookies?

5. A park is 4 times as long as it is wide. If the distance around the park is 12.5 kilometers, what is the area of the park?

Name ________________________________________ Date ____________________

1. Express the missing divisor using a power of 10. Explain your reasoning using a place value model.

   a. $5.2 \div$ __________ $= 0.052$

   b. $7{,}650 \div$ __________ $= 7.65$

2. Estimate the quotient by rounding the expression to relate to a one-digit fact. Explain your thinking in the space below.

   a. $432 \div 73 \approx$ __________

   b. $1{,}275 \div 588 \approx$ __________

3. Generate and solve another division problem with the same quotient and remainder as the two problems below. Explain your strategy for creating the new problem.

$$\begin{array}{r} 3 \\ 17\overline{)63} \\ -\underline{51} \\ 12 \end{array} \qquad \begin{array}{r} 3 \\ 42\overline{)138} \\ -\underline{126} \\ 12 \end{array}$$

4. Sarah says that 26 ÷ 8 equals 14 ÷ 4 because both are "3 R2." Show her mistake using decimal division.

5. A rectangular playground has an area of 3,392 square meters. If the width of the rectangle is 32 meters, find the length.

6. A baker uses 5.5 pounds of flour daily.

   a. How many ounces of flour will he use in two weeks? Use words, numbers, or pictures to explain your thinking. (1 lb = 16 oz)

b. The baker's recipe for a loaf of bread calls for 12 ounces of flour. If he uses all of his flour to make loaves of bread, how many full loaves can he bake in two weeks?

c. The baker sends all his bread to one store. If he can pack up to 15 loaves of bread in a box for shipping, what is the minimum number of boxes required to ship all the loaves baked in two weeks? Explain your reasoning.

d. The baker pays $0.80 per pound for sugar and $1.25 per pound for butter. Write an expression that shows how much the baker will spend if he buys 6 pounds of butter and 20 pounds of sugar.

e. Chocolate sprinkles cost as much per pound as sugar. Find $\frac{1}{10}$ the baker's total cost for 100 pounds of chocolate sprinkles. Explain the number of zeros and the placement of the decimal in your answer using a place value chart.

| End-of-Module Assessment Task Standards Addressed | Topics A–H |
|---|---|

**Write and interpret numerical expressions.**

**5.OA.1** Use parentheses, brackets, or braces in numerical expressions, and evaluate expressions with these symbols.

**5.OA.2** Write simple expressions that record calculations with numbers, and interpret numerical expressions without evaluating them. *For example, express the calculation "add 8 and 7, then multiply by 2" as 2 × (8 + 7). Recognize that 3 × (18932 + 921) is three times as large as 18932 + 921, without having to calculate the indicated sum or product.*

**Understand the place value system.**

**5.NBT.1** Recognize that in a multi-digit number, a digit in one place represents 10 times as much as it represents in the place to its right and 1/10 of what it represents in the place to its left.

**5.NBT.2** Explain patterns in the number of zeros of the product when multiplying a number by powers of 10, and explain patterns in the placement of the decimal point when a decimal is multiplied or divided by a power of 10. Use whole-number exponents to denote powers of 10.

**Perform operations with multi-digit whole numbers and with decimals to hundredths.**

**5.NBT.5** Fluently multiply multi-digit whole numbers using the standard algorithm.

**5.NBT.6** Find whole-number quotients of whole numbers with up to four-digit dividends and two-digit divisors, using strategies based on place value, the properties of operations, and/or the relationship between multiplication and division. Illustrate and explain the calculation by using equations, rectangular arrays, and/or area models.

**5.NBT.7** Add, subtract, multiply, and divide decimals to hundredths, using concrete models or drawings and strategies based on place value, properties of operations, and/or the relationship between addition and subtraction; relate the strategy to a written method and explain the reasoning used.

**Convert like measurement units within a given measurement system.**

**5.MD.1** Convert among different-sized standard measurement units within a given measurement system (e.g., convert 5 cm to 0.05 m), and use these conversions in solving multi-step, real world problems.

## Evaluating Student Learning Outcomes

A Progression Toward Mastery is provided to describe steps that illuminate the gradually increasing understandings that students develop on their way to proficiency. In this chart, this progress is presented from left (Step 1) to right (Step 4). The learning goal for students is to achieve Step 4 mastery. These steps are meant to help teachers and students identify and celebrate what the students CAN do now and what they need to work on next.

| A Progression Toward Mastery | | | | |
|---|---|---|---|---|
| **Assessment Task Item** | **STEP 1 Little evidence of reasoning without a correct answer. (1 Point)** | **STEP 2 Evidence of some reasoning without a correct answer. (2 Points)** | **STEP 3 Evidence of some reasoning with a correct answer or evidence of solid reasoning with an incorrect answer. (3 Points)** | **STEP 4 Evidence of solid reasoning with a correct answer. (4 Points)** |
| **1** 5.NBT.1 5.NBT.2 5.NBT.7 | Student is unable to express the divisors as powers of 10 either as multiples of 10 or as exponents and produces a place value chart with errors. | Student either shows the divisors as powers of 10 (as multiples of 10 or exponents) or uses correct reasoning on the place value chart. | Student correctly expresses the divisors as powers of 10 either as multiples of 10 or exponents and uses correct reasoning on the place value chart for either Part (a) or Part (b). | Student correctly expresses the divisors as powers of 10 either as multiples of 10 or exponents. Student also shows correct reasoning on the place value chart for both Part (a) and Part (b).<br>a. 100 and/or $10^2$<br>b. 1000 and/or $10^3$ |
| **2** 5.NBT.1 5.NBT.2 5.NBT.6 | Student is unable to round either the dividend or the divisor to a one-digit fact. | Student rounds the dividend and divisor but not to a one-digit fact. | Student correctly rounds to a one-digit fact for either Part (a) or Part (b) or rounds both parts correctly without a clear explanation. | Student correctly rounds both Part (a) and Part (b) to a one-digit fact and clearly explains thinking.<br>a. 420 ÷ 70 = 6<br>b. 1,200 ÷ 600 = 2 |
| **3** 5.OA.1 5.NBT.6 | Student is unable to generate a division problem with a quotient of 3 and remainder of 12. | Student generates a division problem with either a quotient of 3 or a remainder of 12 but is unable to explain reasoning used. | Student generates a division problem with both a quotient of 3 and a remainder of 12 but shows no evidence of a strategy other than guess and check. | Student generates a division problem with a quotient of 3 and remainder of 12 and describes a sound strategy (e.g., writes a checking equation ____ = 3 × _____ + 12). |

| A Progression Toward Mastery | | | | |
|---|---|---|---|---|
| **4**<br><br>**5.NBT.7** | Student is unable to perform the decimal division necessary to show non-equivalence of quotients. | Student is able to perform the division necessary to produce the whole number portion of the quotient but is unable to continue dividing the decimal places to show non-equivalence of quotients. | Student is able to explain the non-equivalence of the quotients but with errors in the division calculation. | Student divides accurately and shows the non-equivalence of the quotients.<br>26 ÷ 8 = 3.25<br>14 ÷ 4 = 3.5 |
| **5**<br><br>**5.NBT.6** | Student does not divide to find the length of the playground. | Student makes two errors in division that lead to an incorrect length of the playground. | Student makes one error in division that leads to an incorrect length of the playground. | Student correctly divides and finds the length of the rectangle to be 106 m. |
| **6**<br><br>**5.OA.1**<br>**5.OA.2**<br>**5.NBT.1**<br>**5.NBT.2**<br>**5.NBT.5**<br>**5.NBT.6**<br>**5.NBT.7**<br>**5.MD.1** | Student uses incorrect reasoning for all parts of the task. | Student uses correct reasoning for at least two parts of the task but makes errors in calculation. | Student uses correct reasoning for all parts of the task but makes errors in calculation. | Student describes correct reasoning using words, numbers, or pictures and correctly calculates for all parts of the task.<br>a. 1,232 oz<br>b. 102 loaves<br>c. 7 boxes<br>d. (20 × 0.80) + (6 × $1.25)<br>e. $8.00 |

Name Garrett Date ____________

1. Express the missing divisor using an exponent. Explain your reasoning using a place value chart.

a. $5.2 \div \underline{10^2} = 0.052$

b. $7{,}650 \div \underline{10^3} = 7.65$

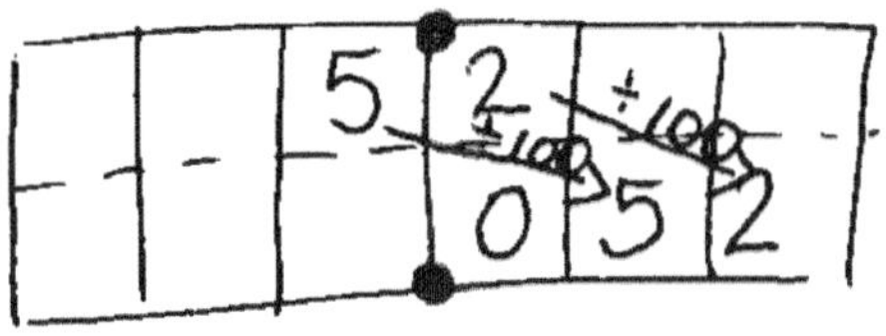


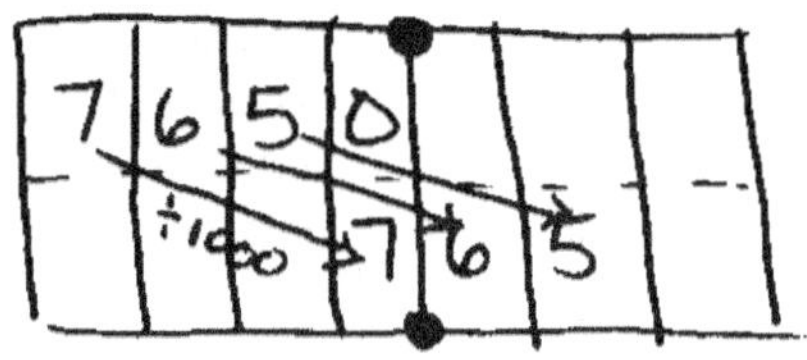


2. Estimate the quotient by rounding the equation to relate to a 1-digit fact. Explain your thinking in the space below.

a. $432 \div 73 \approx \underline{6}$

$420 \div 70 = 42 \div 7 = 6$

73 is close to 7 tens. The nearest multiple of 7 that's like 432 is 42 tens. So $42 \div 7 = 6$

b. $1275 \div 588 \approx \underline{2}$

$1200 \div 600 = 12 \div 6 = 2$

588 is close to 600. The nearest multiple of 600 that is close to 1275 is 12 hundreds. So $12 \div 6 = 2$

3. Generate and solve another division problem with the same quotient and remainder as the two problems below. Explain your strategy for creating the new problem.

```
      3            3           3
17 ) 63     42 ) 138     27 ) 93
     51          126         -81
     12           12          12
```

To check division, I can multiply the answer and the divisor, then add the remainder. So I multiplied 3 X my number which was 27 and got 81 and then I added 12. So my dividend must be 93.

```
  27
x  3
  81
+ 12
  93
```

4. Sarah says that 26 ÷ 8 equals 14 ÷ 4 because both are "3 R2." Show her mistake using decimal division.

$26 \div 8 = 3.25$

$14 \div 4 = 3.5$

5. A rectangular playground has an area of 3,392 square meters. If the width of the rectangle is 32 meters, find the length.

$32 \times ? = 3{,}392$

The length of the rectangle is 106 meters.

6. A baker uses 5.5 pounds of flour daily.

 a. How many ounces of flour will he use in two weeks? Use words, numbers, or pictures to explain your thinking. (1 lb = 16 oz.)

5.5 lbs = ___ oz
5.5 × (1 lb) = ___ oz
5.5 × (16 oz) = ___ oz

55 tenths × 16 = 330 + 550 = 880 tenths = 88

88 oz × 14 = 352 + 880 = 1,232 oz

First, I found the ounces he uses every day. Then I multiplied by 14 days.

The baker uses 1,232 oz of flour in 2 weeks.

b. The baker's recipe for a loaf of bread calls for 12 ounces of flour. If he uses all of his flour to make loaves of bread, how many full loaves can he bake in two weeks?

12 ) 1,232 = 102 R8
- 12
03
- 00
32
- 24
8

The baker can bake 102 full loaves in two weeks.

c. The baker sends all his bread to one store. If he can pack up to 15 loaves of bread in a box for shipping, what is the minimum number of boxes required to ship all the loaves baked in two weeks. Explain your reasoning.

15 ) 102 = 6
- 90
12

He needs 7 boxes to ship all the bread. The last box won't be full. It will only have 12 loaves in it.

d. The baker pays \$0.80 per pound for sugar and \$1.25 per pound for butter. Write an expression that shows how much the baker will spend if he buys 6 pounds of butter and 20 pounds of sugar.

(6 × \$1.25) + (20 × \$0.80)

e. Chocolate sprinkles cost as much per pound as sugar. Find $\frac{1}{10}$ the baker's total cost for 100 pounds of chocolate sprinkles. Explain the number of zeros and the placement of the decimal in your answer using a place value chart.

\$0.80 ÷ 10 = \$0.08

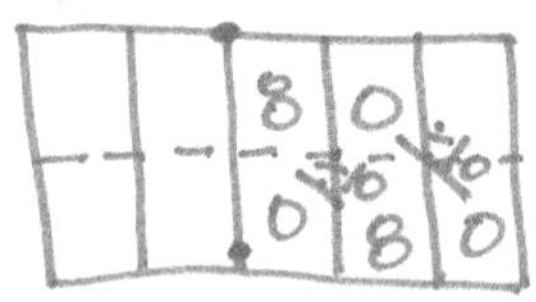

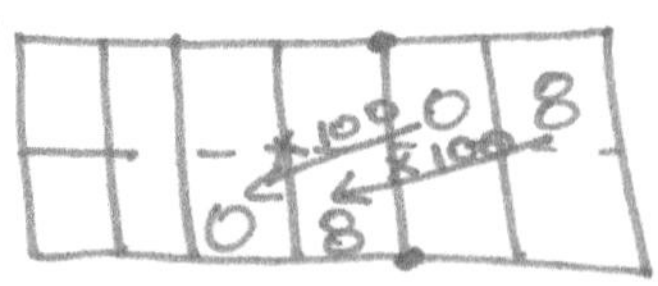

The baker pays \$8.00 for 100 lbs of sprinkles.

## Answer Key

# Eureka Math Grade 5 Module 2

Special thanks go to the Gordon A. Cain Center and to the Department of Mathematics at Louisiana State University for their support in the development of *Eureka Math*.

For a free *Eureka Math* Teacher Resource Pack, Parent Tip Sheets, and more please visit www.Eureka.tools

**Published by the non-profit Great Minds**



Printed in the U.S.A.
This book may be purchased from the publisher at eureka-math.org
10 9 8 7 6 5 4 3 2

Answer Key

# GRADE 5 • MODULE 2

## Multi-Digit Whole Number and Decimal Fraction Operations

# Lesson 1

## Problem Set

1. a. 46; 460
   b. 46 hundreds; 4,600
   c. Ones; 164
   d. Thousands; 164,000
   e. 331; 3; thousands; 993,000
   f. 5; 6; ten thousands; 300,000
2. a. False; answers will vary.
   b. False; answers will vary.
   c. True; answers will vary.
   d. True; answers will vary.
   e. False; answers will vary.
3. a. Answers provided
   b. 135; 1,350; 13,500; 135,000
   c. 200; 2,000; 20,000; 2,000,000
   d. 1,436; 143,600; 1,436,000; 143,600,000
4. Answers will vary.
5. 36,000,000 gal
6. $9,900

## Exit Ticket

1. a. 38,000
   b. 300,000
   c. 75,000
2. Answers will vary.

## Homework

1. a. 129; 1,290
   b. 129; 12,900
   c. Hundreds; 16,600
   d. 44; 4; ten thousands; 1,760,000
   e. 8; 5; ten thousands; 400,000
2. a. True; answers will vary.
   b. False; answers will vary.
   c. True; answers will vary.
   d. False; answers will vary.
3. a. Answers provided
   b. 400; 4,000; 400,000; 400,000
   c. 1,911; 191,100; 1,911,000; 19,110,000
4. 600 $in^2$
5. 2,115,000

# Lesson 2

## Sprint

### Side A

1. 90
2. 900
3. 9,000
4. 80
5. 800
6. 8,000
7. 80,000
8. 70
9. 700
10. 7,000
11. 70,000
12. 700,000
13. 20
14. 300
15. 320
16. 40
17. 500
18. 540
19. 370
20. 840
21. 8,400
22. 84,000
23. 73,000
24. 600
25. 6,000
26. 60,000
27. 6,500
28. 65,200
29. 34,200
30. 80,000
31. 800,000
32. 860,000
33. 867,000
34. 492,000
35. 340
36. 6,290
37. 9,400
38. 23,800
39. 47,000
40. 294,000
41. 17,400
42. 285,000
43. 95,100
44. 129,000

### Side B

1. 80
2. 800
3. 8,000
4. 70
5. 700
6. 7,000
7. 70,000
8. 60
9. 600
10. 6,000
11. 60,000
12. 600,000
13. 30
14. 200
15. 230
16. 50
17. 400
18. 450
19. 730
20. 480
21. 4,800
22. 48,000
23. 37,000
24. 500
25. 5,000
26. 50,000
27. 5,600
28. 56,200
29. 43,200
30. 70,000
31. 700,000
32. 760,000
33. 765,000
34. 942,000
35. 740
36. 2,690
37. 4,900
38. 32,800
39. 37,000
40. 924,000
41. 14,700
42. 825,000
43. 65,100
44. 192,000

## Problem Set

Estimates will vary.

1. a. 600; 50; 30,000; 30,000
   b. 1,000; 60; 60,000; 60,000
   c. 6,000; 25; 150,000; 150,000
2. a. Answers provided
   b. 30,000 × 400; 12,000,000
   c. 9,000 × 60; 540,000
   d. 900 × 600; 540,000
   e. 40,000 × 500; 20,000,000
3. 2,146 × 121; explanations will vary.
4. a. 600; 40
   b. 700; 700
   c. 8,000; 500
5. 800,000
6. a. $20,000
   b. Answers will vary.

## Exit Ticket

a. 700 × 100; 70,000
b. 3,000 × 8,000; 24,000,000
c. 400 × 9,000; 3,600,000
d. 9,000 × 60,000; 540,000,000

## Homework

1. a. 700; 80; 56,000; 56,000
   b. 6,000; 70; 420,000; 420,000
   c. 9,000; 50; 450,000; 450,000
2. a. Answers provided
   b. 80,000 × 200; 16,000,000
   c. 10,000 × 40; 400,000
   d. 1,000 × 400; 400,000
   e. 90,000 × 900; 81,000,000
3. 8,146 × 121; explanations will vary.
4. a. 800; 30
   b. 600; 700
   c. 8,000; 500
5. 3,200,000
6. a. 10
   b. $18,000

# Lesson 3

## Problem Set

1.
   a. Accurate model drawn; (8 + 7) × 2
   b. Accurate model drawn; 4 × (14 + 26)
   c. Accurate model drawn; 3 × (37.5 – 24.5)
   d. Accurate model drawn; (3 × 16) + (2 × 9)
   e. Accurate model drawn; (4 × 25) – (3 × 25)
   f. Accurate model drawn; 3 × (33 + 27)
2.
   a. Answers will vary; 360
   b. Answers will vary; 550
   c. Answers will vary; 2,300
   d. Answers will vary; 76
3.
   a. >; answers will vary.
   b. <; answers will vary.
   c. =; answers will vary.
4. Model drawn correctly; (7 × 15) + (38 × 15)
5. MeiLing's is equivalent; answers will vary.
6.
   a. 24 × (8 + 12)
   b. 2 × (24 × 20)
   c. 1,440

## Exit Ticket

1.
   a. Model drawn correctly; 8 × 47 – 7 × 47
   b. Model drawn correctly; 6 × (12 + 8)
2. >

## Homework

1.
   a. Accurate model drawn; (21 + 4) × 2
   b. Accurate model drawn; 5 × (7 + 23)
   c. Accurate model drawn; 2 × (49.5 – 37.5)
   d. Accurate model drawn; 3 × 15 + 4 × 2
   e. Accurate model drawn; 9 × 37 – 8 × 37
   f. Accurate model drawn; 3 × (45 + 55)
2.
   a. Answers will vary; 160
   b. Answers will vary; 220
   c. Answers will vary; 2,600
   d. Answers will vary; 130
3.
   a. >; answers will vary.
   b. >; answers will vary.
4.
   a. No; answers will vary.
   b. (14 + 12) × (8 + 12); 256

# Lesson 4

## Problem Set

1. a. $(15 - 1) \times 29$ and $(10 \times 29) - (6 \times 29)$
   b. $(38 + 40) \times (38 + 5)$ and $45 \times (40 + 2)$
   c. $(74 \times 5) + (74 \times 9)$
2. a. 19; 20, 1; 500, 25; 475
   b. 11; 10, 1; 10, 1; 240, 24; 264
   c. 79; 80; 80, 1; 1,120, 14; 1,106
   d. 21; 20, 1; 20, 1; 1500, 75; 1,575
3. a. Fifteens; fifteens, fifteen; 15, 15; 300, 15; 285
   b. Fifteens; fifteens, fifteens; 15, 15; 150, 60; 210
   c. Twenty-fives; twenty-fives, twenty-fives; 25, 25; 250, 50; 300
   d. Seventeens; seventeens, seventeens; 17, 17; 340, 34; 306
4. Answers will vary.
5. a. 1,515
   b. 1,782
6. Answers will vary.
7. a. Answers will vary; $(174 + 126) \times 29$
   b. Answers will vary; 8,700

## Exit Ticket

a. 49; 50, 1; 550, 11; 539
b. 13; 10, 3; 10, 3; 250, 75; 325

## Homework

1. a. $(30 \times 19) - (7 \times 29)$ and $(40 - 2) \times 19$
   b. $(26 + 30) \times (26 + 5)$ and $35 \times (20 + 60)$
   c. $(34 \times 8) + (34 \times 9)$
2. a. 19; 20, 1; 1,000, 50; 950
   b. 11; 10, 1; 10, 1; 260, 26; 286
   c. 49; 50; 50, 1; 600, 12; 588
   d. 12; 10, 2; 10, 2; 250, 50; 300
3. a. Twelves; twelves, twelve; 12, 12; 360, 12; 348
   b. Elevens; elevens, eleven; 11, 11; 330, 11; 341
   c. Elevens; elevens, eleven; 11, 11; 220, 11; 209
   d. Fifties; fifties, fifties; 500, 150; 650
4. Answers will vary.
5. a. 1,584
   b. 2,020
6. 266 $ft^2$; answer will vary.
7. 2,323; answers will vary.

# Lesson 5

## Pattern Sheet

Estimates may vary.

| | | | | | | | |
|---|---|---|---|---|---|---|---|
| 1. | 300 | 12. | 450 | 23. | 24,000 | 34. | 2,400 |
| 2. | 600 | 13. | 900 | 24. | 24,000 | 35. | 24,000 |
| 3. | 900 | 14. | 8,000 | 25. | 21,000 | 36. | 1,800 |
| 4. | 250 | 15. | 6,000 | 26. | 1,200 | 37. | 18,000 |
| 5. | 500 | 16. | 4,000 | 27. | 1,500 | 38. | 24,000 |
| 6. | 750 | 17. | 400 | 28. | 1,500 | 39. | 24,000 |
| 7. | 1,000 | 18. | 800 | 29. | 2,500 | 40. | 28,000 |
| 8. | 400 | 19. | 1,200 | 30. | 2,700 | 41. | 32,000 |
| 9. | 800 | 20. | 12,000 | 31. | 27,000 | 42. | 49,000 |
| 10. | 1,000 | 21. | 15,000 | 32. | 2,100 | 43. | 64,000 |
| 11. | 1,000 | 22. | 18,000 | 33. | 21,000 | 44. | 72,000 |

## Problem Set

1. a. 714
   b. 9,114
2. a. 5,172
   b. 2,829
   c. 9,984
3. $7,248
4. 508.2 kg

## Exit Ticket

a. 483

b. 1,716

## Homework

1. a. 504
   b. 5,082
2. a. 6,908
   b. 9,086
   c. 6,816
3. 2.99 m
4. $1,704

# Lesson 6

## Pattern Sheet

| | | | | | | | |
|---|---|---|---|---|---|---|---|
| 1. | 500 | 12. | 594 | 23. | 4,950 | 34. | 2,079 |
| 2. | 495 | 13. | 400 | 24. | 4,950 | 35. | 3,100 |
| 3. | 495 | 14. | 396 | 25. | 8,000 | 36. | 3,069 |
| 4. | 300 | 15. | 700 | 26. | 7,920 | 37. | 7,100 |
| 5. | 297 | 16. | 693 | 27. | 6,000 | 38. | 7,029 |
| 6. | 297 | 17. | 900 | 28. | 5,940 | 39. | 4,200 |
| 7. | 200 | 18. | 891 | 29. | 1,100 | 40. | 4,158 |
| 8. | 198 | 19. | 800 | 30. | 1,089 | 41. | 5,247 |
| 9. | 198 | 20. | 792 | 31. | 1,089 | 42. | 6,336 |
| 10. | 600 | 21. | 500 | 32. | 2,100 | 43. | 7,425 |
| 11. | 594 | 22. | 5,000 | 33. | 2,079 | 44. | 9,603 |

## Problem Set

1. a. 1,680
   b. 22,680
2. a. 69,736
   b. 90,052
   c. 30,940
   d. 35,008
3. $3,584
4. a. $2,375
   b. $19 \times 125 + 4 \times 125 = \$2,875$

## Exit Ticket

a. 3,276

b. 32,886

## Homework

1. a. 972
   b. 18,972
2. a. 34,397
   b. 26,288
   c. 34,868
   d. 25,392
3. $6,000
4. $28,466

# Lesson 7

## Sprint

### Side A

1. 20
2. 120
3. 1,200
4. 40
5. 340
6. 3,400
7. 70
8. 270
9. 2,700
10. 30
11. 6
12. 60
13. 130
14. 26
15. 260
16. 1,300
17. 2,600
18. 8
19. 88
20. 880
21. 8,800
22. 66
23. 660
24. 6,600
25. 240
26. 480
27. 2,400
28. 4,800
29. 690
30. 6,900
31. 142
32. 1,420
33. 28
34. 42
35. 420
36. 4,200
37. 1,640
38. 4,500
39. 42,600
40. 720
41. 2,250
42. 25,200
43. 5,220
44. 63,200

### Side B

1. 30
2. 130
3. 1,300
4. 50
5. 350
6. 3,500
7. 80
8. 280
9. 2,800
10. 40
11. 8
12. 80
13. 140
14. 28
15. 280
16. 1,400
17. 2,800
18. 6
19. 66
20. 660
21. 6,600
22. 88
23. 880
24. 8,800
25. 420
26. 840
27. 4,200
28. 8,400
29. 960
30. 9,600
31. 162
32. 1,620
33. 39
34. 52
35. 520
36. 5,200
37. 2,160
38. 4,500
39. 48,600
40. 640
41. 1,950
42. 14,400
43. 5,340
44. 60,800

## Problem Set

1. a. 169,312
   b. 145,262
   c. Answers will vary.
2. a. 2,562,305
   b. 2,618,350
3. a. 9,342
   b. 399,762
   c. 71,622
   d. 278,622
4. $251,535
5. 461,160
6. 2,494,920

## Exit Ticket

a. 164,944
b. 132,894

## Homework

1. a. 94,458
   b. 83,538
   c. Answers will vary.
2. a. 2,169,490
   b. 1,466,762
3. a. 152,082
   b. 509,082
   c. 130,782
   d. 437,782
4. 986,595
5. $47,188

# Lesson 8

## Problem Set

1. a. 69,864
   b. 280,000; 246,264
   c. 280,000; 326,638
   d. 400,000; 396,237
   e. 2,800,000; 2,297,312
   f. 2,800,000; 3,252,990
   g. 1,000,000; 1,099,750
   h. 2,400,000; 2,414,412
   i. 3,100,000; 3,064,635
2. 776 L 475 mL; answers will vary.
3. $71,713
4. a. No; answers will vary.
   b. 4,410
   c. 9

## Exit Ticket

a. 117,728; 300; 400; 120,000
b. 1,138,018; 2,800; 400; 1,120,000

## Homework

1. a. 46,488
   b. 210,000; 219,185
   c. 240,000; 272,636
   d. 210,000; 210,755
   e. 2,520,000; 2,550,048
   f. 1,500,000; 1,604,564
   g. 1,200,000; 1,317,520
   h. 1,500,000; 1,509,514
   i. 1,800,000; 1,550,416
2. No; answers will vary.
3. 600,000; answers will vary.

# Lesson 9

## Problem Set

1. \$1,518,720
2. a. 26,104
   b. Leah; \$1,996
3. $26 \times \$1,334 + 19 \times \$749 = \$48,915$
4. \$10,822
5. $2 \times (2 \times 9 + 2 \times 6) \times 12 \times 3 = 2,160$
6. a. 3,874
   b. 46,488

## Exit Ticket

a. \$103.50
b. \$89.70
c. \$193.20
d. $15 \times (30 + 26) = 840$

## Homework

1. 1,529
2. a. 4,832 yd
   b. 67,648 yd
3. \$13,608
4. 40,500 $cm^2$
5. a. (120 yd × 80 yd) – (100 yd × 55 yd) = 4,100 $yd^2$
   b. No; length is too short; answers will vary.
   c. \$249,600

# Lesson 10

## Problem Set

Estimates may vary.

1. a. 20, 2, 40; 52.8
   b. 3, 30, 90; 102.3
2. a. 3, 50, 150; 150.4
   b. 3, 90, 270; 300.8
   c. 6, 40, 240; 277.2
   d. 10, 20, 200; 248.2
   e. 8, 30, 240; 278.8
   f. 200, 20, 4000; 2726.8
3. Answers will vary.
4. 4.1 m

## Exit Ticket

1. a. 30, 20, 600; 697.2
   b. 2, 60, 120; 93.5
2. 1,697.5; answers will vary.

## Homework

1. a. 50, 1, 50; 63.6
   b. 2, 80, 160; 172.2
2. a. 4, 30, 120; 142.8
   b. 70, 6, 420; 377
   c. 3 × 20 = 60; 52.8
   d. 15 × 20 = 300; 265.2
   e. 75, 2, 150; 175.2
   f. 200, 60, 12,000; 11,029.5
3. 184.8 $m^2$
4. 114 mi

# Lesson 11

## Sprint

### Side A

1. 9
2. 0.9
3. 0.09
4. 6
5. 0.6
6. 0.06
7. 4
8. 0.4
9. 0.04
10. 15
11. 1.5
12. 0.15
13. 0.12
14. 1.2
15. 12
16. 25
17. 2.5
18. 0.25
19. 28
20. 2.8
21. 0.28
22. 3.6
23. 40
24. 4
25. 0.4
26. 0.3
27. 0.18
28. 3
29. 0.12
30. 0.42
31. 5.4
32. 0.54
33. 0.81
34. 6.4
35. 0.49
36. 3.6
37. 0.25
38. 4.8
39. 0.63
40. 2.4
41. 0.54
42. 3.5
43. 0.48
44. 1.08

### Side B

1. 4
2. 0.4
3. 0.04
4. 8
5. 0.8
6. 0.08
7. 9
8. 0.9
9. 0.09
10. 12
11. 1.2
12. 0.12
13. 0.15
14. 1.5
15. 15
16. 16
17. 1.6
18. 0.16
19. 32
20. 3.2
21. 0.32
22. 2.4
23. 30
24. 3
25. 0.3
26. 0.4
27. 0.24
28. 4
29. 0.16
30. 0.56
31. 7.2
32. 0.72
33. 8.1
34. 0.64
35. 4.9
36. 0.36
37. 2.5
38. 0.48
39. 6.3
40. 0.24
41. 5.4
42. 0.35
43. 0.72
44. 0.96

## Problem Set

Estimates may vary.

1. a. 1, 30, 30; 44.16
   b. 4, 90, 360; answers will vary; 315.95
2. a. 40.32
   b. 9,904.61
   c. 42,038.64
   d. 241.92
3. a. 752.64
   b. 1,140.99
   c. 5,699.4
4. $56.05
5. $3,692.42

## Exit Ticket

1. 407.61
2. a. 85.82
   b. 3,547.9

## Homework

1. a. 2, 12, 24; 29.04
   b. 4, 40, 160; answers will vary; 152.81
2. a. 26.39
   b. 1,807.44
   c. 19,675.19
   d. 678.24
3. a. 48.24
   b. 2,246.16
   c. 6,100.9
4. $42.39
5. a. $283.50
   b. 65.21 m

# Lesson 12

## Problem Set

Estimates may vary.

1. a. 1, 14, 14; 16.94
   b. 2, 300, 600; 747.25
2. a. 1, 12, 12; 14.76
   b. 1, 30, 30; 33.8
   c. 0.2, 10, 2; 3.22
   d. 0.5, 30, 15; 11.7
   e. 7, 30, 210; 197.68
   f. 6, 200, 1,200; 1,409.36
   g. 7, 200, 1,400; 1,468.48
   h. 200, 600, 120,000; 84,060.3
3. 106.95 mi
4. Greg will need $2.08 more.
5. 17.85 g

## Exit Ticket

a. 3, 400, 1,200; 1,218.06
b. 700, 1, 700; 833.75

## Homework

1. a. 25, 2, 50; 55.44
   b. 5, 300, 1,500; 1,653.1
2. a. 1, 21, 21; 25.83
   b. 3, 40, 120; 131.2
   c. 0.3, 40, 12; 13.12
   d. 0.5, 60, 30; 33.48
   e. 6, 30, 180; 170.52
   f. 7, 700, 4,900; 4,664.89
   g. 6, 200, 1,200; 1,266.72
   h. 200, 600, 120,000; 95,326.8
3. 2,007.5 mi
4. $3,174.75
5. $13.60

# Lesson 13

## Problem Set

1. a. Answer provided
   b. 4, 1; 4, 365; 1,460
   c. 9.2, 1; 9.2, 100; 920
   d. 17.1 ft
   e. 6,080 g
   f. 200 oz

2. a. Answer provided
   b. 68,000 g
   c. 170 cm
   d. 80 mL
   e. 180.8 oz
   f. 82.8 in

## Exit Ticket

a. 14; 14, 16; 224
b. 18.2, 1 kilogram; 18.2, 1,000 grams; 18,200

## Homework

1. a. Answer provided
   b. 7, 1; 7, 365; 2,555
   c. 4.5, 1; 4.5, 100; 450
   d. 201.6 oz
   e. 3,090 g
   f. 8,820 in

2. a. Answer provided
   b. 65,000 g
   c. 460 cm
   d. 100 mL
   e. 156.8 oz
   f. 130 mm

# Lesson 14

## Problem Set

1. a. Answer provided
   b. 20; 20; $\frac{20}{4}$; 5
   c. 920, 1; 920, 0.01; 9.2
   d. 1,578, 1; 1,578; 1.578
   e. 6.08 kg
   f. 0.509 L

2. a. Answer provided
   b. 6 pints
   c. 3.78 m
   d. 1.478 L
   e. 3,900 kg
   f. 264.04 km

## Exit Ticket

1. 35, 1; 35, $\frac{1}{7}$; $\frac{35}{7}$ weeks; 5 weeks
2. 4,567, 1 g; 4,567, 0.001 kg; 4.567 kg

## Homework

1. a. Answer provided
   b. 36; 36; $\frac{36}{4}$; 9
   c. 760, 1; 760, 0.01; 7.6
   d. 2,485, 1; 2,485; 2.485
   e. 3.09 kg
   f. 0.205 L

2. a. Answer provided
   b. 4 pints
   c. 5.29 m
   d. 2.06 L
   e. 1,560 kg
   f. 372.06 km

# Lesson 15

## Sprint

### Side A

1. 1; 0
2. 1; 1
3. 1; 2
4. 1; 3
5. 1; 10
6. 1; 8
7. 2; 0
8. 2; 1
9. 2; 2
10. 2; 6
11. 2; 10
12. 2; 11
13. 3; 0
14. 3; 1
15. 3; 10
16. 3; 4
17. 4; 0
18. 4; 10
19. 4; 1
20. 3; 11
21. 4; 2
22. 1; 0
23. 1; 5
24. 2; 0
25. 2; 4
26. 3; 0
27. 3; 9
28. 4; 0
29. 4; 11
30. 5; 0
31. 5; 4
32. 5; 8
33. 5; 11
34. 6; 1
35. 6; 0
36. 6; 8
37. 7; 0
38. 7; 6
39. 8; 0
40. 8; 4
41. 9; 0
42. 9; 9
43. 8; 8
44. 7; 9

### Side B

1. 10; 0
2. 1; 0
3. 1; 1
4. 1; 2
5. 1; 8
6. 1; 10
7. 2; 0
8. 2; 1
9. 2; 2
10. 2; 10
11. 2; 6
12. 2; 11
13. 3; 0
14. 3; 10
15. 3; 1
16. 3; 4
17. 4; 0
18. 4; 1
19. 4; 10
20. 3; 11
21. 4; 2
22. 1; 0
23. 1; 4
24. 2; 0
25. 2; 5
26. 3; 0
27. 3; 6
28. 4; 0
29. 4; 11
30. 5; 0
31. 5; 3
32. 5; 7
33. 5; 10
34. 6; 1
35. 6; 0
36. 6; 5
37. 7; 0
38. 7; 5
39. 8; 0
40. 8; 3
41. 9; 0
42. 9; 7
43. 8; 7
44. 7; 11

## Problem Set

1. 11 oz
2. 1,888 bags of peppercorns
3. 117.76 m
4. a. 48 cups of juice
   b. 12 containers

## Exit Ticket

18,060 m

## Homework

1. 19.2 cm
2. 11 gal
3. 35.7 m
4. a. 1,160 g
   b. 1.16 kg

# Lesson 16

## Sprint

### Side A

1. 3
2. 43
3. 430
4. 43
5. 430
6. 5
7. 85
8. 850
9. 85
10. 850
11. 60
12. 20
13. 20
14. 40
15. 20
16. 20
17. 24
18. 12
19. 12
20. 36
21. 12
22. 12
23. 120
24. 12
25. 2,100
26. 210
27. 21
28. 4,200
29. 420
30. 42
31. 32,000
32. 320
33. 3,200
34. 30
35. 40
36. 43
37. 6
38. 40
39. 60
40. 64
41. 32
42. 23
43. 240
44. 42

### Side B

1. 2
2. 42
3. 420
4. 42
5. 420
6. 4
7. 84
8. 840
9. 84
10. 840
11. 90
12. 30
13. 30
14. 60
15. 30
16. 30
17. 24
18. 12
19. 12
20. 63
21. 21
22. 21
23. 210
24. 21
25. 1,200
26. 120
27. 12
28. 2,400
29. 240
30. 24
31. 23,000
32. 230
33. 2,300
34. 20
35. 30
36. 32
37. 4
38. 40
39. 70
40. 73
41. 32
42. 23
43. 240
44. 42

## Problem Set

1. a. 50
   b. 36
   c. 120
   d. 4,500
   e. 700
   f. 5,300

2. a. Answer provided
   b. 40
   c. 4
   d. 8,000
   e. 800
   f. 80
   g. 700
   h. 900
   i. 90

3. a. 40 m
   b. 60 m
   c. 300 m

4. a. Carter; explanations will vary.
   b. 500

## Exit Ticket

a. 170
b. 59
c. 300
d. 800

## Homework

1. a. 30
   b. 45
   c. 180
   d. 7,300
   e. 900
   f. 680

2. a. Answer provided
   b. 90
   c. 9
   d. 7,000
   e. 700
   f. 70
   g. 800
   h. 800
   i. 50

3. 200

4. a. 4
   b. 16

# Lesson 17

## Problem Set

Estimates may vary.

1. a. Answers provided
   b. 900, 30; 30
   c. 800, 40; 20
   d. 140, 70; 2
   e. 240, 60; 4
   f. 480, 60; 8
   g. 640, 80; 8
   h. 800, 100; 8
   i. 600, 100; 6
   j. 320, 80; 4
   k. 640, 80; 8
   l. 160, 40; 4
   m. 600, 30; 20
   n. 540, 90; 6
   o. 600, 10; 60

2. 800 ÷ 40 = 20; answers will vary.

3. No; answers will vary.

## Exit Ticket

Estimates may vary.

a. 600, 20; 30
b. 900, 30; 30
c. 150, 30; 5
d. 480, 60; 8

## Homework

Estimates may vary.

1. a. Answers provided
   b. 600, 20; 30
   c. 800, 40; 20
   d. 500, 50; 10
   e. 500, 50; 10
   f. 140, 70; 2
   g. 480, 80; 6
   h. 630, 70; 9
   i. 600, 100; 6
   j. 300, 30; 10
   k. 720, 80; 9
   l. 630, 30; 21
   m. 810, 90; 9
   n. 350, 70; 5
   o. 550, 11; 50

2. $8

3. 160 ÷ 40 = 4 barrels; answers will vary.

# Lesson 18

## Problem Set

Estimates may vary.

1. a. Answers provided
   b. 3,000, 30; 100
   c. 9,000, 30; 300
   d. 1,500, 50; 30
   e. 2,400, 60; 40
   f. 2,100, 70; 30
   g. 4,800, 80; 60
   h. 8,000, 80; 100
   i. 8,100, 90; 90
   j. 4,200, 70; 60
   k. 4,800, 80; 60
   l. 4,500, 90; 50
   m. 5,400, 90; 60
   n. 7,200, 90; 80
   o. 3,000, 15; 200

2. 20 ft; answers will vary.

3. a. 40 MB
   b. $28

4. 40 qt

5. 200 hr

## Exit Ticket

Estimates may vary.

a. 6,600, 20; 330
b. 8,400, 40; 210
c. 4,000, 50; 80
d. 4,900, 70; 70

## Homework

Estimates may vary.

1.
   a. Answers provided
   b. 2,000, 20; 100
   c. 8,000, 40; 200
   d. 3,600, 60; 60
   e. 2,800, 70; 40
   f. 5,400, 90; 60
   g. 4,900, 70; 70
   h. 8,100, 81; 100
   i. 9,000, 90; 100
   j. 3,000, 50; 60
   k. 8,100, 90; 90
   l. 4,200, 60; 70
   m. 6,300, 70; 90
   n. 5,000, 50; 100
   o. 6,000, 30; 200

2. 26

3. 30; 11; 3; 40; answers will vary.

# Lesson 19

## Problem Set

1. a. Answer provided
   b. 2 R20; accurate check shown
   c. 1 R21; accurate check shown
   d. 9; accurate check shown
   e. 8 R3; accurate check shown
   f. 2 R35; accurate check shown
2. 7 R19; explanations may vary.
3. 564
4. 5

## Exit Ticket

a. 3 R13
b. 9 R21

## Homework

1. a. Answer provided
   b. 2 R10; accurate check shown
   c. 1 R35; accurate check shown
   d. 9 R10; accurate check shown
   e. 7 R17; accurate check shown
   f. 4 R26; accurate check shown
2. 256
3. 18
4. 4

# Lesson 20

## Problem Set

1. a. Answer provided
   b. 2 R7
   c. 2
   d. 2 R20
   e. 3 R2
   f. 4
2. 1 R39; explanations may vary.
3. 157
4. Answers may vary.
5. $105

## Exit Ticket

a. 3 R15
b. 2 R15

## Homework

1. a. Answer provided
   b. 4 R5
   c. 2 R28
   d. 3 R10
   e. 3 R4
   f. 3 R20
2. 7 ft
3. a. 6 full plates with 12 remaining
   b. 2 bagels

# Lesson 21

## Problem Set

1. a. Answer provided
   b. 2 R14
   c. 8 R7
   d. 8 R7
   e. 8 R61
   f. 7 R5
2. Answers and explanations may vary.
3. 19
4. a. 4
   b. More pencils.

## Exit Ticket

a. 6 R8
b. 5 R2

## Homework

1. a. Answer provided
   b. 4 R10
   c. 5 R16
   d. 7
   e. 8
   f. 8 R8
2. 4
3. 8

# Lesson 22

## Problem Set

1. a. Answer provided
   b. 22 R26
   c. 33 R16
   d. 24 R1
   e. 15 R14
   f. 40 R14
2. 13 R35; explanations may vary.
3. a. 10
   b. 17; explanations may vary.

## Exit Ticket

a. 21 R14
b. 10 R38

## Homework

1. a. Answer provided
   b. 32 R5
   c. 33 R7
   d. 12 R27
   e. 19 R22
   f. 16 R12
2. Explanations may vary.
3. Explanations may vary.

# Lesson 23

## Problem Set

1. a. 211 R6
   b. 84
   c. 213
   d. 70 R14
   e. 315 R17
   f. 70 R14
2. $376
3. 5,460
4. 58 days

## Exit Ticket

a. 435 R18
b. 28 R20

## Homework

1. a. 242 R40
   b. 33 R10
   c. 238 R27
   d. 82 R1
   e. 126 R15
   f. 93 R55
2. 565
3. 57

# Lesson 24

## Problem Set

1. a. Answer provided
   b. Answer provided
   c. 0.6
   d. $(2.4 \div 4) \div 10 = 0.6 \div 10 = 0.06$
   e. 2.1
   f. $(14.7 \div 7) \div 10 = 2.1 \div 10 = 0.21$
   g. 0.17
   h. $(0.34 \div 2) \div 10 = 1.7 \div 10 = 0.17$
   i. 0.05
   j. $(0.45 \div 9) \div 10 = 0.05 \div 10 = 0.005$
   k. 1.15
   l. $(34.5 \div 3) \div 100 = 11.5 \div 100 = 0.115$
2. a. 0.93; explanations may vary.
   b. 0.017; explanations may vary.
   c. 4.2; explanations may vary.
   d. 3.4; explanations may vary.
3. About 122 lb
4. 1.32 kg

## Exit Ticket

1. a. 9.1
   b. 0.091
   c. 0.91
2. 0.081; explanations may vary.

## Homework

1. a. Answer provided
   b. Answer provided
   c. 0.3
   d. (2.4 ÷ 80) ÷ 10 = 0.3 ÷ 10 = 0.03
   e. 7.3
   f. (14.6 ÷ 20) ÷ 10 = 7.3 ÷ 10 = 0.73
   g. 0.2
   h. (80 ÷ 4) ÷ 100 = 20 ÷ 100 = 0.2
   i. 0.08
   j. (0.56 ÷ 7) ÷ 10 = 0.08 ÷10 = 0.008
   k. 1.05
   l. (9.45 ÷ 9) ÷ 100 = 1.05 ÷ 100 = 0.0105

2. a. 8.2; explanations may vary.
   b. 0.5; explanations may vary.
   c. 4.8; explanations may vary.
   d. 0.66; explanations may vary.

3. a. 2.13 mi; explanations may vary.
   b. 29.82 mi

4. 70 mL

# Lesson 25

## Problem Set

1. a. $\approx 0.04$
   b. $\approx 6$
   c. $\approx 0.2$
   d. $\approx 3$
   e. $\approx 1$

2. a. $\approx 0.2$
   b. $\approx 20$
   c. $\approx 2$

3. a. About 13 miles
   b. About 40 days

4. a. About \$3.90
   b. No

## Exit Ticket

a. $\approx 0.08$
b. $\approx 2$
c. $\approx 0.2$

## Homework

1. a. $\approx 0.07$
   b. $\approx 0.6$
   c. $\approx 0.40$
   d. $\approx 2$
   e. $\approx 0.12$

2. a. $\approx 0.2$
   b. $\approx 20$
   c. $\approx 2$

3. $\approx$ \$0.75; answers may vary.

4. a. $\approx$ 0.5 gal; answers may vary.
   b. $\approx$ 3.5 gal
   c. $\approx$ 40 days

# Lesson 26

## Problem Set

1. a. No; 156 ÷ 24 = 6.5, 102 ÷ 15 = 6.8
   b. Answers will vary.
2. All answers should include a multiplication sentence to check division.
   a. 2.78
   b. 2.73
   c. 1.11
   d. 0.53
   e. 4.8
   f. 0.48
   g. 5.1
   h. 0.51
3. 2.55 g
4. 0.25 m
5. 80.25 m

## Exit Ticket

1. All answers should include an estimate and check.
   a. 2.15
   b. 0.23
2. Answers will vary.

## Homework

1. Answers will vary.
2. All answers should include a multiplication sentence to check division.
   a. 3.45
   c. 5.12
   b. 2.03
   c. 0.42
3. a. 2.87
   b. 1.23
4. a. 1.21; answers will vary
   b. 5.3; answers will vary
5. a. 14.2 ft
   b. 3.35 ft taller

# Lesson 27

## Problem Set

1. All answers should include a multiplication sentence to check division.
   a. 0.35
   b. 1.5
   c. 0.5
   d. 1.5
   e. 1.5
   f. 1.04
   g. 0.36
   h. 0.31
   i. 0.175
2. 1.27 kg
3. 34.25 in
4. $33.75
5. $0.81

## Exit Ticket

a. 0.875
b. 1.05

## Homework

1. All answers should include a multiplication sentence to check division.
   a. 0.25
   b. 2.04
   c. 0.5
   d. 8.26
   e. 20.06
   f. 16.8
2. 0.85 L; 850 mL
3. 6.75 kg
4. 24.15 L

# Lesson 28

## Sprint

### Side A

| | | | | | | | |
|---|---|---|---|---|---|---|---|
| 1. | 0.6 | 12. | 4 | 23. | 0.5 | 34. | 0.2 |
| 2. | 0.3 | 13. | 0.4 | 24. | 0.05 | 35. | 0.8 |
| 3. | 0.1 | 14. | 0.3 | 25. | 0.09 | 36. | 0.08 |
| 4. | 0.8 | 15. | 0.2 | 26. | 0.05 | 37. | 0.012 |
| 5. | 0.2 | 16. | 0.6 | 27. | 0.005 | 38. | 0.012 |
| 6. | 0.4 | 17. | 5 | 28. | 0.009 | 39. | 0.011 |
| 7. | 0.4 | 18. | 0.5 | 29. | 0.004 | 40. | 0.034 |
| 8. | 0.2 | 19. | 0.3 | 30. | 0.007 | 41. | 0.034 |
| 9. | 0.1 | 20. | 0.6 | 31. | 0.006 | 42. | 0.064 |
| 10. | 3 | 21. | 0.8 | 32. | 0.007 | 43. | 0.077 |
| 11. | 0.3 | 22. | 0.4 | 33. | 0.2 | 44. | 0.058 |

### Side B

| | | | | | | | |
|---|---|---|---|---|---|---|---|
| 1. | 0.4 | 12. | 6 | 23. | 0.5 | 34. | 0.2 |
| 2. | 0.2 | 13. | 0.6 | 24. | 0.05 | 35. | 0.9 |
| 3. | 0.1 | 14. | 0.4 | 25. | 0.07 | 36. | 0.06 |
| 4. | 0.8 | 15. | 0.3 | 26. | 0.05 | 37. | 0.012 |
| 5. | 0.4 | 16. | 0.2 | 27. | 0.005 | 38. | 0.012 |
| 6. | 0.2 | 17. | 3 | 28. | 0.007 | 39. | 0.013 |
| 7. | 0.9 | 18. | 0.3 | 29. | 0.007 | 40. | 0.046 |
| 8. | 0.3 | 19. | 0.5 | 30. | 0.006 | 41. | 0.044 |
| 9. | 0.1 | 20. | 0.7 | 31. | 0.007 | 42. | 0.074 |
| 10. | 3 | 21. | 0.9 | 32. | 0.009 | 43. | 0.087 |
| 11. | 0.3 | 22. | 0.6 | 33. | 0.2 | 44. | 0.068 |

## Problem Set

1. 43.5 hr
2. $16.25
3. 6.75 in
4. 321.25 kg

## Exit Ticket

$23.20

## Homework

1. 9
2. 2.22 oz
3. 0.67 kg
4. $5.75
5. 38

# Lesson 29

## Problem Set

1. 903 kg
2. 6 weeks; explanations will vary.
3. 39.12 m
4. 10,800 $m^2$

## Exit Ticket

$618

## Homework

1. 13
2. $178
3. 64.5 m
4. $3.50
5. 6.25 $km^2$



EUREKA MATH™